MULTIAXIAL FATIGUE AND FRACTURE

Other titles in the ESIS Series

EGF 1	*The Behaviour of Short Fatigue Cracks* Edited by K.J. Miller and E.R. de los Rios
EGF 2	*The Fracture Mechanics of Welds* Edited by J.G. Blauel and K.-H. Schwalbe
EGF 3	*Biaxial and Multiaxial Fatigue* Edited by M.W. Brown and K.J. Miller
EGF 4	*The Assessment of Cracked Components by Fracture Mechanics* Edited by L.H. Larsson
EGF 5	*Yielding, Damage, and Failure of Anisotropic Solids* Edited by J.P. Boehler
EGF 6	*High Temperature Fracture Mechanisms and Mechanics* Edited by P. Bensussan and J.P. Mascarell
EGF 7	*Environment Assisted Fatigue* Edited by P. Scott and R.A. Cottis
EGF/ESIS 8	*Fracture Mechanics Verification by Large Scale Testing* Edited by K. Kussmaul
ESIS/EGF 9	*Defect Assessment in Components –* *Fundamentals and Applications* Edited by J.G. Blauel and K.-H. Schwalbe
ESIS 10	*Fatigue under Biaxial and Multiaxial Loading* Edited by K. Kussmaul, D.L. McDiarmid, and D.F. Socie
ESIS 11	*Mechanics and Mechanisms of Damage in Composites and Multi-Materials* Edited by D. Baptiste
ESIS 12	*High Temperature Structural Design* Edited by L.H. Larsson
ESIS 13	*Short Fatigue Cracks* Edited by K.J. Miller and E.R. de los Rios
ESIS 14	*Mixed-Mode Fatigue and Fracture* Edited by H.P. Rossmanith and K.J. Miller
ESIS 15	*Behaviour of Defects at High Temperatures* Edited by R.A. Ainsworth and R.P. Skelton
ESIS 16	*Fatigue Design* Edited by J. Solin, G. Marquis, A. Siljander, and S. Sipilä
ESIS 17	*Mis-Matching of Welds* K.-H. Schwalbe and M. Koçak
ESIS 18	*Fretting Fatigue* Edited by R.B. Waterhouse and T.C. Lindley
ESIS 19	*Impact and Dynamic Fracture of Polymers and Composites* Edited by J.G. Williams and A. Pavan
ESIS 20	*Evaluating Material Properties by Dynamic Testing* Edited by E. van Walle
ESIS 21	*Multiaxial Fatigue & Design* Edited by A. Pinian, G. Cailletand and T.C. Lindley
ESIS 22	*Fatigue Design of Components.* ISBN 008-043318-9 Edited by G. Marquis and J. Solin
ESIS 23	*Fatigue Design and Reliability.* ISBN 008-043329-4 Edited by G. Marquis and J. Solin
ESIS 24	*Minimum Reinforcement in Concrete Members.* ISBN 008-043022-8 Edited by Alberto Carpinteri

MULTIAXIAL FATIGUE AND FRACTURE

Editors: E. Macha, W. Będkowski and T. Łagoda

ESIS Publication 25

This volume contains 18 papers selected from 90 presented at the Fifth International Conference on Biaxial/Multiaxial Fatigue and Fracture held in Cracow, Poland, 8–12 September 1997. The meeting was organised by the Department of Mechanics and Machine Design, Technical University of Opole, the Institute of Fundamental Technological Research, Polish Academy of Sciences and the Committee of Machine Design, Polish Academy of Sciences and sponsored by the European Structural Integrity Society (ESIS).

1999

Elsevier

Amsterdam • Lausanne • New York • Oxford • Shannon • Singapore • Tokyo

ELSEVIER SCIENCE Ltd
The Boulevard, Langford Lane
Kidlington, Oxford OX5 1GB, UK

First edition 1999

Library of Congress Cataloging-in-Publication Data
International Conference on Biaxial/Multiaxial Fatigue (5th : 1997 : Kraków, Poland)
Multiaxial fatigue and fracture / editors, E. Macha, W. Będkowski, and T. Łagoda ; sponsored by ESIS.
p. cm.
ISBN 0-08-043336-7
1. Materials--Fatigue--Congresses. I. Macha, Ewald. II. Będkowski, W. III. Łagoda, T. IV. European Structural Integrity Society. V. Title.
TA460.I53185 1999
620.1'126--dc21 99-31215
CIP

British Library Cataloguing in Publication Data
Multiaxial fatigue and fracture. - (ESIS publication ; 25)
1.Materials - Fatigue - Congresses 2.Fracture mechanics - Congresses 3.Structural failures - Congresses 4.Structural engineering - Congresses
I.Macha, E. II.Bedkowski, W III.Lagoda, T.
620.1'126
ISBN 0080433367

ISBN: 0 08 043336 7

∞ The paper used in this publication meets the requirements of ANSI/NISO Z39.48-1992 (Permanence of Paper).

Transferred to digital printing 2005

CONFERENCE ORGANISERS

International Scientific Committee

A. ABEL	Australia	
S. CURIONI	Italy	
M. DE FREITAS	Portugal	
F. ELLYIN	Canada	
D. FRANÇOIS	France	
T. INOUE	Japan	
S. KOCAŃDA	Poland	
E. KREMPL	USA	
K. KUSSMAUL	Germany	
S.B. LEE	Korea	
P. LUKAS	Czech Republic	
E. MACHA	Poland	*Co-chairman*
D.L. McDOWELL	USA	
K.J. MILLER	UK	
Z. MRÓZ	Poland	*Co-chairman*
M. OHNAMI	Japan	
J. PETIT	France	
A. PINEAU	France	
V.N. SHLYANNIKOV	Russia	
D. SOCIE	USA	
C.M. SONSINO	Germany	
V. TROSHCHENKO	Ukraine	
E.K. TSCHEGG	Austria	

National Committee

W. BĘDKOWSKI	Opole	*Secretary*
J. DZIUBIŃSKI	Katowice	
L. GOŁASKI	Kielce	
K. GOŁOŚ	Warszawa	
A. JAKOWLUK	Białystok	
S. KOCAŃDA	Warszawa	
E. MACHA	Opole	*Co-chairman*
Z. MRÓZ	Warszawa	*Co-chairman*
A. NEIMITZ	Kielce	
A. SKORUPA	Kraków	
K. SOBCZYK	Warszawa	
J. SZALA	Bydgoszcz	
R. ŻUCHOWSKI	Wrocław	

Organising Committee

H.Achtelik, R.Bryś, B.Chrobak, J.Dembicka, E.Helleńska, C.Lachowicz, T.Łagoda, J.Marynowski, R.Pawliczek, P.Piaseczny, J.Sołtysek

Elsevier Titles of Related Interest

Books

ABE & TSUTA
AEPA '96: Proceedings of the 3rd Asia-Pacific Symposium on Advances in Engineering Plasticity and its Applications (Hiroshima, August 1996). ISBN 008-042824-X

JONES
Failure Analysis Case Studies. ISBN 008-043338-3

KARIHALOO *ET AL.*
Advances in Fracture Research: Proceedings of the 9th International Conference on Fracture (Sydney, April 1997). ISBN 008-042820-7

KISHIMOTO *ET AL.*
Cyclic Fatigue in Ceramics. ISBN 0-444-82154-6

KLESNIL & LUKÁŠ
Fatigue of Metallic Materials. 2nd Edn. ISBN 0-444-98723-1

LADEVÈZE & ODEN
Advances in Adaptive Computational Methods in Mechanics. ISBN 008-043327-8

LIEW *ET AL.*
Vibration of Mindlin Plates. ISBN 008-043341-3

LÜTJERING & NOWACK
Fatigue '96: Proceedings of the 6th International Fatigue Congress (Berlin, May 1996). ISBN 008-042268-3

RAVICHANDRAN *ET AL.*
Small Fatigue Cracks: Mechanics, Mechanisms & Applications. ISBN 008-043011-2

RIE & PORTELLA
Low Cycle Fatigue and Elasto-Plastic Behaviour of Materials. ISBN 008-043326-X

VOYIADJIS & KATTAN
Advances in Damage Mechanics: Metals and Metal Matrix Composites. ISBN 008-043601-3

VOYIADJIS *ET AL.*
Damage Mechanics in Engineering Materials. ISBN 008-043322-7

Journals

Acta Metallurgica et Materialia
Composite Structures
Computers and Structures
Corrosion Science
Engineering Failure Analysis
Engineering Fracture Mechanics
International Journal of Fatigue
International Journal of Impact Engineering
International Journal of Mechanical Sciences
International Journal of Non-Linear Mechanics
International Journal of Pressure Vessels & Piping
International Journal of Solids and Structures
Journal of Applied Mathematics and Mechanics
Journal of the Mechanics and Physics of Solids
Materials Research Bulletin
Mechanics of Materials
Mechanics Research Communications
NDT&E International
Scripta Metallurgica et Materialia
Theoretical and Applied Fracture Mechanics
Tribology International
Wear

For more information Elsevier's catalogue can be accessed via the internet on http://www.elsevier.com

CONTENTS

PREFACE

This volume contains 18 papers selected from 90 presented at the Fifth International Conference on Biaxial/Multiaxial Fatigue and Fracture held in Cracow, Poland, 8-12 September 1997. The meeting was organised by the Department of Mechanics and Machine Design, Technical University of Opole, the Institute of Fundamental Technological Research, Polish Academy of Sciences and the Committee of Machine Design, Polish Academy of Sciences and sponsored by the European Structural Integrity Society (ESIS).

The First International Conference on Biaxial/Multiaxial Fatigue was in San Francisco in 1982. The next conferences were held in Sheffield (2^{nd}ICBMF, 1995), Stuttgart (3^{rd}ICBMF, 1989) and in St Germain en Laye (4^{th}ICBMF, 1994). Three other important meetings connected with biaxial/multiaxial fatigue and fracture should be also mentioned here. The first was the International Conference on Mixed-Mode Fracture and Fatigue, Vienna, 1991 and the other two – ASTM Symposia in San Diego, 1991 and in Denver, 1995.

The papers in this book deal with theoretical, computational and experimental aspects of the multiaxial fatigue and fracture of engineering materials and structures.

The paper are divided into the following four categories:

1. Proportional cyclic loading (3 papers),
2. Non-proportional cyclic loading (7 papers),
3. Variable amplitude and random loading (5 papers),
4. Crack growth (3 papers).

Most papers in this publication talk about the behaviour of constructional materials and elements of machines under non-proportional loading and under variable amplitude and random loading, which are more realistic load histories met in industrial practice. Here we classify variable amplitude loading under cyclic load with basic frequency and random loading under load with a continuous band of frequency.

The book gives a review of the latest world successes and directions of investigations on multiaxial fatigue and fracture. More and more often publications are results of the co-operation of researchers from different laboratories and countries. Seven out of eighteen papers included here were worked out by international authors teams. This is a symptom of our times, when science and investigations know no borders.

A large number of people contributed to the issue of this publication. The editors particularly wish to thank the authors who have invested many hours, both in the laboratory and for preparing their papers. The editors also gratefully acknowledge the scientific roles played by 57 manuscript reviewers from 12 countries, as well as the staff members at ESIS and Elsevier who have made this publication possible.

E.Macha, W.Będkowski and T.Łagoda, Editors

Opole, March 1999

I. PROPORTIONAL CYCLIC LOADING

MODELLING THRESHOLD CONDITIONS FOR CRACKS UNDER TENSION/TORSION LOADING

Yury G. MATVIENKO*, Mike W. BROWN** and Keith J. MILLER**

* Mechanical Engineering Research Institute of the Russian Academy of Sciences, 4 M.Kharitonievsky Per., 101830 Moscow, Russia
** SIRIUS, University of Sheffield, Mappin Street, Sheffield, S1 3JD, U.K.

ABSTRACT

A model of microstructurally and physically short crack growth, with the threshold criterion for combined tension/torsion loading, has been proposed to analyse short crack propagation in medium carbon steel. Tresca's equivalent strain criterion for the high strain range regime and Rankine's criterion for fatigue limit strain range were employed for an analysis of Stage II crack growth. As a result, the threshold conditions and equations for Stage II physically short crack growth under combined loading were derived from the equation for push-pull loading. The parameters and exponents of the short crack growth equations are dependent on microstructure as well as on the type of loading. The influence of biaxial strain ratio on non-propagating crack length is discussed, with regard to the distance between major microstructural barriers.

KEY WORDS

Fatigue, threshold, short crack, combined loading

NOTATION

a	crack length
a_f	non-propagating fatigue crack length
a_{tr}	transitional crack length
A, B, D_{th}	constants of crack growth equations
d	microstructural parameter
d_a	average ferrite grain size
da/dN	growth rates of short fatigue cracks
m	grain number
α, β	exponents of crack growth equations
$\lambda = \Delta\gamma / \Delta\varepsilon$	strain ratio
ν	Poisson's ratio
σ_y	yield stress
$\Delta\gamma$, $\Delta\varepsilon$	shear and normal strain ranges
$\Delta\gamma_f$, $\Delta\varepsilon_f$	fatigue limit strain ranges

INTRODUCTION

The consideration of two dominant phases of the fatigue failure process, Stage I and Stage II fatigue crack growth, proposed by Forsyth **(1)** plays a significant role in the understanding of damage accumulation. Certainly, advances in the mechanics of fatigue crack propagation have been connected with an analytical description of microstructurally short and physically small crack growth **(2)-(4)**. Thus, damage accumulation has been interpreted in the failure process as the physical propagation of cracks. The general fatigue lifetime of a solid may be determined by integration of these crack propagation stages.

Equations, which describe the propagation of microstructurally and physically short cracks **(2) (3)**, have been derived from experimental analyses of fatigue crack behaviour in medium carbon steel. Microstructurally short crack growth for Stage I can be represented by the function

$$\frac{da}{dN} = A\Delta\varepsilon^{\alpha}(d-a) \tag{1}$$

which will be equal to zero when a crack reaches a microstructural barrier. Here a is the crack length, A and α are material constants depending on the type of loading and $\Delta\varepsilon$ is the applied normal strain range. The microstructural parameter d refers to the possible distances between microstructural barriers.

Stage II short fatigue crack propagation follows and, being normal to the maximum principal strain, it exhibits mode I crack opening. A general equation for physically short crack growth may be written in the form

$$\frac{da}{dN} = B\Delta\varepsilon^{\beta}a - D_{th} \tag{2}$$

where B and β are material constants depending on the type of loading.

Parameter D_{th} represents the mechanical threshold for Stage II short cracks, providing a crack length a_{th} below which Stage II cracks cannot propagate. The microstructural barrier d represents a microstructural threshold, above which Stage I cracks do not grow.

The functions (1) and (2) for Stage I and Stage II short crack growth rates will be equal to zero at microstructural and mechanical threshold conditions respectively. There are two threshold conditions for short cracks. The first threshold condition of Stage I is determined by microstructural barriers in the material. The second mechanical threshold for Stage II physically short crack growth is dependent on crack length as well as the applied strain range.

Apparently, the methodology of damage accumulation analysis based on crack propagation can be more complex for multiaxial modes of loading. Therefore, the issue addressed in this paper is to work out a two-stage model of crack growth and a threshold criterion, which will allow analytical equations for short crack propagation under torsion and combined tension/torsion fatigue loading to be obtained.

MODIFIED EQUATIONS FOR STAGE I AND STAGE II SHORT CRACK GROWTH

Equations (1) and (2), which describe Stage I and Stage II short crack growth in materials under a uniaxial stress state, for example, push-pull tests, can be transferred to the general case of multiaxial loading. The transformation is based on the reduction of a multiaxial cyclic strain state to an equivalent uniaxial strain range. Various equivalent strain criteria could be employed for this purpose.
A widely used equivalent strain formula employed in plasticity theory is based on the Tresca or maximum shear criterion

$$\Delta\varepsilon_{eqT} = \frac{1}{(1+\nu)}(\Delta\varepsilon_1 - \Delta\varepsilon_3) \tag{3}$$

where ν is the elasto-plastic value of Poisson's ratio, and $\Delta\varepsilon_1 \geq \Delta\varepsilon_2 \geq \Delta\varepsilon_3$ are principal strain ranges. In the low stress regime $(\sigma_{max} << \sigma_y)$ the Rankine failure criterion of maximum principal stress has been applied to biaxial fatigue studies **(5)**. For proportional loading conditions this criterion may be extended to elastic-plastic conditions, i.e.

$$\Delta\varepsilon_{eqR} = \frac{1}{(1+\nu)}[\Delta\varepsilon_1 + (\Delta\varepsilon_1 + \Delta\varepsilon_2 + \Delta\varepsilon_3)\,\nu/(1-2\nu)] \tag{4}$$

For combined tension-torsion loading the Tresca equivalent strain range formula can be written as

$$\Delta\varepsilon_{eqT} = \sqrt{\left[\Delta\varepsilon^2 + \Delta\gamma^2/(1+\nu)^2\right]} \tag{5}$$

and Rankine's equivalent strain range formula becomes

$$\Delta\varepsilon_{eqR} = (\Delta\varepsilon_{eqT} + \Delta\varepsilon)/2 \tag{6}$$

where $\Delta\varepsilon$ and $\Delta\gamma$ are the axial and torsional strain range components, and ν is the elasto-plastic value of Poisson's ratio. Thus, Eqs. (1) and (2) of short crack propagation (Stage I and Stage II) may be re-written for combined loading in terms of the torsion strain range components taking into account Eqs. (5) and (6) and employing the strain ratio $\lambda(=\Delta\gamma/\Delta\varepsilon)$.

$$\frac{da}{dN} = A_{eq}\Delta\gamma^{\alpha}(d-a) \quad \text{where} \quad A_{eq} = A\left(\frac{1}{\lambda^2} + \frac{1}{(1+\nu)^2}\right)^{\alpha/2} \tag{7}$$

$$\frac{da}{dN} = B_{eq}\Delta\gamma^{\beta}a - D_{th} \tag{8}$$

where

$$B_{eq} = B\left(\frac{1}{\lambda^2} + \frac{1}{(1+\nu)^2}\right)^{\beta/2} \qquad \text{for Tresca}$$

$$B_{eq} = B\left[\frac{1}{2}\left(\frac{1}{\lambda^2} + \frac{1}{(1+\nu)^2}\right)^{1/2} + \frac{1}{2\lambda}\right]^{\beta} \qquad \text{for Rankine}$$

Here the parameter D_{th} is assumed to be a material constant. New constants A_{eq} and B_{eq} are determined by the parameters and exponents of Eqs. (1) and (2), using the equivalent strain formula.

A MODEL OF MICROSTRUCTURALLY SHORT CRACK GROWTH

Since Stage I fatigue crack propagation is a shear mode, the employment of Tresca's equivalent strain criterion and Eq. (1) for microstructurally short crack growth under push-pull loading allows a description of the observed behaviour of microstructurally short cracks under torsion and combined loading. Microstructurally short cracks are influenced in their behaviour by the texture of the metal. Textural effects include crystallographic orientation, the size and shape of grains as well as size and distribution of second phases and inclusions. From this point of view the following mechanism of short crack growth and overcoming of microstructural barriers can be proposed for a carbon medium steel. Since this steel contains ferrite and pearlite grains, let us postulate that there are two types of microstructural barrier. Weak barriers are connected with boundaries of ferrite grains, and strong barriers are boundaries of pearlite grains.

Microstructurally short cracks can grow at a constant strain range below the fatigue limit of a material but will be arrested at weak barriers, i.e. ferrite grain boundaries, or at strong barriers. A microstructural fatigue limit barrier is a material threshold and represents the distance a crack must grow to meet the major microstructural barrier in the metal under fatigue limit conditions.

For example, Table 1 reflects threshold conditions derived from the experimental results of Zhang **(6)** and Eqs. (1) and (2) for carbon medium steel under push-pull loading. The maximum non-propagating fatigue crack represents the fatigue limit strain range. It can be seen that the non-propagating crack length a_f exceeds the average ferrite grain size d_a (37 μm) in medium carbon steel in the transverse direction, with a standard deviation of 18 μm. It is clear that a physically short fatigue crack will propagate at a strain range that is above the fatigue limit $\Delta\varepsilon_f$. However, there is a significant difference between microstructural barriers d and the threshold short crack length of Stage II for strain ranges close to the fatigue limit. What is the mechanism of microstructural short crack propagation beyond the first grain to reach the Stage II threshold? To answer this question it is necessary to consider basic microstructural aspects of fatigue crack growth and fatigue resistance of metals and develop the representation of microstructural barriers.

Table 1. Constants and exponents of equations (1) and (2) and microstructural fatigue limit parameters for a medium carbon steel under push-pull loading

A	α	B	β	D_{th} [μm / cycle]	$\Delta\varepsilon_f$	a_f [μm]
4.68×10^5	4.5	2.96×10^{-1}	1.771	3.74×10^{-3}	4.1×10^{-3}	213

The differences in sizes of weak microstructural barriers and the mechanical threshold is due to the crack propagation process, connected with overcoming weak barriers (boundaries of ferrite grains), in accordance with the equation for Stage I crack growth. The transcrystalline crack that becomes the short fatigue failure crack is the one located in the largest ferrite grain, because it grows fastest. It is clear that the microstructural parameter d will change for each crack step. Parameter d represents an above-average ferrite grain size d_1 = 55μm for the first grain that is likely to initiate a Stage I crack. For the subsequent m crack steps of Stage I the microstructural parameter $d = d_m$ will be of the form

$$\left.\begin{aligned} d_m &= d_1 + 2(m-1)\ d_a, \quad d_m < D \\ d_m &= D \end{aligned}\right\} \tag{9}$$

where D is the spacing of strong barriers, between pearlite grains. Thus D, the strong barrier gives an absolute upper bound to weak barrier values d_m. For a crack in grain m, d is assumed to change from d_m to d_{m+1} when the crack length achieves a critical value of 0.95 d_m **(7)**. The transition of short crack propagation from Stage I to Stage II is determined by an equality of crack rates from Eqs. (7) and (8)

$$A_{eq}\Delta\gamma^{\alpha}(d_m - a_{tr}) = B_{eq}\Delta\gamma^{\beta}a_{tr} - D_{th}\ , \tag{10}$$

giving the transitional crack length a_{tr}.

The number of microstructurally short crack steps depends on a material's microstructure, and also on the applied strain range because Stage I is limited by the transition of crack propagation to Stage II after the mechanical threshold has been exceeded.

THRESHOLD CRITERION FOR A STAGE II SHORT CRACK

Stage II short crack growth in medium carbon steel subjected to torsion cyclic loading has been analysed, using the experimental results of Zhang **(6)**. The experimental results lie between two curves described by the Tresca and Rankine criteria. The Rankine and Tresca curves were derived from Zhang's push-pull crack growth results, using Eq. (8). The location of the experimental curve is dependent on the applied strain range and can be determined by the threshold conditions. Eq. (8) allows the analysis of mechanical threshold conditions for Stage II. Rankine's criterion correlates with experimental results for the low strain range regime. The use of Tresca's criterion is more justified in the high strain range regime of loading. It is clear that it is necessary to modify these criteria and the parameters of Eq. (8) to describe threshold conditions for any possible strain range regime. The following procedure is proposed to establish the modified criterion and parameters of Eq. (8).

It is considered that two points will create a modified threshold criterion. The first point is determined by Rankine's criterion at the fatigue limit strain range $\Delta\gamma_f$. The second point is determined by the intersection of the Tresca's criterion threshold line with the experimental threshold in the high strain range regime of Fig.1. Thus, modified equations of Stage II fatigue short crack growth under torsion may be determined by the following threshold conditions

$$a_{th}\Delta\gamma_{th}^{\beta_{eq}} = constant\ ,\ B_{eq}\Delta\gamma_f^{\beta_{eq}}a_f = D_{th} \tag{11}$$

which gives a new modified constant and exponent for medium carbon steel.

The "modified" parameter β_{eq} is found as follows:

i) taking the first fixed point at the fatigue limit, derive $a_{th} = a_f$ from the Rankine, Eq. (8),

ii) taking the second fixed point at $a = 7\mu m$, calculate the threshold strain range $\Delta\gamma_{th}$ from the Tresca Eq. (8),

iii) assuming a straight line of slope $-1/\beta_{eq}$ in Fig.1, evaluate β_{eq} from a line shown between these fixed points,

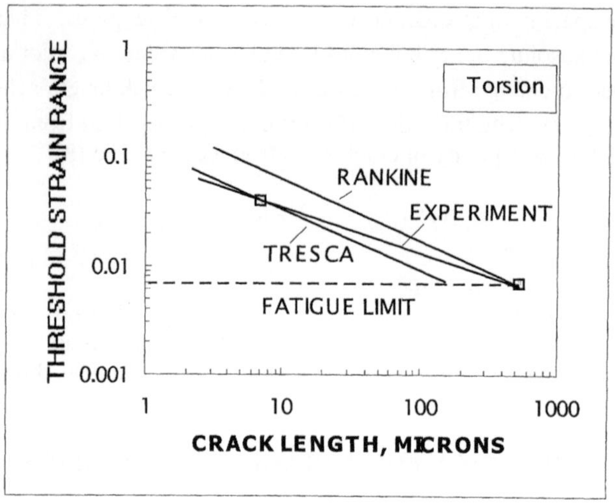

Fig.1 Threshold conditions for medium carbon steel under torsion loading.

Table 2 shows excellent agreement with the experimental results because the critical value of 7 μm was selected for the torsion data in Fig.1.

Table 2 Constants and exponents of equations (7) and (8) and microstructural fatigue limit parameters for medium carbon steel under torsion loading.

Criterion	A_{eq}	α	B_{eq}	β_{eq}	$\Delta\gamma_f$	a_f μm
Tresca	1.03×10^5	4.5	1.63×10^{-1}	1.771	6.9×10^{-3}	154
Rankine	-	-	4.78×10^{-2}	1.771	6.9×10^{-3}	526
Experiment	8.66×10^3	4.02	1.64	2.483	6.9×10^{-3}	531
Modified	1.03×10^5	4.5	1.605	2.477	6.9×10^{-3}	526

Let us use the previous procedure with two selected boundary threshold points to establish a modified equation for Stage II short crack propagation under combined loading at strain ratio λ. Obviously, the fatigue limit condition for combined loading differs from the fatigue limit under torsion and push-pull. To estimate the fatigue limit condition under combined loading, an approach based on the Γ-plane **(8)** can be used. It was shown that lifetime is controlled by the maximum shear strain γ_{max} and the normal strain ε_n on that plane of maximum shear. Contours of constant endurance plotted on a graph of $\frac{1}{2}\Delta\gamma_{max}$ versus $\Delta\varepsilon_n$ may be represented by the elliptical equation

$$\left(\frac{\frac{1}{2}\Delta\gamma_{max}}{g}\right)^2+\left(\frac{\Delta\varepsilon_n}{l}\right)^2=1 \tag{12}$$

where g and l are fitted constants. In the case considered here, constant endurance is referred to the fatigue limit. So, the maximum shear strain and the normal strain are written in the usual form

$$\frac{1}{2}\Delta\gamma_{max}=\sqrt{\left(\frac{\Delta\gamma_f}{2}\right)^2+\left[\frac{(1+\nu)\Delta\varepsilon_f}{2}\right]^2} \tag{13}$$

$$\Delta\varepsilon_n=\left(\frac{1-\nu}{2}\right)\Delta\varepsilon_f \tag{14}$$

where $\Delta\gamma_f=\lambda\Delta\varepsilon_f$ for combined loading. Constants g and l can be obtained from Eq. (12) using fatigue limit strain ranges for torsion and push-pull loading. For example, for a medium carbon steel for a given $\lambda = 1.5$, $g = 3.45\times10^{-3}$ *and* $l = 2.22\times10^{-3}$. Thus, the Γ-plane approach allows us to estimate fatigue limit strain range components for combined loading (Table 3).

Table 3. Constants and exponents of equations (7) and (8) and microstructural fatigue limit parameters for a medium carbon steel under combined loading at λ=1.5

Criterion	A_{eq}	α	B_{eq}	β_{eq}	$\Delta\gamma_f$	a_f [μm]
Tresca	4.21×10^5	4.5	2.84×10^{-1}	1.771	4.6×10^{-3}	182
Rankine	-	-	2.09×10^{-1}	1.771	4.6×10^{-3}	247
Modified	4.21×10^5	4.5	5.14×10^{-1}	1.938	4.6×10^{-3}	247

The use of two selected boundary threshold points, transferred to combined axial and torsion loading is illustrated in Fig.2. The modified threshold line described by Eq. (11) allows us to obtain the real constant B_{eq} and exponent β_{eq} for the crack growth Eq. (8) for Stage II (Table 3). We have assumed that the critical length 7 μm used to obtain the second fixed point is a constant value, irrespective of the strain state λ.

A modified threshold condition and an equation for Stage II short crack growth under combined loading have been derived from Tresca's and Rankine's equivalent strain criteria and two selected boundary threshold points, which are fitted to empirical torsion fatigue threshold results. It is clear that the parameters and exponents of short crack growth equations are dependent on microstructure as well as on the type of loading.

The dependence on microstructure is illustrated by the maximum non-propagating fatigue crack lengths a_f (or distances between strong microstructural barriers), which have been derived from equations for Stage II with a growth rate equal to zero under fatigue limit strain ranges (Table 4). Obviously, the change of size of non-propagating cracks for various loading conditions is connected with a change of crack growth directions between major microstructural barriers. The high value for a_f in torsion reflects the anisotropy of the material, with greater microstructural dimensions for cracks growing along the longitudinal axis of the bar.

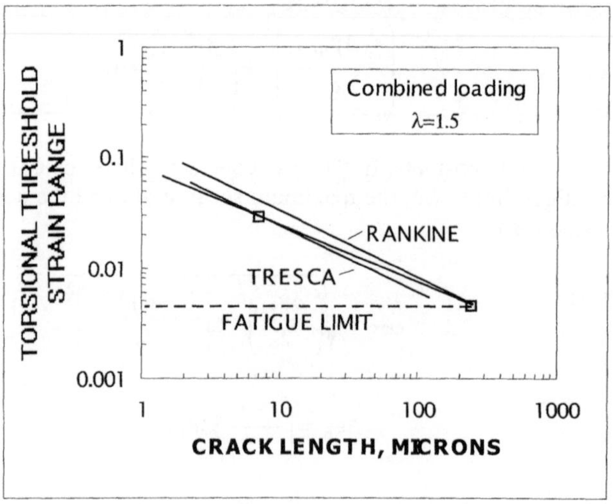

Fig.2 Threshold conditions for a medium carbon steel under combined loading at λ=1.5.

Table 4 The influence of strain ratio λ on constants and exponent of Stage II short crack growth equations and non-propagating crack length.

	λ	B_{eq}	β	a_f [μm]	$\Delta\gamma_f$	$\Delta\varepsilon_f$
Uniaxial	0	0.296	1.771	213	0	0.0041
Combined	1.5	0.514	1.938	247	0.0046	0.0031
Pure shear	∞	1.605	2.477	526	0.0069	0

VERIFICATION OF SHORT CRACK GROWTH MODEL

The proposed model and modified equations for Stage I and Stage II short crack growth have been employed to predict the lifetime of smooth specimens under combined loading. A general fatigue lifetime N_c is determined by summation of lifetimes for each crack propagation stage. Fatigue lifetime N_I for Stage I has been obtained by integrating Eq. (7) from $a = 1$μm to the transitional crack length a_{tr}. The number of crack steps for Stage I was taken into account in Eq. (9). Integrating Eq. (8) from a_{tr} to the failure crack length gives Stage II lifetime N_{II}. The transition length a_{tr} is given by Eq. (10). It is assumed that after transition cracks will remain in Stage II without reverting to Stage I under constant amplitude loading.

The predicted lifetime for an equivalent regime of combined loading is compared with experimental life for regimes of torsion and push-pull loading of solid specimens (Table5). The equivalent regime of combined loading was calculated by employing the Γ-plane approach with Eq. (12) and applied strain ranges for torsion and push-pull loading at constant life. Theoretical results for endurance under combined load closely correlate with experimental lives for equivalent regimes of torsion and push-pull loading.

Table 5 Prediction of life for medium carbon steel under combined loading at λ =1.5.

Push-pull (6)		Torsion (6)		Tension/torsion	
$\Delta\varepsilon$, %	N_c^{exp}	$\Delta\gamma$, %	N_c^{exp}	$\Delta\gamma$, %	N_c^{pred}
0.55	3.5×10^5	1.0	3.5×10^5	0.65	3.7×10^5
0.77	1.3×10^5	1.4	1.7×10^5	0.89	1.0×10^5
1.3	2.4×10^4	2.4	2.4×10^4	1.5	3.3×10^4

CONCLUSIONS

Equations and modified threshold conditions for short crack growth are developed on the basis of Tresca and Rankine equivalent strain criteria and experimental push-pull data on short crack propagation in a medium carbon steel. The following conclusions can be drawn from the present study:

(1) Satisfactory correlation of Stage II calculated short crack behaviour and experimental data for torsion is ensured by fitting the formula for threshold condition to two limited cases, which are connected with the use of Rankine's equivalent strain criterion for fatigue limit strain range and Tresca's criterion for the high strain range regime.

(2) A modified threshold condition and an equation for Stage II short crack growth under tension/torsion loading have been derived from these equivalent strain criteria and two selected boundary threshold points, which are fitted to empirical torsion fatigue threshold results.

(3) The parameters, exponents of short crack growth equations and the maximum non-propagating physically short fatigue crack length are dependent on distance between major microstructural barriers as well as on the type of loading.

REFERENCES

(1) Forsyth P.J.E., (1961), A two stage process of fatigue crack growth, Proc. Crack Propagation Symposium, Cranfield, UK, pp. 76-94.

(2) Carbonell E. Perez and Brown M.W., (1985), A study of short crack growth in torsional low cycle fatigue for a medium carbon steel, Fatigue Fracture Engng Mater. Structures, vol. 9, pp. 15-33.

(3) Hobson P.D., Brown M.W. and de los Rios E.R., (1986), Two phases of short crack in a medium carbon steel, The Behaviour of Short Fatigue Cracks (EGF), (Edited by K.J. Miller and E.R. de los Rios), pp. 441-459.

(4) Brown M.W., Miller K.J., Fernando U.S., Yates J.R. and Suker D.K., (1994), Aspects of multiaxial fatigue crack propagation, Fourth International Conference on Biaxial/Multiaxial Fatigue, vol. I, pp. 3-16.

(5) Brown M.W. and Buckthorpe D.E., (1989), A crack propagation based effective strain criterion, Biaxial and Multiaxial Fatigue (EDF 3), (Edited by M.W. Brown and K.J. Miller), pp. 499-510.

(6) Zhang W., (1991), Short fatigue crack behaviour under different loading systems, Ph.D. Thesis, University of Sheffield, UK.

(7) Yates J.R. and Grabowski L., (1990), Fatigue life assessment using a short crack growth model, Fatigue'90, vol. 4, pp. 2369-2376.

(8) Brown M.W. and Miller K.J., (1973), A theory for fatigue under multiaxial stress-strain conditions, Proc. Instn Mech. Engrs, vol. 187, pp. 745-755.

Acknowledgements

Prof. Yu.G. Matvienko would like to thank the University of Sheffield and the Russian Academy of Sciences for funding a one-year visit to the Department of Mechanical Engineering, University of Sheffield.

FATIGUE AND FRACTURE OF PLANE ELEMENTS WITH SHARP NOTCHES UNDER BIAXIAL LOADING

Krzysztof L. MOLSKI* and Andrzej SEWERYN**

* Warsaw University of Technology, Narbutta 84, 02-524 Warsaw, Poland

**Białystok University of Technology, Wiejska 45C, 15-351 Białystok, Poland

ABSTRACT

Theoretical, numerical and experimental investigations on fatigue and fracture of plane rectangular elements with opposite V-notches of various opening angles are presented. Two materials - a ductile aluminium alloy and brittle PMMA - were used to analyse damage processes under biaxial loading conditions, as tension and shear. Experiments have shown that material properties strongly affect the damage process in both monotonic and variable loading. Fracture mode, fatigue crack direction and shape depend on material properties, notch angle and tension/shear ratio. The explanation of this behaviour is based on an elastic-plastic FEM analysis for the aluminium alloy and a non-local damage accumulation criterion for the PMMA.

KEY WORDS

Sharp notch, fatigue and fracture, multiaxial loading

NOTATION

d_0	damage zone length (characteristic parameter)
K_I^λ, K_{II}^λ	generalised stress intensity factors for Mode I and Mode II, respectively
$\Delta K_I^\lambda, \Delta K_{II}^\lambda$	range of generalised stress intensity factors for Mode I and Mode II
n_σ	damage accumulation exponent
N_f	number of cycles to crack initiation
f	ratio of fatigue limit to critical normal stress
R_d	damage accumulation factor
P, T, F	tensile, shearing and resultant forces
ΔF	range of variable load
F_{max}	maximum peak force of cyclic load
Δu	range of displacements corresponding to ΔF
2β	notch angle
r,ϑ	polar coordinates
$\overline{R}_\sigma$	non-local brittle failure function
q_i	component of displacement vector
ψ	angle of the resultant force

σ_n	normal stress on the physical plane
σ_c	critical normal stress
σ_o	fatigue stress limit
$\Delta\sigma_{az}$	equivalent stress range along the damage zone
λ_I, λ_{II}	eigen exponents of the stress field for Mode I and Mode II
ϑ_0	direction of crack initiation
ω_n	damage accumulation measure

INTRODUCTION

Results of many investigations have indicated that stresses and strains distributed near a notch root strongly affect the damage accumulation process, and control early stages of fatigue crack growth. High stress and strain gradients are associated with the presence of sharp corners and may be represented by various asymptotic relations depending on the opening angle, loading modes and boundary conditions. Such corners frequently appear as weak points in engineering structures reducing their strength and durability. They can also serve as interesting objects for theoretical and experimental studies related to fatigue and fracture processes. In such cases elastic stress fields based on the solutions of the theory of elasticity are of great importance. They are also useful for local plasticity near the notch root, due to the fact that the differences between the elastic and elastoplastic strain fields some distance from the notch are small.

This paper deals with theoretical, numerical and experimental investigations of fatigue and fracture of plane rectangular plates with two opposite V-notches of different opening angles, subjected to biaxial loading conditions with various tension/shear ratios. Two materials, one ductile and one brittle, were used. For each material and V-notch considered, the following experimental data have been collected and analysed: fracture mode during monotonic and variable loading, fatigue crack (or cracks) direction and shape, number of cycles to crack initiation and resultant forces applied to the element.

THE STRESS FIELD AROUND A SHARP CORNER

Let us consider a sharp corner in a polar coordinate system (r,ϑ), as is shown in Fig.1. If the notch faces are free from loading, the asymptotic stress fields near the notch are given by Williams solution **(1)**, or may be also described, together with the displacement field, by Eqs (1) recently published in **(2)**:

$$\begin{aligned} \sigma_{ij} &= (2\pi r)^{\lambda_I - 1} K_I^{\lambda} a_{ij}(\vartheta) + (2\pi r)^{\lambda_{II}-1} K_{II}^{\lambda} b_{ij}(\vartheta) \\ q_i &= (2\pi r)^{\lambda_I} K_I^{\lambda} c_i(\vartheta) + (2\pi r)^{\lambda_{II}} K_{II}^{\lambda} d_i(\vartheta) \end{aligned} \tag{1}$$

where $a_{ij}(\vartheta)$, $b_{ij}(\vartheta)$, $c_i(\vartheta)$ and $d_i(\vartheta)$ represent combinations of trigonometric functions of ϑ, and λ_I and λ_{II} are eigen exponents for pure tension and in-plane shearing of the corner respectively. Their values are obtained from the characteristic Eqs (2):

$$\begin{aligned} \lambda_I \sin 2\alpha + \sin 2\lambda_I \alpha = 0 \\ \lambda_{II} \sin 2\alpha - \sin 2\lambda_{II} \alpha = 0 \end{aligned} \tag{2}$$

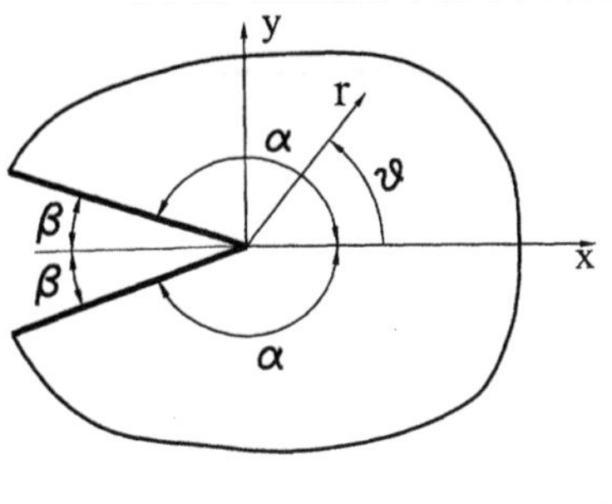

Fig.1 Sharp corner with vertex angle 2α and a polar coordinate system (r,ϑ).

The numerical solutions of the characteristic Eqs.(2) are shown in Fig.2. Generalised stress intensity factors, K_I^λ and K_{II}^λ, that appear in Eqs. (1) for opening and in-plane shearing modes respectively, are defined as follows **(2)**:

$$K_I^2 = \lim_{\vartheta=0,\gamma\to 0}\left[(2\pi r)^{1-\lambda_I}\sigma_{\vartheta\vartheta}(r,\vartheta)\right]$$
$$K_{II}^2 = \lim_{\vartheta=0,\gamma\to 0}\left[(2\pi r)^{1-\lambda_{II}}\tau_{\gamma\vartheta}(r,\vartheta)\right] \quad (3)$$

and present the advantage of being simple stress for $\lambda = 1$. When $\lambda = 0.5$ the classic definition of stress intensity factor for a crack is obtained.

Asymptotic values of the stress field near the corner strongly depend on the notch opening angle 2β, the element shape, the loading mode and the displacement conditions imposed on the body. Therefore approximate values of the generalised stress intensity factors are usually based on FEM results, where appropriate special finite elements fill the core region of the apex. Such numerical procedure has been applied to determine K values for plates with V-notches experimentally tested as described below.

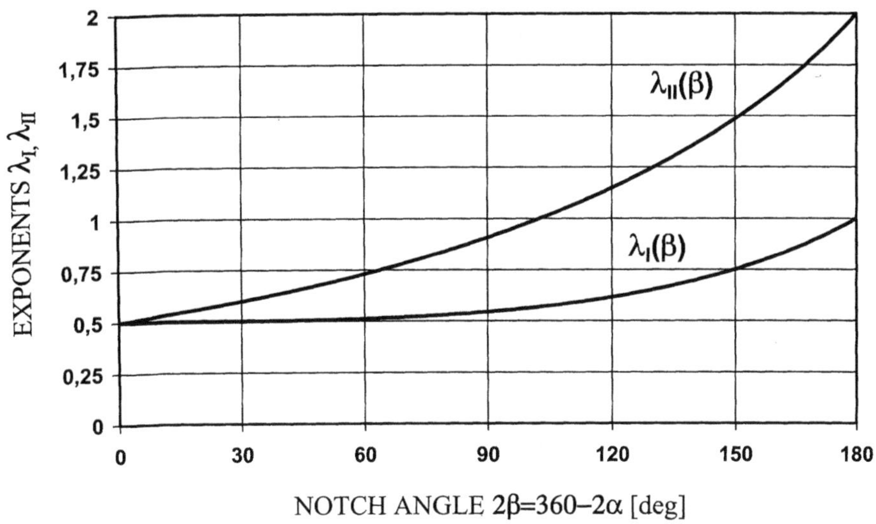

Fig.2 Solution of the characteristic Eqs (2) (first asymptotic term).

EXPERIMENTAL PROCEDURE - MATERIALS AND METHODS

Experimental tests were carried out using 100 mm wide rectangular plates with two opposite 25 mm deep sharp V-notches and with two semi-circular edge notches, as is shown in Fig.3.

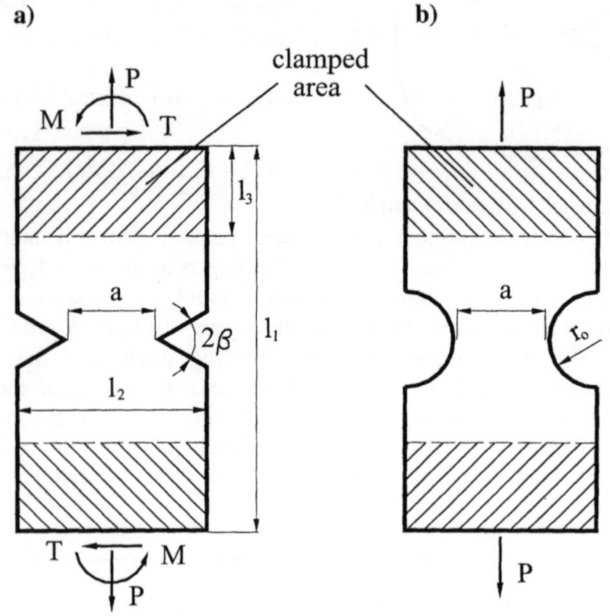

Fig.3 Rectangular plates with a) sharp V-notches and b) semi-circular notches, where l_1=200mm, l_2=100mm, l_3=47mm, a=50mm and r_o=25mm.

An aluminium alloy with ultimate strength equal to 130 MPa and an elongation about 50% has been used as a representative ductile material. Sharp notches with total opening angles 2β = 20 and 80 degrees were cut from 3 mm thick plates.
Brittle materials were represented by 5 mm thick PMMA with opening angles 2β equal to 40 and 80 degrees. All notches were made on a vertical milling machine for a set of PMMA specimens fixed between external plates of aluminium alloy. This prevented PMMA from cracking during the cutting process. Final machining of notches was executed by means of a special cutting tool with the profile corresponding to the notch angle.
Microscopic measurements have shown that the notch root radii were much lower than 0.02 mm, while the opening angles varied about 0.5^o from the theoretical ones.
In order to impose biaxial loading conditions, a special device has been used together with a standard uniaxial fatigue machine. As is shown in Fig.4, such a device consists of two parts (*1* and 2) with two slide-ways (*3*) that provide mutual separation of both parts (*1*) and (2), and can also withstand internal bending reactions appearing in the structure in skew positions of the tested plate. Four washers (*4*) and (*5*) with increased coefficient of friction are placed on both sides of the tested plate (*6*) and fix it firmly with four bolts (*7*) and nuts. In this way both extremes of the plate are always kept always parallel and the external load applied by means of this device generates a biaxial stress state.

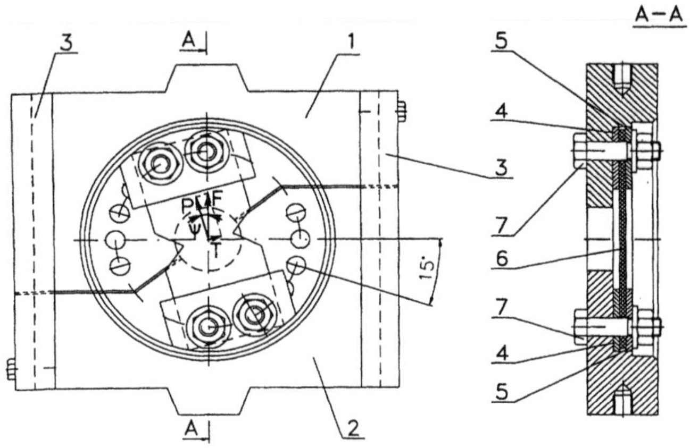

Fig.4 Special device for biaxial loading of rectangular plates with two equal and opposite notches.

Various biaxial stress states along the central cross section of the plate (collinear with the notch plane of symmetry) are represented by different values of the tension/shear ratio, realised by changing the orientation of the plate (*6*) with respect to the device. Rotational angle ψ can be changed by 15°, and is determined by a set of holes made in both parts of the device. For $\psi = 0°$ (vertical position of the plate), the plate is subjected to simple tension, and for $\psi = 90°$, pure shear appears in the middle cross section. The resultant tensile and shearing forces in the element (*6*) can be calculated from Eqs. (4):

$$T = F \sin \Psi, \qquad P = F \cos \Psi \tag{4}$$

where *F* is the load generated by the fatigue machine.
All tests have been carried out on a hydraulic machine with MTS controlling system. A frequency of 3 Hz was chosen for all fatigue tests run with constant displacement (aluminium) and loading (PMMA) amplitudes applied to the device. First of all some experiments under monotonic loading were made, in order to find the critical forces (limiting load) causing fracture of the element for various notch angles and tension/shear ratios. After that, basic parameters of fatigue tests were established.

EXPERIMENTAL RESULTS

Critical load values for monotonic tests of aluminium alloy and PMMA are shown in Fig.5 and Fig.6, respectively.
In the case of the aluminium alloy, the limit load of the plate depends on the shape of the far field plastic zones and their mutual interaction in the central cross section, showing the important role of shearing stresses. The strength varies with the tension/shear ratio, represented by the angle ψ. Maximum strength, obtained for ψ between 30 and 45 degrees, corresponds to the maximal force necessary to develop and join together two plastic zones growing from the opposite notches. This phenomenon has been confirmed by an elastoplastic FEM analysis. Notch angle 2β, equal to 20 and 80 degrees, had no significant influence on the limit load value. However, sharper notches slightly reduced the strength of the plate.

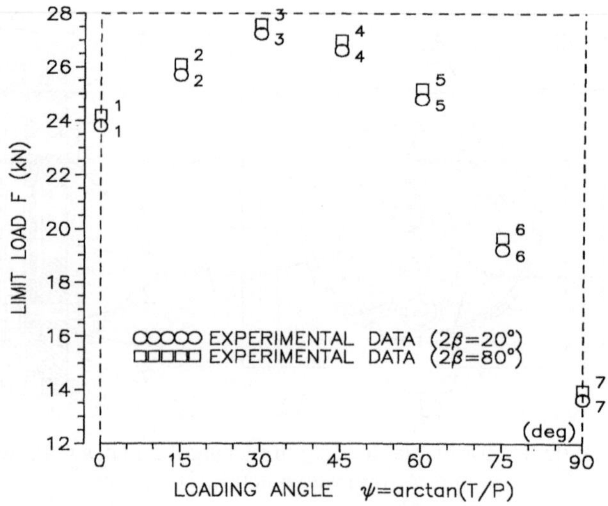

Fig.5 Limit load for monotonic biaxial tests of aluminium alloy.

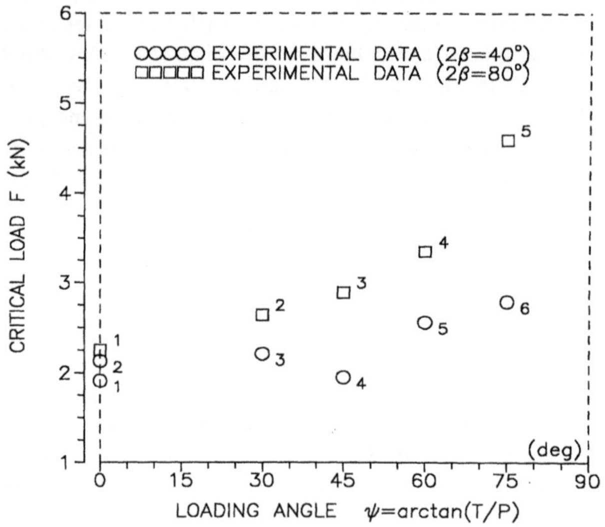

Fig.6 The critical load for monotonic biaxial tests of PMMA.

The critical load for the monotonic test of PMMA was quite different to that for the aluminium alloy. The strength of the element depends on the notch angle 2β and the orientation of the plate with respect to the device. Sharper notches produced lower values of the critical load, which increased with the angle ψ, as is shown in Fig.6.
Fatigue behaviour of specimens made of the aluminium alloy mainly depends on the plastic zone size and its orientation. Fatigue cracks appeared on the specimen surface as two small curvilinear cracks, in both sides of the plastic zone. After a certain number of cycles one crack dominated and developed across the plate. The crack plane was inclined 45 degrees to the lateral surface of the specimen, and 0 degrees in the middle plane, showing 3-dimensional fatigue behaviour of the damage process. Some experimental results for the aluminium alloy are shown in Table 1.

Table 1 Experimental results of fatigue tests for the aluminium alloy.

No.	2β [deg]	ψ [deg]	Δu [mm]	N_f	F_{max} [kN]
1	20	0	0.4	770	12.7
2	20	0	0.5	376	-
3	20	0	0.3	1421	11.4
4	20	0	0.25	8300	7.8
5	20	30	0.5	4200	9.0
6	20	30	0.6	1903	11.9
7	20	30	0.7	1130	15.7
8	20	45	0.8	2100	14.5
9	20	45	1.3	810	-
10	20	60	0.8	4386	8.9
11	20	60	1.0	2800	8.9
12	20	60	1.3	850	13.0
13	20	60	1.6	650	16.0
14	20	75	1,3	820	8.3
15	20	75	1,6	717	9.5
16	20	90	1.6	1500	7.4
17	20	90	1.8	610	8.8

In the case of fatigue tests of PMMA, crack shape and directions were similar to those under monotonic loading. Similar behaviour of PMMA was also reported in **(8)**, where fracture process for cruciform specimens with an artificial crack under biaxial loading was investigated.

Some experimental results related to the present study are presented in Table 2 and Fig.7.

Table 2 Experimental results of fatigue tests for PMMA.

No.	2β [deg]	ψ [deg]	ΔF [kN]	N_f	ϑ_0 [deg]
1	40	0	1,95	1	-2
2	40	30	2,20	1	-25
3	40	60	2,56	1	-41
4	40	75	2,78	1	-47
5	40	30	1,80	4	-23
6	40	75	2,20	15	-42
7	40	30	1,70	26	-24
8	40	60	2,00	59	-42
9	40	75	2,00	122	-44
10	40	60	1,80	169	-40
11	40	0	1,60	310	-1
12	40	30	1,50	395	-29
13	40	0	1,40	1832	-4
14	80	0	2,25	1	-0
15	80	30	2,64	1	-21
16	80	60	3,35	1	-33
17	80	75	5,03	1	-42
18	80	75	4,02	3	-50
19	80	0	1,90	19	-5
20	80	30	2,10	48	-24
21	80	60	2,50	53	-40
22	80	75	2,75	126	-44
23	80	30	1,80	528	-24
24	80	0	1,70	579	-3
25	180	0	11,50	1	0
26	180	0	8,00	266	2
27	180	0	6,50	2226	0

EXPERIMENTAL VERIFICATION OF A NON-LOCAL STRESS CRITERION OF FATIGUE CRACK INITIATION FOR PMMA

In order to estimate the fatigue initiation period for cracks emanating from V-notches, a non-local criterion, published by Seweryn and Mróz in **(6, 7)**, has been applied. This fracture criterion, in the case of brittle materials, is based on the assumption that brittle failure occurs at the notch tip when the mean value of the normal stress σ_n distributed over a characteristic damage zone d_0 on a physical plane defined by angle ϑ, reaches its critical value **(3)**:

$$R_f = \max_{(\vartheta)} \overline{R}_\sigma(\sigma_n/\sigma_c) = \max_{(\vartheta)} \left[\frac{1}{d_0} \int_0^{d_0} \frac{\sigma_n}{\sigma_c} dr \right] = 1 \tag{5}$$

where R_f is the failure factor and $\overline{R}_\sigma$ is the non-local failure function represented by the normal stress condition. For multiaxial loading conditions, the orientation of the critical plane is not known in advance, so first we need to search for the maximum value of the non-local failure function $\overline{R}_\sigma$.

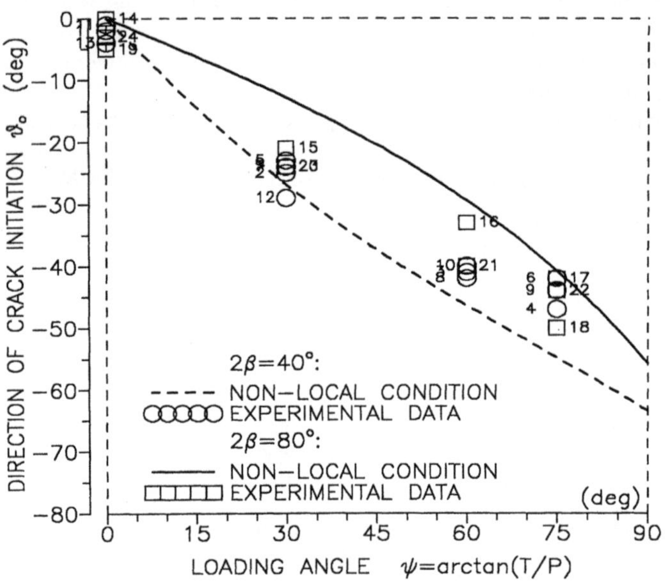

Fig.7 Direction of fatigue crack initiation vs. loading angle (PMMA).

The damage zone length can be found from Eq (6) **(4)**,

$$d_0 = \frac{2}{\pi}\left(\frac{K_{Ic}}{\sigma_c}\right)^2 \tag{6}$$

where K_{Ic} represents the critical value of the stress intensity factor.

In a similar way the fatigue damage initiation condition can be defined. The fatigue damage accumulation occurs at the notch tip, when the averaged value of the normal stress σ_n distributed over the damage zone d_0 exceeds the threshold value *f*:

$$R_f = \max_{(\vartheta)} \overline{R}_\sigma(\sigma_n/\sigma_c) = \max_{(\vartheta)}\left[\frac{1}{d_0}\int_0^{d_0}\frac{\sigma_n}{\sigma_c}dr\right] = f \qquad (7)$$

where $f=\sigma_o/\sigma_c$ and σ_o is the permanent fatigue strength. In such a case the maximum value of the non-local failure function $\overline{R}_\sigma$ should lie in the following range: $f<\overline{R}_\sigma(\sigma_n/\sigma_c)<1$. If we apply the incremental accumulation damage measure $d\omega_n$, it should depend on the positive incremental values of the non-local failure function $d\overline{R}_\sigma$, given in the following form:

$$d\omega_n(\vartheta) = A_1\left(\frac{\overline{R}_\sigma - f}{1-f}\right)^{n_\sigma}\frac{d\overline{R}_\sigma}{1-f} \qquad (8)$$

where A_1 and n_σ denote material constants to be determined experimentally.
In this approach the fatigue crack initiates when the accumulation damage measure ω_n on any of the physical planes reaches the critical value:

$$R_d = \max_{(\vartheta)} \omega_n = 1 \qquad (9)$$

In the case of plates with V-notches subjected to proportional loading starting from zero, the condition of fatigue crack initiation has the following form:

$$R_d = N_f \max_{(\vartheta)}\left[\left(\frac{\Delta\overline{\sigma}_\vartheta(\vartheta)-\sigma_o}{\sigma_c-\sigma_o}\right)^{n_\sigma+1}\right] = 1, \qquad (10)$$

where r and ϑ are polar coordinates originated at the notch tip.
The range of averaged stresses $\Delta\overline{\sigma}_\vartheta$ along the characteristic damage zone d_0, is then given by Eq. (11)

$$\Delta\overline{\sigma}_\vartheta = \frac{1}{d_o}\int_0^{d_o}\Delta\sigma_\vartheta(\vartheta)dr. \qquad (11)$$

By using Eqs (1) - (3), it may be also represented by Eq (12):

$$\Delta\overline{\sigma}_\vartheta = \frac{\Delta K_I^\lambda f_I(\vartheta)}{\lambda_I(2\pi d_o)^{1-\lambda_I}} + \frac{\Delta K_{II}^\lambda f_{II}(\vartheta)}{\lambda_{II}(2\pi d_o)^{1-\lambda_{II}}}, \qquad (12)$$

Relations between generalised stress intensity factors K_I^λ and K_{II}^λ and loading force F, are given by Eq (13):

$$K_I^\lambda = \zeta_I(\beta)F\cos\psi, \qquad K_{II}^\lambda = \zeta_{II}(\beta)F\sin\psi, \qquad (13)$$

Combining Eqs (12) and (13), the range of averaged stresses $\Delta\overline{\sigma}_\vartheta$ has the following form:

$$\Delta\overline{\sigma}_\vartheta = \Delta F\left(\frac{\xi_I(\beta)f_I(\vartheta)}{\lambda_I(2\pi d_o)^{1-\lambda_I}}\cos\psi + \frac{\xi_{II}(\beta)f_{II}(\vartheta)}{\lambda_{II}(2\pi d_o)^{1-\lambda_{II}}}\sin\psi\right), \qquad (14)$$

In the presented case, normalised parameters ζ_I and ζ_{II} relating generalised stress intensity factors with external load F, have been obtained by means of FEM tautology using special asymptotic finite elements around the notch tip. Their numerical values are shown in Table 3.

Table 3 FEM results of generalised stress intensity factors for the tested specimens.

2β [deg]	λ_I	λ_{II}	$\zeta_I = K_I^{\lambda} / P$ [MPa $m^{1-\lambda_I}$ / kN]	$\zeta_{II} = K_{II}^{\lambda} / T$ [MPa $m^{1-\lambda_{II}}$ / kN]
20	0,5004	0,5620	0,6709	0,7103
40	0,5035	0,6382	0,6865	1,0416
60	0,4122	0,7309	0,7276	1,5994
80	0,5304	0,8434	0,8007	2,2531
180	0,0000	-	7,0628	-

By introducing an equivalent normal stress range $\Delta\sigma_{az}$, defined by Eq (15)

$$\Delta\sigma_{az} = \max_{(\vartheta)} \Delta\bar{\sigma}_{\vartheta}(\vartheta), \tag{15}$$

into the above formulas, fatigue initiation period N_f is obtained:

$$\log_{10} N_f = -(n_\sigma + 1)\log_{10}\left(\frac{\Delta\sigma_{az} - \sigma_o}{\sigma_c - \sigma_o}\right). \tag{16}$$

Making use of the above procedure, the non-local criterion expressed by Eqs (5) - (9) makes it possible to transform any multiaxial and singular stress problem into a simple uniaxial one.

Theoretical results of fatigue initiation cycle number N_f, obtained from the above model, are compared to the experimental data for PMMA in Fig.8.

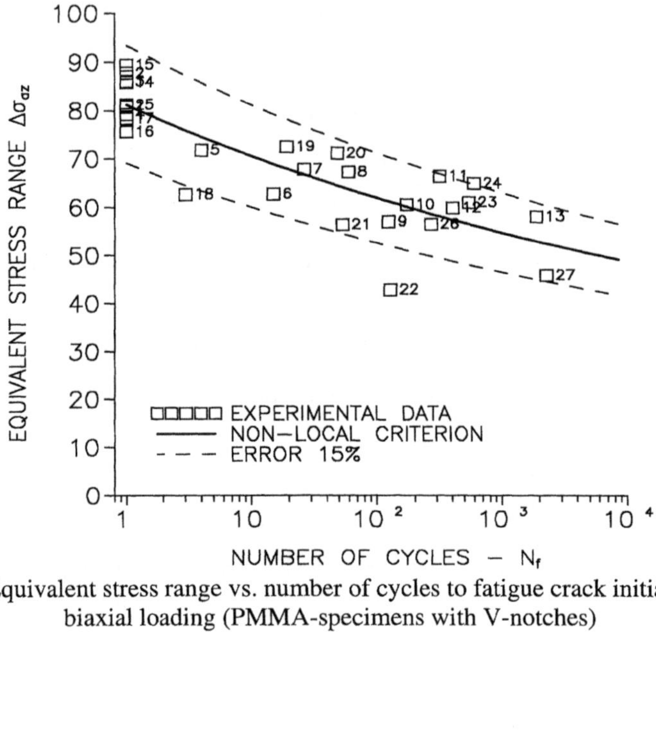

Fig.8. Equivalent stress range vs. number of cycles to fatigue crack initiation for biaxial loading (PMMA-specimens with V-notches)

The critical stress σ_c = 81.2 MPa, derived from monotonic tests for specimens with two semicircular notches, and the characteristic length of the damage zone d_0 = 0.164 mm, calculated from Eq. (6), served as basic input values for the model. The fatigue limit σ_o and the exponent value n_σ have been determined by minimising the standard deviation of experimental results. A non-linear system of equations, solved by means of a gradient method, gave the following values: n_σ=10.7 and σ_o=21.6 MPa.

Both dashed lines shown in Fig.8 represent ±15% of stress deviation. Only two experimental points (for 2β = 80 degrees and ψ = 75 degrees) are placed outside of this range. This is probably due to the small singularity effect for in-plane shearing, compared with the higher order terms of the stress field near the corner. The agreement between the theoretically estimated fatigue initiation period and the experimental results for PMMA with sharp notches may be considered as quite satisfactory, since the discrepancy of different tests on monotonically loaded specimens made of PMMA is about 12% (see Williams (**9**)).

CONCLUSIONS

Experimental tests carried out using rectangular elements with opposite V-notches have indicated that material damage depends on many factors, such as: notch geometry, loading mode (tension/shear ratio), monotonic or variable loading and properties of the material (ductile or brittle).

In the case of an aluminium alloy subjected to monotonic load (representative for ductile materials) the strength of the element depends on the shape of the far field plastic zones expanding from the opposite notches and joining together. Thus, notch angles 2β equal to 20 and 80 degrees had no significant influence on the limit load value. This phenomenon has been confirmed qualitatively and quantitatively by an elastic-plastic FEM analysis.

The results of fatigue tests for the same material have shown the importance of the plastic zone sizes on the fracture process, which inside the material was initially different from that on the surfaces of the plate. Local plasticity and corresponding hardening of the material have made the early stage of crack propagation more complicated for analysis, due to the 3-dimentional nature of the damage process. Under pure shear, only one fatigue crack was initiated very easily and propagated from the notch apex through the centre of the plastic zone lying in the central plane of the element.

In the damage process of the brittle material, represented by PMMA, the elastic stress and strain fields appeared to be the most important factor. In both monotonic and fatigue tests, the crack paths were almost identical and depended on the notch opening angle 2β and the ratios between tension and shear, showing the most important role of normal stresses in the fracture process. Thus, the critical load depended also on these parameters.

The non-local stress damage accumulation criterion by Seweryn and Mróz, which has appeared useful in predicting critical monotonic loads (**3**), also remains in satisfactory agreement with experimental results of fatigue tests for PMMA and may be applied in estimating the fatigue crack initiation period as well as initial fracture angle.

Fatigue tests of brittle materials are generally difficult to carry out due to the very narrow range between the threshold and critical values of the stress intensity factor.

REFERENCES

(1) Williams M.L. (1952), Stress singularities resulting from various boundary conditions in angular corners of plates in extension, *Trans. ASME, J. Appl. Mech.*, Vol. 19, pp. 526-528.

(2) Seweryn A. and Molski K. (1996), Elastic stress singularities and corresponding generalised stress intensity factors for angular corners under various boundary conditions, *Eng. Fract. Mech.*, Vol. 55, pp. 529-556.

(3) Seweryn A., Poskrobko S. and Mróz Z. (1997), Brittle fracture in plane elements with sharp notches under mixed-mode loading, *J. Eng. Mech. ASCE*, Vol. 123, pp. 535-543.

(4) Seweryn A. (1994), Brittle fracture criterion for structures with sharp notches, *Eng. Fract. Mech.*, Vol. 47, pp. 673-681.

(5) Seweryn A. and Mróz Z. (1995), A non-local stress failure condition for structural elements under multiaxial loading, *Eng. Fract. Mech.*, Vol. 51, pp. 955-973.

(6) Seweryn A. and Mróz Z. (1996), A non-local stress failure and fatigue damage accumulation condition, In. *Multiaxial Fatigue and Design* (eds Pineau A., Cailletaud G., Lindley T.C.), Mech. Eng. Publ., London, pp.259-280.

(7) Seweryn A. and Mróz Z. (1998), On the criterion of damage evolution for variable multiaxial stress state, *Int. J. Solids Struct.*, Vol. 35, pp. 1599-1616.

(8) Molski K. and Będkowski W. (1997), Investigation of the crack growth on the cruciform specimens with sharp notches under biaxial loading, *Proc. 5th Int. Conf. on Biaxial/Multiaxial Fatigue & Fracture*, Eds E.Macha and Z.Mroz, TU Opole, Poland, vol.I, pp. 481-490.

(9) Williams J.G. (1984), *Fracture Mechanics of Polymers*, Ellis Horwood, Chichester.

Acknowledgements

The investigation described in this paper is a part of research project No. 7 T07C 006 12 sponsored by the Polish State Committee for Scientific Research.

ASSESSMENT OF THE CYCLIC LIFE OF AN ELEMENT WITH A CONCENTRATOR AND RESIDUAL STRESSES TAKING INTO ACCOUNT THE LOCAL COMPLEX STRESS STATE OF THE MATERIAL

Valery T. TROSHCHENKO, Georgy V. TSYBAN'OV and Alexander V. STEPURA

Institute for Problems of Strength, National Academy of Sciences of Ukraine

ABSTRACT

The known criteria for the material ultimate state under a complex stress state have been verified using the results of an experimental investigation into fatigue of steel specimens with a stress concentrator in the shape of a central hole and with residual stresses therein. The computation of the stress-strain state in the concentrator has been done and the applicability of the modified criterion of shear octahedral stresses to the assessment of the cyclic life of an element under test has been shown. The modification of the criterion lies in the proposed method for taking into account the residual stresses as a non-uniaxial cycle asymmetry. Based on the criterion applied and taking into account the dept of the material damaged layer, a procedure has been proposed for computing the cyclic lives of the tested elements

KEY WORDS

Complex stress state, concentrator, residual stress, cyclic life, ultimate state criteria

NOTATION

σ_{oX}, σ_{oY}; σ_{oZ};	components of residual stresses along the X, Y and Z axes, respectively
α	theoretical stress intensity factor
σ_n	nominal stress in the specimen with a concentrator calculated by the net section
σ_{aX}, σ_{aY}, σ_{aZ}	components of the stress amplitude in the X, Y and Z - directions, respectively, under symmetrical loading
X_{max}	distance along the X -axis from the concentrator surface to the point on the stress intensity amplitude diagram where the stress intensity amplitude reaches its maximum
X_{eq}	distance along the X -axis from the concentrator surface to the point on the diagram of stress intensity amplitudes where this amplitude is equal to that for a smooth specimen for lives which are the same as in a notched specimen
N	cyclic life of the specimen
τ_{oct}	shear octahedral stress

P_{max}	maximum normal octahedral stress
A , B	constants
τ_a , σ_a	amplitudes of cyclic stresses in tests for symmetrical torsion and tension-compression, respectively
σ_{oj}	at $j = X, Y, Z$, it corresponds to the residual stress components along their respective axes σ_{oX}, σ_{oY}, σ_{oZ}
σ_{aj}	at $j = X, Y, Z$, it corresponds to the stress amplitude components along their respective axes σ_{aX}, σ_{aY}, σ_{aZ}
σ_{ajr}	at $j = X, Y, Z$, it corresponds to the stress amplitude components along the corresponding axes σ_{aXr}, σ_{aYr}, σ_{aZr} for the asymmetrical cycle of loading with the σ_{oj} asymmetry which are equivalent in the life attained to a fully-reversed cycle with the amplitude σ_{aj}
ψ	coefficient of sensitivity to the cycle asymmetry
$\tau_{oct}(r)$	shear octahedral stress amplitude calculated for the asymmetrical cycle of loading taking into account the coefficient of sensitivity of the cycle asymmetry
$P_{max}(r)$	maximum normal octahedral stress for the loading cycle asymmetry taking into account the coefficient of sensitivity to the cycle asymmetry.

INTRODUCTION

Most fatigue failures of machine parts and structural elements are associated with the initiation and subsequent growth of cracks from the zones of local high stresses which can be caused by various defects, structural stress concentrators, sites of fretting corrosion, manufacturing tensile residual stresses, etc. In the above zones, the material is under conditions of nonuniform non-uniaxial stress state, which requires the assessment of its lifetime, based on the established ultimate state criteria. These criteria are most frequently established and verified given the uniform field of stresses and macroscopic volumes of the material in this field, and in this connection their formal application in the case of a local nonuniform complex stress state is not well substantiated, although in many practical problems it is this question that needs to be settled. A structural stress concentrator with residual stresses induced in the material by machining is one of the most typical cases for structural elements. The loading of such elements results in a complex stress state which, being established in the concentrator, interacts with the volumetric field of residual stresses relaxing in the process of cyclic loading. The complexity of resolving this problem on the prediction of the cyclic life has led, in this case, to the appearance of a great number of works in this line of investigation **(1 - 4)**. However, the question has not yet been settled, which is effects the many-sided nature of the problem to be resolved.

In the present work an attempt has been undertaken by the authors to develop the calculation-and-experimental procedure for assessing the cyclic life of an element, with a stress concentrator and residual stresses based on the material complex stress state, which is taken into account by means of calculations and the experimental justification of the applicability of the material ultimate state criteria.

EXPERIMENTAL PROCEDURE

A low-alloy type 10GN2MFA steel was employed in the tests. It is used for manufacturing steam generator collectors during the operation of which both concentrators in the form of holes and residual stresses induced by pressing-in of pipelines are present therein. The steel under study has the following mechanical characteristics under static loading: $\sigma_u = 620$ MPa, $\sigma_{0.2} = 540$ MPa.

Test specimens were prepared in the form of a plate of rectangular section with dimensions of 14×2.5 mm. A central circular hole 3 mm in diameter was made in the specimens. The theoretical stress concentration factor, α, was equal to 2.6. Cyclic loading at a frequency of 36 Hz was realised under fully-reversed tension-compression.

The application of various methods of machining result in inducing non-uniaxial residual stresses in an element subjected to machining. In this case, the ratios of residual stress components along different axes can be different. Therefore, when preparing the test specimens with a stress concentrator, the residual stresses were induced in the concentrators by various methods providing different ratios of residual stress components along the axes. The methods for inducing residual stress are presented schematically in Fig.1. The residual stresses were induced immediately prior to fatigue order to reduce their possible decrease with time as a result of the relaxation processes.

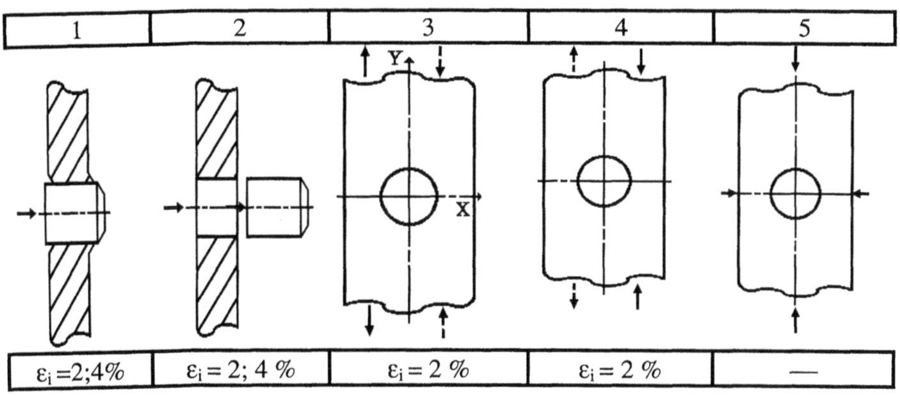

Fig.1 Methods for inducing the residual stresses and generating the static ones.

Referring to Fig.1, the residual stresses were generated using the following operations:

- scheme 1 – pressing in pins; in this case a hardened pin with a negative allowance of 2% is pressed into the concentrator hole;
- scheme 2 – pulling a pin through with a negative allowance of 2% and 4%; residual stresses are induced by pulling the same pin as in scheme 1 through the hole with a certain negative allowance as shown in Fig.1;
- scheme 3: preliminary compression of the specimen with a concentrator along the axis of the subsequent application of the cyclic load; the maximum strain in the concentrator is 2%;
- scheme 4: preliminary tension of the specimen with a concentrator; the maximum strain in the concentrator is 2%;

– scheme 5 in Fig.1 corresponds to the biaxial compression of the specimen by static loads to induce a static compressive stress in the minimal cross section of the specimen with a concentrator in the process of fatigue testing. This mode of loading was realised using a testing machine (along the axis of the cyclic loading) and a clamping device specially designed for this purpose (perpendicular to the axis of cyclic loading, i.e., the *X*-axis).

RESULTS OF THE RESIDUAL STRESS CALCULATION

The diagram of the residual stress distribution diagram was obtained by computation based on the strain model whose equations were solved by the finite element method **(5, 6)**. The software applied enables one to make an elastoplastic calculation of the material stress-strain state in the stress concentrator. The initial data for the calculation were the geometrical characteristics of the specimen, the material static stress-strain diagram and the specimen loading parameters.
Figure 2 presents the results of the calculation of residual stresses in the stress concentrator induced according to the schemes under study.

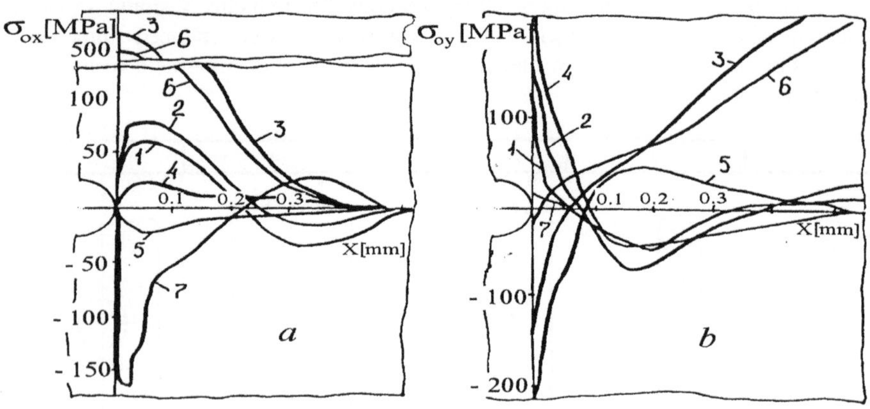

Fig.2. Distribution of residual stress components through the thickness of the material in the concentrator: 1 - pulling a pin through the hole with an interference of 2%; 2 - pulling a pin through the hole with an interference of 4%; 3 - pressing in a pin with an interference of 2% (α=2.6); 4 - prestressing by compression; 5 - prestressing by tension; 6 - pressing in a pin (α=2); 7 - biaxial compression

It follows from the above diagrams that with scheme 2 being realised; the value of the residual stress component near the concentrator surface σ_{oX} is comparable with the value of σ_{oY}. In the case when the residual stresses are generated according to schemes 3 and 4, the stress components σ_{oY} far exceed the σ_{oX} components. When the concentrator is loaded according to scheme 1, the residual stresses along the X-axis reach in fact the yield strength (σ_{oX} = 520 MPa) on the concentrator surface and thus are nearly 4 times higher than those along the y-axis (σ_{oY} = 132 MPa). If, with the same scheme of generating the residual stresses, the theoretical stress concentration factor is decreased to α = 2 by increasing the hole diameter, then, almost for the same σ_{oX}-value, the longitudinal component σ_{oY} on the concentrator surface is virtually absent (curve 6 in Fig.2). Almost the same ratio of static stresses on the concentrator surface can also be obtained for the hole with $\alpha = 2.6$ with the use of the loading scheme where, in the process of cyclic loading, the specimen is additionally subjected to biaxial loading along the longitudinal and transverse axes. In doing so, the calculated variation in the forces applied has been made so as to induce the static stress on the concentrator surface in the direction of the x-axis with the subsequent investigation of the influence of this static component directed perpendicular to the axis of the varying loading on the cyclic life of specimens with a concentrator.

From the results given it follows that by changing the method of generating the residual stresses and applying biaxial static load, one can obtain various cases of residual stress combinations (or static ones under loading according to scheme 5) along the X and Y axes including the limiting cases where the stress component along one of the axes exceeds the component along the other axis to an extent that the latter can be neglected. The residual stress diagrams given in Fig.2 are the initial induced prior to fatigue testing.

Therefore, later on, in the development of the model of calculation for assessing the cyclic life of an element under test, one should be familiar with the method of determining the level of residual stresses taking into account their relaxation under cyclic loading. For this purpose, the literature data on the direct measurement of the residual stress kinetics resulting from the relaxation under the action of a cyclic load in steels of similar grade have been analysed **(7)**. It follows from this analysis that, in the case of high-cycle fatigue the stage of stabilisation of residual stresses is no more than 10% of the life prior to crack initiation. For this reason, in the subsequent assessment of the cyclic life for specimens with a concentrator and residual stresses, the fatigue damage to the material at the stage where the residual stresses reach the level of stabilisation was not taken into account. As a result of the analysis of the data on the relaxation of residual stresses under high-cycle loading of steels, the relationship of a closed form has been obtained **(7)**. It enables one to determine the level of the stabilised residual stress along each axis. The initial data for computation are the material static and cyclic stress-strain diagrams geometrical characteristics of the specimen and the calculation of the distribution of residual and cyclic stress components through the material depth based on these data.

RESULTS OF FATIGUE TESTS AND THEIR ANALYSIS

Fatigue tests were carried out on the specimens prepared according to the schemes given in Fig.1. Fatigue curves were constructed in nominal stresses as a function of lifetime prior to initiation of a 0.1 mm long crack.

Figure 3 illustrates the fatigue curves for smooth specimens (curve 1), specimens with a concentrator without residual stresses (curve 2), specimens with stress concentrators (curves 3 to 7) and specimens with a concentrator with the biaxial static compression being superimposed (curve 8).

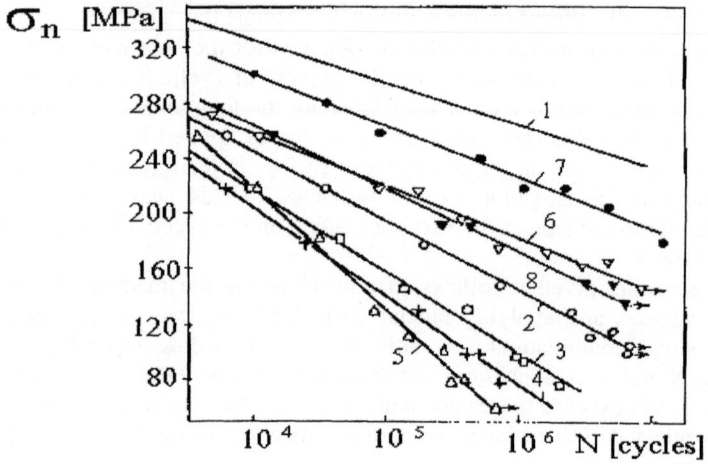

Fig.3. Fatigue curves for the investigated specimens: 1 - smooth specimens; 2 - specimens with a concentrator without residual stresses; 3,4 - pulling a pin through with an interference of 2% and 4%, respectively; 5 - after prestressing in compression; 6 - after prestressing by tension; 7 – pressing in a pin; 8 - in biaxial static compression.

Based on the given experimental data, we consider the possible methods for going from fatigue curves for smooth specimens to those for specimens with concentrators and relaxed residual stresses. In this case the nonuniformity of stress distribution through the material depth should be taken into account. This can be done in a variety of ways, namely, by the recalculation of stress-strain diagrams using the coefficients which take into account the stress gradient **(8)**; by taking into account the depth of action of the stresses which exceed half the fatigue limit with the use of a relative gradient of the first principal stress **(9)**; by establishing the critical depth of the surface layer at which the fatigue failure occurs and which is constant for a given material **(10, 11)**. Since the distributions of stresses through the depth of the material in the concentrator under conditions of elastoplastic deformation are nonuniform, the application of the first two methods is troublesome. For this reason, the possible application of the third method of those listed above has been considered.

Based on the calculation procedure applied for determining residual stress distributions, an elastoplastic analysis has been made of the stress-strain state in a stress concentrator under cyclic loading. In so doing, the cyclic stress-strain diagrams for the steel under study were used (8). Calculations were made for several levels of nominal stresses. Figure 4 presents the distribution of stress intensity amplitudes along the X-axis. The point on the specimen median plane where the stress in the concentrator is maximum for the elastic solution of the problem is taken as the X-axis origin.

Curves in Fig.4 are constructed on the basis of the calculated components of the stress amplitudes σ_{aX}, σ_{aY} and σ_{aZ} along the X, Y and Z-axes, respectively. For all the nominal stresses the maximum stresses are observed along the Y-axis. Here, with distance from the concentrator surface, the values of σ_{aY} change non-monotonically and have a maximum at a distance of 10-20 μm from the concentrator surface. The stress σ_{aZ} are lower by a factor of 3 or 4 than the value of σ_{aY} for the corresponding nominal stress, and decrease with distance from the concentrator surface; stress σ_{aX} is rather small and increases from zero on the concentrator surface to a quantity which is several times lower than the σ_{aX} values on the concentrator surface. In view of the fact that the stress in the concentrator along the Y-axis is significantly higher than those along the two other orthogonal axes, the distribution of stress intensity amplitudes through the material depth given in Fig.4 is representative of the varying nature of stress σ_{aY}.

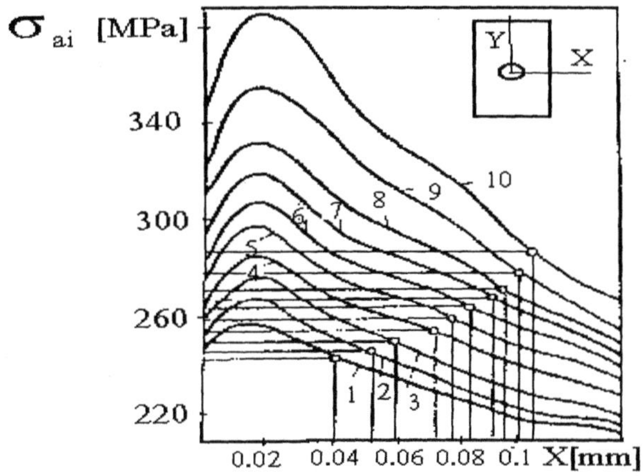

Fig.4. Distribution of stress intensity amplitudes through the depth of the material in the concentrator: σ_n=106 (1), 114 (2), 121 (3), 130 (4), 137 (5), 146 (6), 153 (7), 166 (8), 177 (9), 180 (10) MPa.

The diagrams of stress intensity amplitudes (Fig.4) also include points plotted according to the stress intensity amplitudes in smooth specimens in tension-compression for the same lives as those for the specimens with a stress concentrator. As is evident from those results, the stress amplitude intensity both on the concentrator surface and down to a certain depth exceeds that in a smooth specimen for equal lives. We designated this depth by X_{eq} and note that on the concentrator surface ($X = 0$) and at the maximum point of the stress intensity amplitude ($X = X_{max}$) the values of σ_{ai} characterising the stress-strain state in the stress concentrator correspond to much shorter lives on the fatigue curve for smooth specimens, as compared to point X_{eq}. The difference in stresses at points $X = 0$ and $X = X_{eq}$ increases with the nominal stress applied to a notched specimen.

Considering the analysis made on the stresses in the concentrator, the following may be suggested: when a notched specimen is cyclically loaded to the number of cycles which corresponds to the life of a smooth specimen for the stress intensity amplitude equal to its value at the point X_{eq}, the material microdiscontinuities have already been formed in the stress concentrator down to the depth X_{eq} and there is a nucleated crack which is equivalent to that in the smooth specimen. As is evident from the data in Fig.4, the values of X_{eq} increase with the nominal stresses applied to the specimen with a concentrator, which somewhat differs from the assumption made in Refs.

(**10,11**) about the constancy of the critical depth of the material surface layer at which the stress equivalence for a smooth and notched specimen is observed. The analysis of similar results made for other materials allows one to say that for low characteristics of cyclic plasticity, the X_{eq}-value is close to a constant value, which is in agreement with Refs. (**10, 11**). It has been found that in the region of the high-cycle fatigue lives, the X_{eq}-value varies within 40 to 100 μm for aluminium alloy AMg6, 60 to 200 μm for type 45 steel, and 45 to 140 μm for type 1Kh2M steel.

DEVELOPMENT OF THE METHOD, WHICH TAKES INTO ACCOUNT THE NON-UNIAXIALITY OF CYCLIC AND RESIDUAL STRESSES IN A CONCENTRATOR.

Using the proposed relationship (7), the components of stabilised residual stresses were determined for specimens with a concentrator and with residual stresses taking into account the components of the cyclic stresses, which are acting on the corresponding axes. The residual stresses obtained are thereafter considered to be applied statically. They were determined for the following material depths in the concentrator: $X = 0$, $X = X_{max}$ and $X = X_{eq}$ for the life $N = 2 \cdot 10^5$ cycles. Other lives are not considered in this case in view of the similarity of the results and regularities established at the chosen life level.

In the realisation of various schemes for inducing residual stresses, measures were taken to provide their combination with both different ratios along the two principal axes and a prevailing value along either individual axis. In view of this, the application of the approaches wherein the first principal stress is used for assessing the cyclic lifetime seems to be impossible. Therefore, a search for the equivalent stresses for asymmetrical cycles presenting a superposition of cyclic non-uniaxial stress in the concentrator and non-uniaxial stabilised residual stresses will be made with the use of the stress intensity, as in the case of determining the X_{eq}. In so doing, we take by convention that the intensity of residual compressive stresses is a negative value which is necessary for construction of the diagram of limiting amplitudes in terms of stress intensities referred to as the conventional diagram of limiting amplitudes. Such diagrams presented in Fig.5 were constructed for the stresses acting at the depths $X = 0$, $X = X_{max}$ and $X = X_{eq}$.

To construct them, the data on both the specimens with a concentrator and with residual stresses and smooth specimens (solid symbols) and specimens with a concentrator without residual stresses (open symbols) were used. From the given data the following conclusions can be drawn:

- the best description of the results and their approximation to the result obtained from the smooth specimen is observed when the stresses at the depth $X = X_{eq}$ are used for describing the stress state;
- the scatter of the results with respect to the approximating straight line even for $X = X_{eq}$ is sufficiently high for the agreed description of the stress state to be used for assessing the cyclic life of specimens with a concentrator and residual stresses; particularly poor is the agreement for the results which are indicative of the fatigue with residual stresses of different signs along the axes, e.g. a semi-open symbol for the case of the concentrator with pulling a pin through.

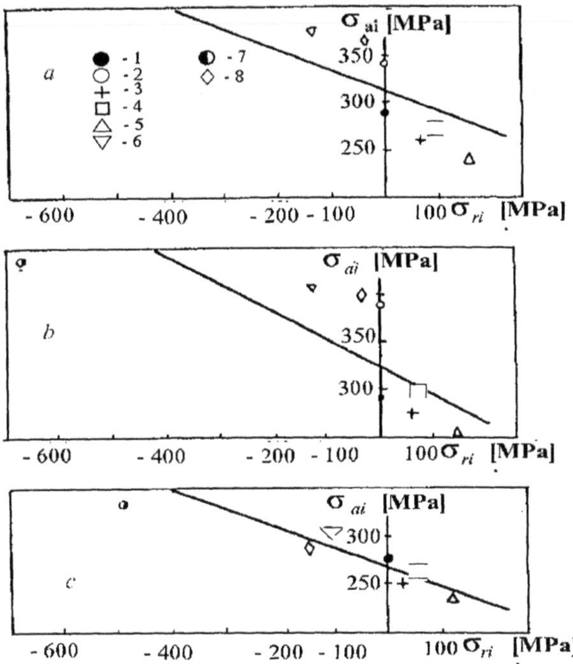

Fig.5 Conventional diagrams of limiting amplitudes in terms of stress intensities for different depths of the material: (*a*) $X = 0$; (*b*) $X = X_{\max}$; (*c*) $X = X_{eq}$; 1 - without residual stresses; 2 - pressing in a pin; 3,4 - pulling a pin through with an interference of 2 and 4%; 5,6 - prestressing by compression and by tension; 7 - biaxial compression; 8 - smooth specimen.

Taking into account these conclusions and also the previously discussed fact of the appearance of microdiscontinuities in the material in the vertex of the concentrator from its surface down to the depth $X = X_{eq}$, the alternatives considered have been to use the material ultimate state criteria wherein the maximum or octahedral shear stresses depend on the stress which is normal to the site with a maximum shear stress or on the hydrostatic stress in a cycle of loading, respectively **(12)**. The calculated and experimental data obtained are in the best agreement with the criterion of octahedral stresses in the form **(13)**:

$$\tau_{oct} + A \cdot P_{max} = B \tag{1}$$

where A, B are constants, P_{max} is the maximum hydrostatic stress in a cycle, τ_{oct} is the shear octahedral stress determined by the following relationships:

$$P_{max} = \frac{1}{3}(\sigma_{oX} + \sigma_{oY} + \sigma_{oZ} + \sigma_{aX} + \sigma_{aY} + \sigma_{aZ}); \tag{2}$$

$$\tau_{oct} = \frac{1}{3}\sqrt{\left(\sigma_{aX} - \sigma_{aY}{}^{2}\right) + \left(\sigma_{aY} - \sigma_{aZ}\right)^{2} + \left(\sigma_{aZ} - \sigma_{aX}\right)^{2}}. \tag{3}$$

Figure 6a illustrates the test results for specimens with a concentrator and residual stresses which are calculated according to criterion (1) for the life of $2 \cdot 10^5$ cycles with the computation of the magnitudes for the material depth in the concentrator $X = X_{eq}$ which are included in this criterion. In this figure the test results for smooth specimens in symmetrical, tension-compression and in torsion were taken as reference points for the construction of criterion curves I (for $N = 2 \cdot 10^5$ cycles) and II (for $N = 10^6$ cycles) which were defined by the following coordinates:

in tension-compression $$\tau_{oct} = \frac{\sqrt{2}}{3}\sigma_a;\ P_{max} = \frac{\sigma_a}{3}; \tag{4}$$

in cyclic torsion $$\tau_{oct} = \sqrt{\frac{2}{3}}\tau_a;\ P_{max} = 0; \tag{5}$$

As is seen from the given data, in this case the experimental points also have deviations from the criterion line plotted according to the cyclic tension-compression and torsion test results for smooth specimens. Analysing those deviations and the nature of residual stresses, the authors have noticed the following regularity: the points corresponding to the test results for specimens with a concentrator and residual (or static components) compressive stresses therein are below the criterion line and, consequently, the assessment of their cyclic lives according to criterion (1) is conservative, whereas the points corresponding to the test results for specimens with a concentrator and tensile residual stresses are above the criterion line, i.e. criterion (1) overestimates their lives.

The indicated regularity can result from the fact that the criterion employed like criteria of similar kind (**12**) does not take into account the material sensitivity to the cycle asymmetry which manifests itself in the change in the amplitude of a varying stress depending on the value and sign of the mean stress: tensile static stresses reduce the amplitude values of stresses whereas the compressive ones increase them for the same cyclic life.

In view of the aforesaid, each stress intensity amplitude in relationships (2), (3) should be changed by taking into account the material sensitivity to the cycle asymmetry rather than by a simple summation of stresses as it was done in relationship (2), or by a total neglect of the asymmetry in the computation of τ_{oct} according to relationship (3). Therefore, the stress components in Eqs. (2) and (3) should be used taking into account the coefficient of sensitivity to the cycle asymmetry ψ in the form

$$\sigma_{ajr} = \sigma_{aj} - \Psi\sigma_{oj} \tag{6}$$

where σ_{ajr} is the stress amplitude of the asymmetrical cycle with the asymmetry σ_{oj} resulting in the same life as the symmetrical cycle with the amplitude σ_{aj}; $j = X, Y, Z$.

The value of ψ was determined from a special dependence and the material experimental data for asymmetrical cycles of loading in accordance with (**14**). In view of the aforesaid, Eqs (2) and (3) will be written in the following form

$$\tau_{oct(r)} = \left(\frac{1}{3}\right)\sqrt{(\sigma_{aXr} - \sigma_{aYr})^2 + (\sigma_{aYr} - \sigma_{aZr})^2 + (\sigma_{aZr} - \sigma_{aXr})^2}\ ; \tag{7}$$

$$P_{oct(r)} = \left(\frac{1}{3}\right)\left[\sigma_{aX} + \sigma_{aY} + \sigma_{aZ} + (1-\Psi)(\sigma_{oX} + \sigma_{oY} + \sigma_{oZ})\right]. \qquad (8)$$

Three variations of relationship (1) were verified using the experimental data obtained which differ in considering or neglecting the cycle asymmetry influence on the cyclic stress components in determinations of τ_{oct} and P_{max}:

$$\tau_{oct(r)} = A\,P_{max} = B; \qquad (9)$$

$$\tau_{oct} = A\,P_{max}(r) = B; \qquad (10)$$

$$\tau_{oct(r)} = A\,P_{max}(r) = B. \qquad (11)$$

Figure 6 illustrates the results of this kind of presentation of the experimental data according to which the results of computation using relationship (11) are in the best agreement with the criterion curve, i.e. a version wherein the influence of the cycle asymmetry on the variation in the stress amplitude is taken into account in the computation of both τ_{oct} and P_{max}. Considering the aforesaid we can say that dependence (11) is the extension of criterion (1).

ASSESSMENT OF THE CYCLIC LIFE OF THE ELEMENT WITH A CONCENTRATOR AND RESIDUAL STRESSES

The series of investigations performed makes it possible to determine the algorithm for calculating cyclic life of a structural element with a concentrator under symmetrical loading with the account taken of relationship (11). For the calculation it is necessary to have the data on smooth specimens: fatigue curves in tension-compression and in torsion, stress-strain diagrams for cyclic and static loadings. The criterion relationship τ_{oct} vs P_{max} for several lives are constructed from the fatigue curves for smooth specimens, whereas the distribution curves for the components of residual and cyclic stresses (for different levels of nominal stresses applied) and octahedral cyclic stresses τ_{oct} are constructed from the concentrator geometry and stress-strain diagrams. Using the τ_{oct} vs X curves and the criterion curves, the X_{eq} value is determined for the given values of σ_{a1}, N_1 by the method of successive approximations. Stabilised values of residual stresses are found for this depth (X_{eq}) and their consideration as a cycle asymmetry is made according to dependencies (7) and (8). By introducing the refinements according to formula (11) and using the criterion curves τ_{oct} vs P_{max}, the N_1 value is found which corresponds to the life of the specimen with a stress concentrator and residual stresses under the action of nominal stress σ_{1n}. By specifying several values of σ_a, N_1, and repeating the described procedure, we get the fatigue curve for an element with a stress concentrator.

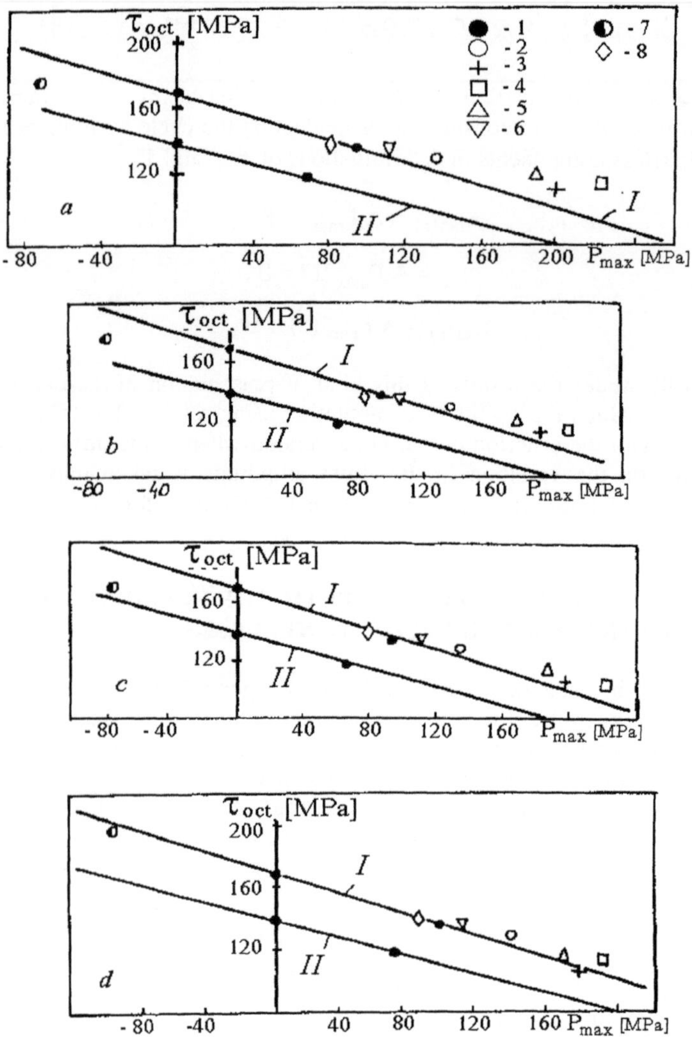

Fig.6 The influence of the negative allowance for the cycle asymmetry on the correspondence to the criterion of shear octahedral stresses: a, b, c, d correspond to Eqs. (1), (9), (10), (11); I, II - criterion lines.

The described algorithm for assessing cyclic life was realised for several variations of tested specimens with concentrators and residual stresses. The results are given in Fig.7 as calculated and experimental fatigue curves.

As follows from the data obtained, the prediction is satisfactory, though an increase in the prediction error is observed with a decrease in cyclic life. We relate this fact to a larger error of the determination of the stabilised residual stresses for those lives.

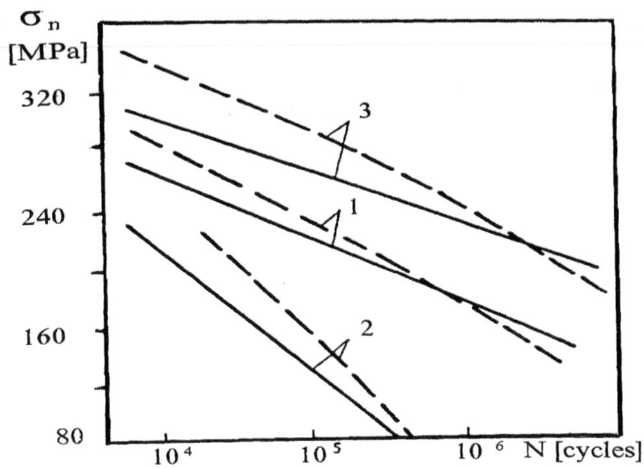

Fig.7 Comparison of the calculated and experimental fatigue curves (dashed lines-calculation, solid lines–experiment): 1-biaxial compression; 2-prestressing by compression; 3-pressing in a pin.

CONCLUSIONS

On the basis of the calculations and experiments performed, the applicability of the criterion of octahedral stresses has been determined taking into account a hydrostatic stress.

However, it has been shown that this criterion should be modified to take into account the components of the cycle asymmetry along different axes. The results have been demonstrated on plane specimens with a central hole and residual stresses induced therein whose stabilised values are taken as the static components of the loading.

On the basis of the modified criterion of the material ultimate state, an algorithm has been proposed for the calculation of high-cycle fatigue life for a typical structural element with a concentrator and residual stresses, and the agreement between the calculated and experimental results has been shown. Here, account is taken of a certain layer of a damaged material in the stress concentrator wherein its ultimate state has been attained.

REFERENCES

(1) Leis D.C. and Topper T.H., (1977), Long-life notch strength reduction due to local biaxial state of stress, Trans. ASME J. Engng. Mater. Technol., Ser. H, pp.215-221

(2) Walker E.K. (1977), Multiaxial stress-strain approximations for notch fatigue behaviour, Journal of Testing and Evaluation, Vol.5, No.2, pp.106-113

(3) Flavenot J.F. and Scalli N.,(1989), A comparison of multiaxial fatigue criteria incorporating residual stress effects, Biaxial and Multiaxial Fatigue, (Edited by M.W.Brown and K.J.Miller), Mech. Engng Public., London, pp.437-457

(4) Dang Van K., Griveau B. and Message O., (1989), On a new multiaxial fatigue limit criterion: theory and application, Biaxial and Multiaxial Fatigue (Edited by M.W. Brown and K.J.Miller) Mech. Engng Public., London, pp.479-496

(5) Umanskiy S., (1983), Optimization of Approximate Methods for the Mechanics Border Problems Solution, Naukova Dumka, Kiev (in Russian)

(6) Malinin N.N., (1975), Applied Theory of Plasticity and Creep, Mashinostroenie, Moscow (in Russian)

(7) Troshchenko V.T., Tsybanyov G.V. and Stepura A.V., (1993), Development of a method for accounting the influence of stress concentration and residual stresses on cyclic longevity of steel 10GN2MFA, Report 1. Analysis of materials stress-strain state, Problemy Prochnosti, No.8, pp.3-13

(8) Troshchenko V.T., (1971), Fatigue and Inelasticity of Metals, Naukova Dumka, Kiev (in Russian)

(9) Serensen S.V., Kogaev V.P., and Shneiderovich R.M., (1975), Load-carrying capacity of machine parts and their strength calculation, Mashinostroenie, Moscow (in Russian)

(10) Panasyuk V.V., Ostash O.P. and Kostyuk E.M., (1986), Relation between characteristics of cyclic crack resistance of materials in the stages of crack initiation and growth, Fiz. -khim. Mekh.Mater., No.6, pp.46-52

(11) Flavenot J.F. and Scalli N., (1989), A critical depth criterion for the evaluation of long-life fatigue strength under multiaxial loading and a stress gradient, Biaxial and Multiaxial Fatigue, (Edited by M.W. Brown and K.J.Miller) Mech. Engng Public., London, pp.459-478

(12) Garud Y.S., (1981), Multiaxial fatigue: A survey of the state of the art, Journal of Testing and Evaluation, Vol. 9, No. 3, pp. 165 – 178

(13) Crossland G., (1956), Effect of large hydrostatic pressures on the torsional fatigue strength of an alloy steel, Proc. of the Int. Conf. of Fatigue of Metals, London, pp.138-149

(14) Kryzhanovsky V.I., (1993), Ultimate state of materials under cyclic loading and a complex stress state, Proc. XVII Sci. Conf. Young Scientists of the Inst. Mechanics Nat. Ac. Sci. Ukraine (in Russian), Kiev, pp. 74-78

II. NON-PROPORTIONAL CYCLIC LOADING

NONPROPORTIONAL LOW CYCLE FATIGUE OF 6061ALUMINUM ALLOY UNDER 14 STRAIN PATHS

Takamoto ITOH*, Takumi NAKATA*, Masao SAKANE**, Masateru OHNAMI**

*Department of Mechanical Engineering, Faculty of Engineering, Fukui University, 9-1, Bunkyo 3-chome, Fukui, 910-8507, Japan
** Department of Mechanical Engineering, Ritsumeikan University, Japan

ABSTRACT

This paper studies the nonproportional low cycle fatigue of 6061 aluminum alloy under 14 strain paths. Tension-torsion low cycle fatigue tests were carried out using hollow cylinder tube specimens (OD 12 mm, ID 9 mm, gage length 6.4 mm) under 14 proportional and nonproportional cyclic strain paths at room temperature. Nonproportional strain written with only strain path and having a material constant correlated nonproportional fatigue lives within a factor of two scatter band. The additional hardening of 6061 aluminum alloy under nonproportional straining was also discussed in relation with fatigue life.

KEY WORDS

Low cycle fatigue, multiaxial stress, nonproportional loading, life prediction, 6061 aluminum alloy

NOTATION

$\varepsilon_1(t)$	Maximum principal strain at time t
$\varepsilon_3(t)$	Minimum principal strain at time t
$\varepsilon_I(t)$	Maximum absolute value of the principal strain at time t : Max $[|\varepsilon_1(t)|, |\varepsilon_3(t)|]$
$\varepsilon_{I\,max}$	Maximum value of $\varepsilon_I(t)$ in a cycle
$\varepsilon^*(t)$	Equivalent strain based on COD at time t
$\phi(t)$	Principal strain ratio at time t
$\Delta\varepsilon_I$	Maximum principal strain range under nonproportional straining
$\Delta\varepsilon_{ASME}$	Equivalent strain range defined in Code Case N-47
$\Delta\varepsilon^*_I$	Equivalent strain range based on COD under nonproportional straining
$\Delta\varepsilon_{NP}$	Nonproportional strain range
$\Delta\varepsilon^*_{NP}$	Nonproportional strain range based on COD
$\sigma_1(t)$	Maximum principal stress at time t
$\sigma_3(t)$	Minimum principal stress at time t
$\sigma_I(t)$	Maximum absolute value of the principal stress at time t : Max $[|\sigma_1(t)|, |\sigma_3(t)|]$

$\Delta\sigma_I$ Maximum principal stress range under nonproportional straining
$\xi(t)$ Angle between the principal strain directions of $\varepsilon_I(t)$ and $\varepsilon_{I\,max}$
f_{NP} Nonproportional factor
f^*_{NP} Nonproportional factor based on COD
α Material constant which expresses the amount of additional hardening
N_f Number of cycles to failure

INTRODUCTION

ASME Code Case N-47 **(1)** has been frequently used as a design criterion for nonproportional low cycle fatigue, but recent studies have shown that the Code Case estimates unconservative lives for nonproportional fatigue. Nonproportional loading reduces the low cycle fatigue life due to the additional hardening depending on strain history, so the nonproportional parameter must take account of the additional hardening. A couple of nonproportional parameters which include the stress range or amplitude have been proposed **(2-4),** and stress terms in the parameters are able to be calculated using the inelastic constitutive equation **(5-8),** but it is not a simple procedure in general and requires many material constants. There is no well established method of estimating nonproportional low cycle fatigue life based on only strain history.

The authors **(9)** carried out nonproportional low cycle fatigue tests using a hollow cylinder specimen of Type 304 stainless steel and proposed a nonproportional low cycle fatigue parameter written with only strain history. Type 304 stainless steel is known as a material, which shows the large additional hardening under nonproportional loadings **(5, 9-11).** Fatigue lives drastically reduced by additional hardening, which depends on strain history. The maximum reduction is a factor of 10 when compared with the proportional fatigue life. However, the degree of additional hardening is material dependent, so that the reduction of nonproportional lives is also material dependent.

The aim of this paper is to examine the nonproportional low cycle fatigue life of 6061

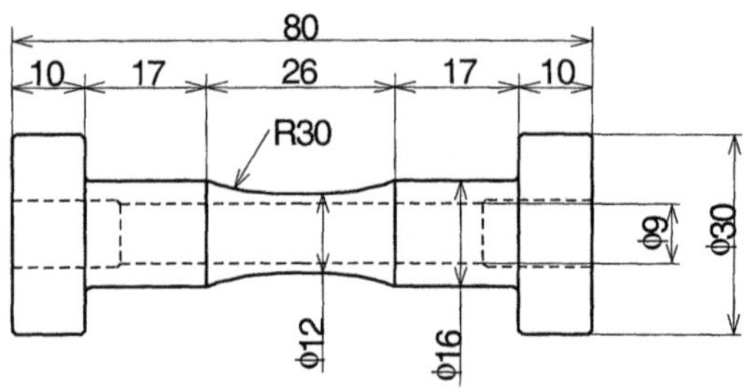

Fig.1 Shape and dimensions of the specimen tested (mm).

aluminum alloy which shows a small additional hardening and to confirm the availability of the nonproportional strain proposed previously to the small additional hardening material, by making extensive nonproportional low cycle fatigue tests using 14 strain paths.

EXPERIMENTAL PROCEDURE

The material tested was 6061 aluminum alloy (6061 Al alloy) which received T6 heat treatment. Mises' equivalent total strain controlled nonproportional low cycle fatigue tests were carried out using hollow cylinder specimens with 9 mm inner diameter, 12 mm outer diameter and 6.4 mm gage length as shown in Fig.1. Test machine used was a tension-compression and reversed torsion electric servo hydraulic low cycle fatigue machine.

Figure 2 shows strain paths employed, where ε and γ are the axial and shear strains, respectively. Case 0 is a push-pull test and is the base data used for the nonproportional life prediction. Total axial strain range was varied from 0.5 % to 1.5 %. Strain paths shown in the figure were determined so as to make clear the various effects in nonproportional straining **(9)**. In strain paths 1-13, the total axial strain range, $\Delta\varepsilon$, had the same strain magnitude as the total shear strain range, $\Delta\gamma$, on Mises' equivalent basis.

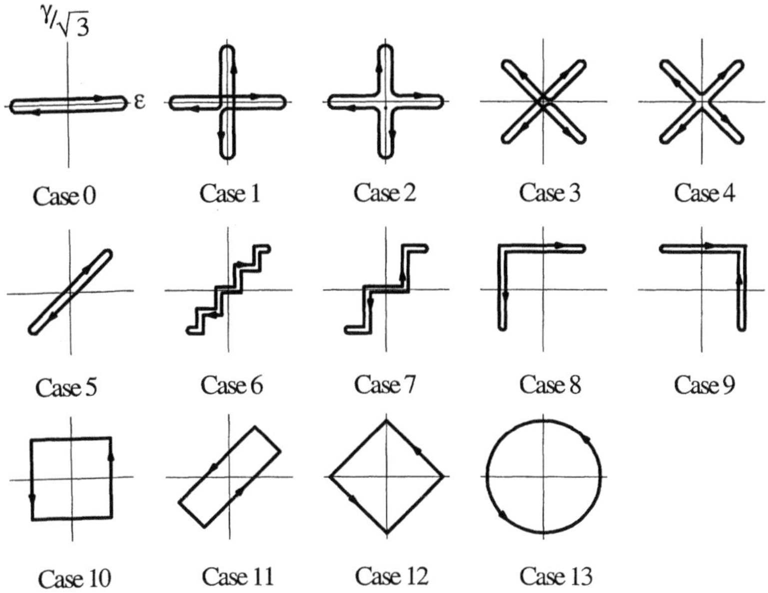

Fig.2 Proportional and nonproportional strain paths.

In this paper, one cycle is defined as full straining for both axial and shear cycles. Thus, a complete straining along the strain paths shown in Fig.2 was counted as one cycle for all the Cases except Case 3 and 4. In Case 3 and 4, a complete cycling was counted as two cycles. The number of cycles to failure (N_f) was defined as the cycle at which the axial stress amplitude was decreased by 5 % from its cyclically stable value.

EXPERIMENTAL RESULTS AND DISCUSSION

Definition of Stain and Stress Ranges

This study defined the maximum principal strain range as

$$\Delta\varepsilon_I = \mathrm{Max}\left[\varepsilon_{Imax} - \cos\xi(t)\cdot\varepsilon_I(t)\right] \tag{1}$$

where $\varepsilon_I(t)$ is the maximum absolute value of principal strain at time t and is given by Eq. (2)

$$\varepsilon_I(t) = \begin{cases} |\varepsilon_1(t)| & \text{for } |\varepsilon_1(t)| \geq |\varepsilon_3(t)| \\ |\varepsilon_3(t)| & \text{for } |\varepsilon_1(t)| < |\varepsilon_3(t)| \end{cases} \tag{2}$$

where $\varepsilon_1(t)$ and $\varepsilon_3(t)$ are the maximum and minimum principal strains at time t, respectively. The maximum value $\varepsilon_{I\,max}$ of $\varepsilon_I(t)$ is expressed as

$$\varepsilon_{Imax} = \mathrm{Max}[\varepsilon_I(t)] \tag{3}$$

$\xi(t)$ is the angle between $\varepsilon_{I\,max}$ and $\varepsilon_I(t)$ directions and expresses the variation angle of the principal strain direction.

Figure 3 schematically shows the relationship between $\varepsilon_I(t)$ and $\xi(t)$ on a polar figure of $\varepsilon_I(t)$. The angle $\xi(t)$ becomes a half value in physical plane, i.e. in the specimen.

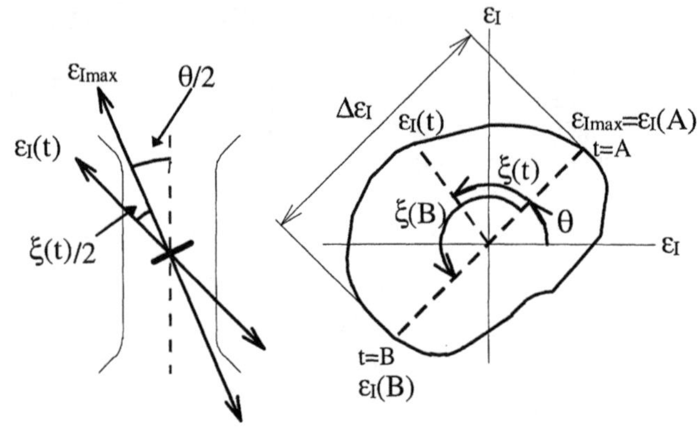

Fig.3 Schematic graph of $\varepsilon_I(t)$, $\xi(t)$ and $\Delta\varepsilon_I$.

The principal strain range, $\Delta\varepsilon_I$, is determined by two strains, $\varepsilon_I(A)$ and $\varepsilon_I(B)$, and the angle between them, where A and B are the times maximising the strain range in bracket in Eqs. (1) and (3). Thus, Eq. (1) is equivalent to finding the largest principal strain range occurred in specimen and is rewritten as,

$$\begin{aligned} \Delta\varepsilon_I &= \varepsilon_I(A) - \cos\xi(B)\cdot\varepsilon_I(B) \\ \varepsilon_I(A) &= \varepsilon_{Imax} \end{aligned} \tag{4}$$

The angle $\xi(B)$ is the angle between the principal strain directions of ε_{Imax} and $\varepsilon_I(B)$.

Nonproportional Low Cycle Fatigue Life

Table 1 lists the test results of 6061 aluminum alloy and Type 304 stainless steel together with the stress and strain parameters. Stress ranges were measured at the half life ($1/2N_f$).

Figure 4 shows the correlation of nonproportional low cycle fatigue (LCF) lives of 6061 Al alloy with the equivalent strain range defined in ASME Code Case N-47 (ASME strain range) **(1)**, which has been used as a design parameter for the nonproportional fatigue. In the figure, a factor of two scatter band is shown by lines based on the push-pull data, i.e. Case 0 data, and attached numbers denote the Case number. ASME strain range correlates fatigue lives unconservatively for some Cases by more than a factor of two. The lowest fatigue lives occurred in Case 13, i.e. circular path. Fatigue lives in that Case are about 1/3 of those in Case 0. The significant reduction in fatigue life also occurred in Case 10 and 12, box paths, as well as circular path.

For the comparison, the data correlation of Type 304 stainless steel with ASME strain is shown in Fig.5, of which tests were made by the authors **(9)**. Specimen shape and strain paths are the same as those in this study. The figure shows the same trend of the data correlation as that in Fig.4, but ASME strain gives a more unconservative estimate for Type 304 steel than for 6061 Al alloy. The minimum lives are found in Case 12 which is about 10 % of the failure cycle in Case 0 at the same strain range, whereas it is about 30 % for 6061 Al alloy.

Comparison of the results between Fig.4 and Fig.5 leads to the conclusion that the nonproportional LCF damage is a function of strain history and material. Thus, nonproportional strain parameter must take account of these two factors.

Stress under Nonproportional Straining

Figures 6 (a) and (b) show the correlation of LCF lives of 6061 Al alloy and Type 304 steel with maximum principal stress range, $\Delta\sigma_I$, defined by Eq.(5) similar to that of $\Delta\varepsilon_I$,

$$\Delta\sigma_I = \sigma_I(A) - \cos\xi(B)\cdot\sigma_I(B)$$

$$\sigma_I(t) = \begin{cases} |\sigma_1(t)| & \text{for} \quad |\sigma_1(t)| \geq |\sigma_3(t)| \\ |\sigma_3(t)| & \text{for} \quad |\sigma_1(t)| < |\sigma_3(t)| \end{cases} \qquad (5)$$

where $\sigma_1(t)$ and $\sigma_3(t)$ are the maximum and minimum principal stresses at time t. Fatigue lives of 6061 Al alloy, Fig.6 (a), are mostly within a factor of two band, while those of Type 304 steel, Fig.6 (b), are correlated too conservatively where most of the data are out of a factor of two scatter band. The results in these two figures indicate that the small reduction in LCF life occurs for small additional hardening material and the large reduction for large additional hardening material. The maximum principal stress range is a suitable parameter for the former material but is not for the latter material.

Reduction in nonproportional LCF life is connected with the degree of additional hardening **(2-4)**. In nonproportional loading, the principal strain direction is changed with proceeding cycles, so the maximum shear stress plane is changed continuously in a cycle. This causes an interaction between slip systems and which results in the formation of small cells **(10,11)** for Type 304 steel. Large additional hardening occurred by the interaction of slip systems for that steel. 6061 Al alloy, on the other hand, is a material of high stacking fault energy and slips of dislocations are wavy. No large interaction occurred in 6061 Al alloy since dislocations change their glide planes easily following the variation of the maximum principal strain direction **(11)**.

Nonproportional LCF Strain Parameter

The authors proposed nonproportional strain range, $\Delta\varepsilon_{NP}$, below.

$$\Delta\varepsilon_{NP} = (1+\alpha \cdot f_{NP}) \cdot \Delta\varepsilon_I \tag{6}$$

where α is a material constant related to the additional hardening. f_{NP} is the nonproportional factor, which expresses the severity of nonproportional straining and is described by only the strain history.

Table 1 Summary of the test results.

6061 aluminum alloy											
Case No.	N_f (Cycles)	Strain range, MPa					Stress range, MPa			NP factor	
		$\Delta\varepsilon_{ASME}$	$\Delta\varepsilon$	$\Delta\gamma$	$\Delta\varepsilon_{NP}$	$\Delta\varepsilon^*_{NP}$	$\Delta\sigma$	$\Delta\tau$	$\Delta\sigma_I$	f_{NP}	f^*_{NP}
0	44500	0.50	0.50	0	0.50	0.50	368	0	368	0	0
13	7500	0.50	0.50	0.87	0.60	0.60	400	185	407	1.00	1.00
0	2900	0.80	0.80	0	0.80	0.80	527	0	527	0	0
1	955	0.80	0.80	1.39	0.85	0.84	519	302	523	0.34	0.32
2	975	0.80	0.80	1.39	0.85	0.84	536	314	551	0.34	0.32
3	1740	0.80	0.57	0.98	0.85	0.83	409	221	533	0.39	0.41
4	2610	0.80	0.57	0.98	0.85	0.83	393	236	530	0.39	0.41
5	2050	0.80	0.57	0.98	0.79	0.76	446	177	543	0	0
6	3370	0.80	0.57	0.98	0.81	0.77	373	238	536	0.10	0.10
7	2800	0.80	0.57	0.98	0.82	0.79	383	228	514	0.20	0.21
8	1310	0.80	0.57	0.98	0.91	0.89	430	235	529	0.77	0.81
9	890	0.80	0.57	0.98	0.91	0.89	454	214	512	0.77	0.81
10	1650	0.80	0.57	0.98	0.91	0.89	411	210	521	0.77	0.81
11	1310	0.80	0.72	0.36	0.82	0.80	454	271	535	0.46	0.46
12	1250	0.80	0.80	1.39	0.92	0.92	514	259	520	0.77	0.77
13	785	0.80	0.80	1.39	0.96	0.96	493	292	506	1.00	1.00
0	740	1.20	1.20	0	1.20	1.20	561	0	561	0	0
5	970	1.20	0.85	1.47	1.18	1.13	435	212	560	0	0
13	220	1.20	1.20	2.08	1.44	1.44	581	362	630	1.00	1.00
0	225	1.80	1.80	0	1.80	1.80	586	0	575	0	0
13	69	1.80	1.80	3.12	2.16	2.16	637	394	692	1.00	1.00

Table 1 (cont.).

304 stainless steel												
Case No.	N_f (Cycles)	Strain range, %					Stress range, MPa			NP factor	
		$\Delta\varepsilon$	$\Delta\gamma$	$\Delta\varepsilon_{ASME}$	$\Delta\varepsilon_{NP}$	$\Delta\varepsilon^*_{NP}$	$\Delta\sigma$	$\Delta\tau$	$\Delta\sigma_I$	f_{NP}	f^*_{NP}
0	49000	0.50	0	0.50	0.50	0.50	530	0	530	0	0
0	23400	0.65	0	0.65	0.65	0.65	580	0	580	0	0
0	7100	0.80	0	0.80	0.80	0.80	630	0	630	0	0
0	1500	1.00	0	1.00	1.00	1.00	730	0	730	0	0
0	1700	1.13	0	1.13	1.13	1.13	730	0	730	0	0
0	690	1.20	0	1.20	1.20	1.20	805	0	805	0	0
0	540	1.50	0	1.50	1.50	1.50	825	0	825	0	0
1	9500	0.50	0.87	0.50	0.65	0.64	685	395	715	0.34	0.32
2	20000	0.50	0.87	0.50	0.65	0.64	670	355	680	0.34	0.32
3	2400	0.50	0.87	0.71	0.95	0.92	670	420	1020	0.39	0.41
4	3400	0.50	0.87	0.71	0.95	0.92	790	395	950	0.39	0.41
5	17500	0.50	0.87	0.71	0.70	0.67	485	185	655	0	0
6	9700	0.50	0.87	0.71	0.76	0.74	500	240	695	0.10	0.10
7	18000	0.50	0.87	0.71	0.82	0.80	530	285	695	0.20	0.21
8	2050	0.50	0.87	0.71	1.18	1.17	760	410	915	0.77	0.81
9	2950	0.50	0.87	0.71	1.18	1.17	780	370	885	0.77	0.81
10	2600	0.50	0.87	0.71	1.18	1.17	765	400	1035	0.77	0.81
11	14400	0.50	0.87	0.56	0.79	0.78	570	280	595	0.46	0.46
12	4750	0.50	0.87	0.50	0.85	0.84	660	360	840	0.77	0.77
1	1400	0.80	1.39	0.80	1.04	1.03	950	530	985	0.34	0.32
2	2100	0.80	1.39	0.80	1.04	1.03	860	490	865	0.34	0.32
3	820	0.80	1.39	1.13	1.50	1.47	975	545	1350	0.39	0.41
4	900	0.80	1.39	1.13	1.50	1.47	1010	520	1220	0.39	0.41
5	3200	0.80	1.39	1.13	1.11	1.07	590	250	820	0	0
6	2600	0.80	1.39	1.13	1.21	1.17	670	320	905	0.10	0.10
7	1700	0.80	1.39	1.13	1.31	1.27	735	390	920	0.20	0.21
8	470	0.80	1.39	1.13	1.88	1.85	1055	560	1220	0.77	0.81
9	660	0.80	1.39	1.13	1.88	1.85	1075	600	1245	0.77	0.81
10	320	0.80	1.39	1.13	1.88	1.85	1060	555	1345	0.77	0.81
11	1200	0.80	1.39	0.89	1.26	1.24	850	500	975	0.46	0.46
12	710	0.80	1.39	0.80	1.36	1.35	940	510	975	0.77	0.77

The results in Fig.6 showed that the degree of additional hardening is material dependent. Other literature reported that aluminum alloys show little or no additional hardening **(3,4,10),** while Type 304 stainless steel usually gives a significant additional hardening **(5,10,11).** Doong et al. **(10)** have reported that little additional hardening has almost no effect on nonproportional fatigue life and significant additional hardening causes a drastic reduction in fatigue life. Thus, the nonproportional LCF parameter must include a parameter, which expresses the amount of additional hardening.

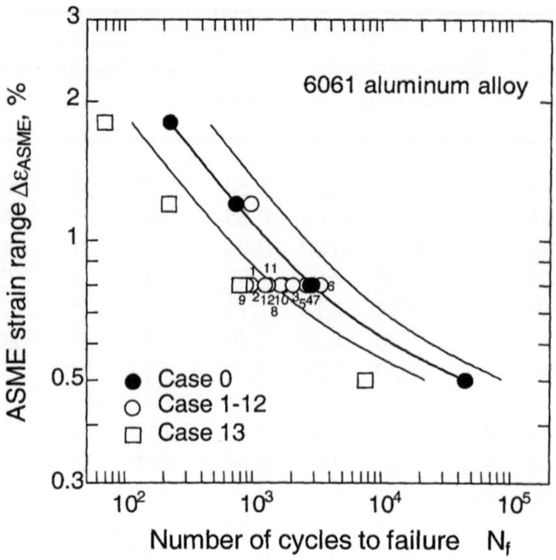

Fig.4 Correlation of nonproportional LCF lives of 6061 aluminum alloy with ASME strain range.

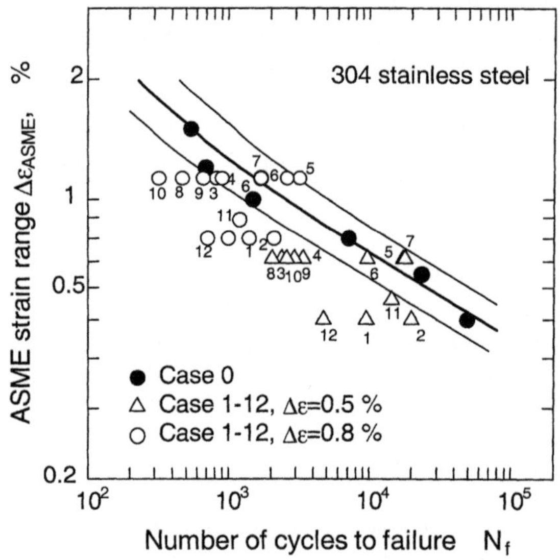

Fig.5 Correlation of nonproportional LCF lives of Type 304 stainless steel with ASME strain range.

Equation 6 takes account of the amount of additional hardening by the material constant, α. The value of α is defined as the ratio of stress amplitude under 90 degrees out-of-phase loading (circular strain path in $\gamma/\sqrt{3}-\varepsilon$ plot) to that under proportional loading. 90 degrees out-of-phase loading shows the maximum additional hardening among all the nonproportional histories **(3,12)**. For 6061 Al alloy, the stress amplitude under 90 degrees out-of-phase loading was increased up to 20 % in comparison with the proportional loading, so the value of α is 0.2. For Type 304 stainless steel, it was 0.9 due to the large additional hardening **(3,9)**.

The nonproportional factor, which expresses the severity of nonproportional straining, is defined as

$$\begin{aligned} f_{NP} &= \frac{k}{T \cdot \varepsilon_{I\max}} \int_0^T \left(|\sin \xi(t)| \cdot \varepsilon_I(t) \right) dt \\ k &= \pi/2 \end{aligned} \tag{7}$$

where $\varepsilon_I(t)$, $\varepsilon_{I\,max}$ and $\xi(t)$ are the parameters defined in Eqs.(1)-(3) and Fig.3. T is time for a cycle and f_{NP} is normalised by T and $\varepsilon_{I\,max}$. k is a constant to make f_{NP} unity under 90 degrees out-of-phase loading. The reason for making f_{NP} integral form is that the experimental results indicate that the nonproportional LCF life is significantly influenced by the degree of principal strain direction change and strain length after the direction change. The value of f_{NP} takes zero for proportional straining. In Eq.(7), f_{NP} is calculated from only the strain path and accounts for the severity of nonproportional loading. The parameter given by Eq.(6) in this study would evaluate the degree of additional hardening due to nonproportional loading.

The authors **(13)** proposed another nonproportional strain on the basis of the equivalent strain based on crack opening displacement (COD strain) to improve the data correlation of proportional LCF lives. The COD strain physically expresses the intensity of COD in multiaxial stress and strain states. The nonproportional strain range based on COD is defined similar to Eq.(6) as,

$$\Delta\varepsilon^*_{NP} = (1 + \alpha \cdot f^*_{NP})\, \Delta\varepsilon^*_I \tag{8}$$

where $\Delta\varepsilon^*_I$ is the COD strain range under nonproportional straining and is given by

$$\begin{aligned} \Delta\varepsilon^*_I &= \varepsilon^*(A) - \cos\xi(B) \cdot \varepsilon^*(B) \\ \varepsilon^*(t) &= \beta \cdot (2 - \phi(t))^{m'} \cdot \varepsilon_I(t) \end{aligned} \tag{9}$$

In this equation, $\varepsilon^*(t)$ and $\phi(t)$ are the COD strain and the principal strain ratio at time t. $\phi(t)$ is defined as

$$\phi(t) = \begin{cases} \varepsilon_3(t)/\varepsilon_I(t) & \text{for} \quad |\varepsilon_I(t)| \geq |\varepsilon_3(t)| \\ \varepsilon_I(t)/\varepsilon_3(t) & \text{for} \quad |\varepsilon_I(t)| < |\varepsilon_3(t)| \end{cases} \tag{10}$$

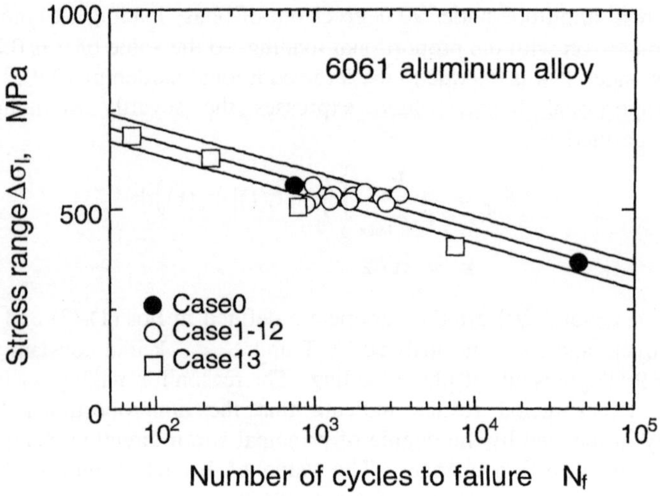

(a) 6061 aluminum alloy

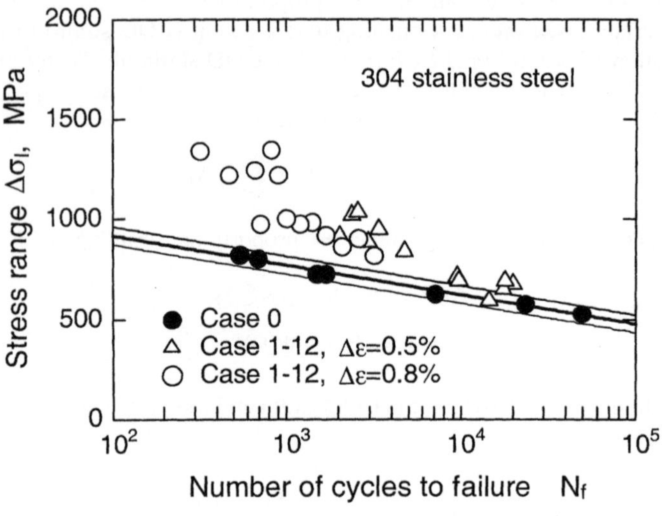

(b) Type 304 stainless steel

Fig.6 Correlation of nonproportional LCF lives with principal stress range.

Constants, β and m', in Eq.(9) take the values of 1.83 and -0.66, respectively, independent of material which was verified by FEM analyses **(13).** The nonproportional factor based on COD strain, f^*_{NP}, is given by

$$f^*_{NP} = \frac{k^*}{T \cdot \varepsilon^*_{Imax}} \int_0^T \left(|\sin \xi(t)| \cdot \varepsilon^*(t) \right) dt \qquad k^* = 1.66 \qquad (11)$$

Figures 7 and 8 correlate the nonproportional fatigue data with the two nonproportional strains shown in Eqs.(6) and (8), respectively. In these figures, (a) and (b) are the correlation of fatigue lives for 6061 Al alloy and Type 304 steel, respectively.

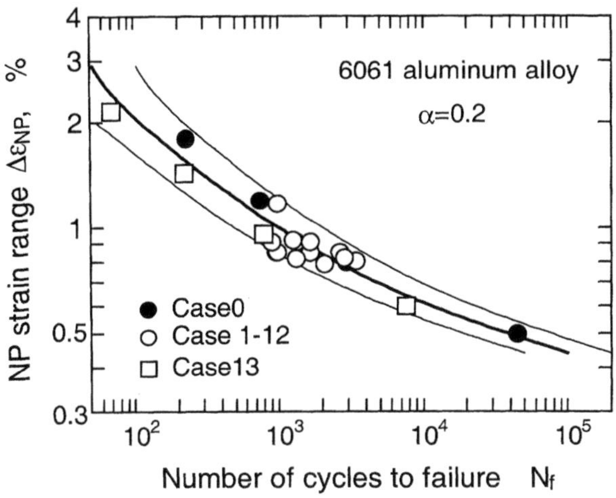

(a) 6061 aluminum alloy

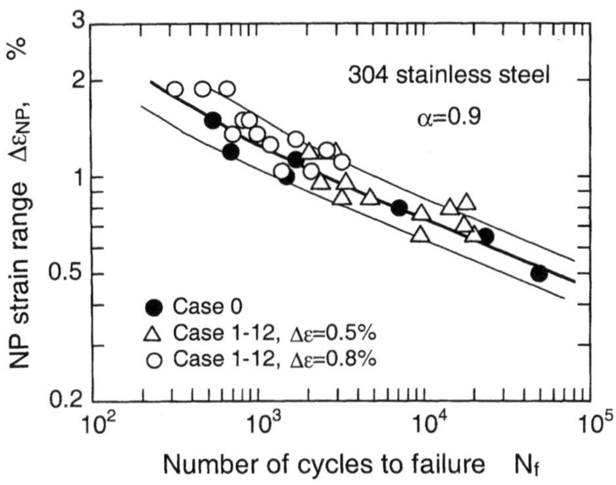

(b) Type 304 stainless steel

Fig.7 Correlation of nonproportional LCF lives with the nonproportional LCF strain range based on the principal strain.

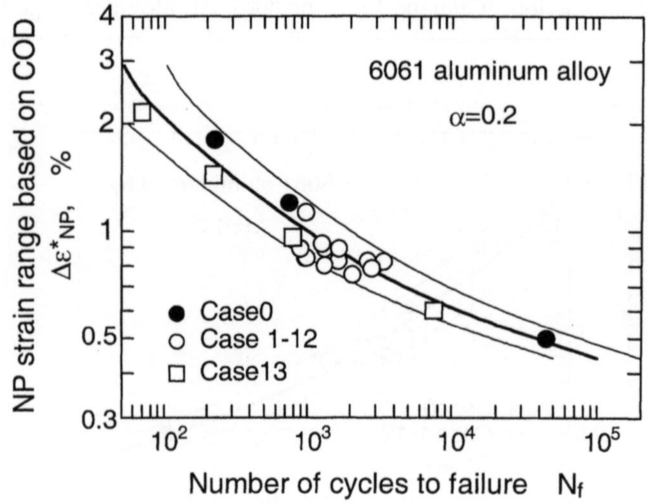

(a) 6061 aluminum alloy

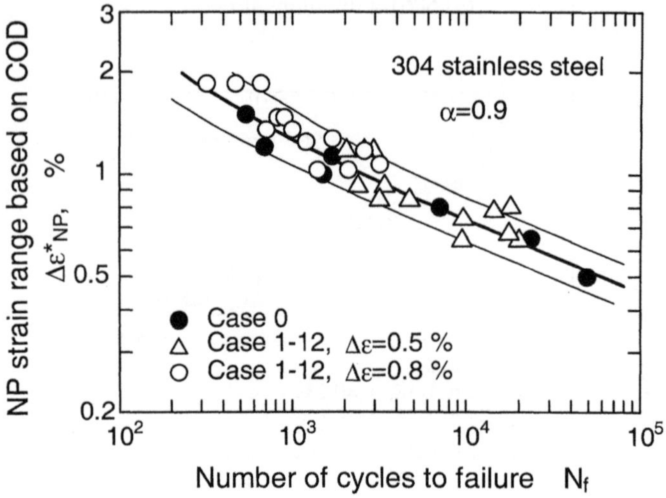

(b) Type 304 stainless steel

Fig.8 Correlation of nonproportional LCF lives with the nonproportional LCF strain range based on COD.

Almost all the nonproportional data are within a factor of two scatter band in the correlation with the two nonproportional strains. The scatter of the data appears to be somewhat smaller in the correlation with $\Delta\varepsilon^*_{NP}$, which arises from the better correlation of proportional data with COD strain range than that with the maximum principal strain range **(13)**. Therefore, the two nonproportional strains proposed by Eqs.(6) and (8) are able to predict nonproportional LCF lives with various strain histories. These equations have only one material constant, which is determined by the stress range ratio under 90 degrees out-of-phase and proportional loadings.

CONCLUSION

Proportional and nonproportional low cycle fatigue tests were carried out using fourteen strain paths for 6061 aluminum alloy hollow cylinder specimens at room temperature. Fatigue lives of 6061 aluminum alloy were reduced by nonproportional loading but the reduction was not so large as that of Type 304 stainless steel. The two nonproportional strains were applied nonproportional LCF lives of the two materials. The scatter of the data was within a factor of two for both the materials and which indicates the two strains are suitable parameter for correlating nonproportional LCF data of small and large additional hardening materials.

REFERENCES

(1) ASME Code Case N-47, (1978), Case of ASME Boiler and Pressure Vessel Code, Case N-47, Class 1, Section 3, Division 1, ASME.

(2) Smith R. N., Watson P. and Topper T. H., (1970), A stress strain function for the fatigue of materials, J. Materials JMLSA, Vol.5, No.4, pp.767-778.

(3) Socie D. F., (1987), Multiaxial fatigue damage models, Trans. ASME J. Engng. Mater. Technol., Vol.109, No.4, pp.293-298.

(4) Fatemi A. and Socie D. F., (1988), A critical plane approach to multiaxial fatigue damage including out-of-phase loading, Fatigue Fract. Engng Mater. Struct., Vol.11, No.3, pp.149-165.

(5) McDowell D. L., (1983), On the path dependence of transient hardening and softening to stable states under complex biaxial cyclic loading. Proc. Int. Conf. on Constitutive Laws for Engng. Mater., Tucson, AZ, (Desai and Gallagher, eds.), p.125.

(6) Krempl E. and Lu H., (1983), Comparison of the stress responses of an aluminum alloy tube to proportional and alternate axial and shear strain paths at room temperature, Mechanics of Materials, Vol.2, pp.183-192.

(7) Benallal A. and Marquis D., (1987), Constitutive equations for nonproportional cyclic elasto-viscoplasticity, Trans. ASME J. Engng. Mater. Technol., Vol.109, No.4, pp.326-336.

(8) Doong S. H. and Socie D. F., (1991), Constitutive modelling of metals under nonproportional loading, Trans. ASME J. Engng. Mater. Technol., Vol.113, No.1, pp.23-30.

(9) Itoh T, Sakane M, Ohnami M. and Socie D. F., (1995), Nonproportional low cycle fatigue criterion for type 304 stainless steel, Trans. ASME J. Engng. Mater. Technol., Vol.117, No.3, pp.285-292.

(10) Doong S. H., Socie D. F. and Robertson, I. M. (1990), Dislocation substructures and nonproportional hardening, Trans. ASME J. Engng. Mater.

Technol., Vol.112, No.4, pp.456-465.

(11) Itoh T., Sakane M., Ohnami M. and Ameyama K., (1992), Additional hardening due to nonproportional loading (a contribution of stacking fault energy), MECAMAT'92, Proc. Int. Seminar on Multiaxial Plasticity, France, (Benallal et al. eds.), pp.43-50.

(12) Kanazawa K., Miller K. J. and Brown M. W., (1979), Cyclic deformation of 1%Cr-Mo-V steel under out-of-phase loads, Fatigue Eng. Mater. Struct., Vol.2, No.2, pp.217-228.

(13) Itoh T., Sakane M. and Ohnami M., (1994), High temperature multiaxial low cycle fatigue of cruciform specimen, Trans. ASME J. Engng. Mater. Technol., Vol.116, No.1, pp.90-98.

WEAKEST LINK THEORY AND MULTIAXIAL CRITERIA

Jiping LIU

Volkswagen AG, Abt. EGPF/1712, 38436 Wolfsburg, Germany,
Formerly Technical University of Clausthal, Institut fur Maschinelle Anlagentechnik und Betriebsfestigkeit, Leibnizstrasse 32, 38678 Clausthal-Zellerfeld, Germany

ABSTRACT

From the weakest link theory, the classical multiaxial criteria, the principal normal stress criterion, the maximum shear stress criterion and the von Mises criterion, have been derived as special cases. On the basis of this analysis, a general fatigue criterion is formulated for multiaxial stress. The existing multiaxial criteria of integral approach and of the critical plane approach can be derived as special cases from the general fatigue criterion. On this basis, a new modification of shear stress intensity hypothesis SIH that provides satisfactory agreement between experimental and calculated results is proposed.

KEY WORDS

Multiaxial stress state, fatigue strength, weakest link theory, multiaxial criteria

INTRODUCTION

A multiaxial stress state, which varies with time, is generally present at the most severely stressed point in a structural component. As a rule, the multiaxial stress state is of a very complex nature. The individual stress components may vary in a mutually independent manner or at different frequencies, for instance, if the flexural and torsional stresses on shaft are derived from two vibrational systems with different natural frequencies.

For assessing this multiaxial stress, the classical multiaxial criteria, such as the von Mises criterion or the maximum shear stress criterion, are not directly applicable. This is illustrated in Fig.1 for the example of two load cases. In the first case, an alternating normal stress occurs in combination with an alternating shear stress with a phase shift of 90°, Fig.1a. The second case involves a pulsating tensile normal stress, σ_x, and a compressively pulsating normal stress σ_y, Fig.1b. In both load cases, the principal stresses exhibit the same variation with time. In accordance with the classical multiaxial criteria, the same equivalent stresses are calculated in both cases. The endurance limits are very different, however, as shown by experiments **(1)**. This is explained by the fact that the principal direction can vary in the case of multiaxial fatigue stress. A variable principal direction is not taken into account by the classical fatigue criteria.

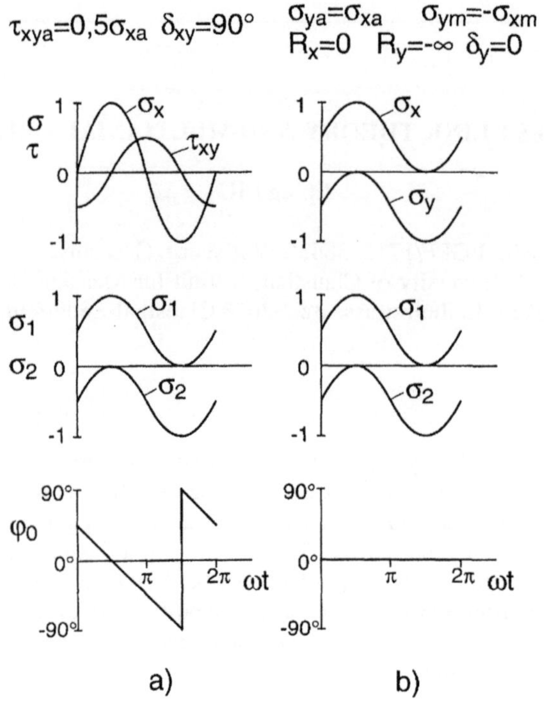

Fig.1 Two examples of multiaxial stresses and the time histories of principal stresses and principal directions.

For calculating the endurance limit in the case of multiaxial stresses, a number of multiaxial criteria have been developed during the past decades **(2, 12)**. These developments are even more comprehensive, as indicated by recent studies **(13, 17)**. The multiaxial criteria differ considerably in formulation, in the range of applicability, and in the reliability of prediction. Furthermore, they also involve highly different physical interpretations, if such an interpretation has been considered and indicated at all in formulating the hypothesis.

As a matter of principle, the known multiaxial fatigue criteria can be subdivided into hypotheses of the critical plane approach, hypotheses of integral approach, as well as empirical criteria. In the case of integral approach, the equivalent stress is calculated as an integral of the stresses over all intersection planes of a volume element; compare with the hypothesis of the effective shear stress **(2)** and the shear stress intensity hypothesis SIH **(3, 11)**. In the case of the critical intersection plane approach, only the intersection plane with the critical stress combination is considered, for instance, with the modified shear stress hypothesis proposed by McDiarmid. **(5)**.

In the present publication, the weakest link theory is first analysed. Subsequently, the relationship between the weakest link theory and the classical multiaxial criteria is explained. The multiaxial fatigue criteria of the critical intersection plane approach and of integral approach prove to be limiting cases of the weakest link theory. On the basis of this analysis, a general multiaxial criterion is formulated for arbitrary multiaxial stresses. From the general criterion, the known multiaxial criteria can be derived as special cases. Finally, the fatigue behaviour under multiaxial stress is described for a few load cases as examples. The further developed shear stress intensity hypothesis is verified on the basis of test results.

WEAKEST LINK THEORY AND CLASSICAL MULTIAXIAL CRITERIA

The weakest link theory was originally developed by Weibull for describing the static strength of brittle materials **(18)**. It was extended by Batdorf for considering the probability of failure of ceramic materials under multiaxial static load **(19-21)**. The weakest link theory is frequently applied for calculating the probability of failure of ceramic components under multiaxial load **(22-24)**.

In accordance with the weakest link theory, the probability of survival of a component can be described as follows:

$$P_{\ddot{U}} = \exp\left[-\frac{1}{4\pi}\int\limits_{V}\int\limits_{\Omega}\left(\frac{\sigma_{\gamma\varphi e}}{\sigma_0}\right)^{\kappa}\cdot d\Omega\cdot dV\right] \tag{1}$$

where denotes:

$\sigma_{\gamma\varphi e}$ - local equivalent stress in the intersection plane of the defect

κ - Weibull's exponent

Ω - the spherical surface area, Fig.2

V - the volume of the machine component.

The equivalent stress can be calculated as follows:

$$\sigma_V = \left[\frac{1}{4\pi}\int\limits_{\Omega}\left(\sigma_{\gamma\varphi e}\right)^{\kappa}\cdot d\Omega\right]^{\frac{1}{\kappa}} \tag{2}$$

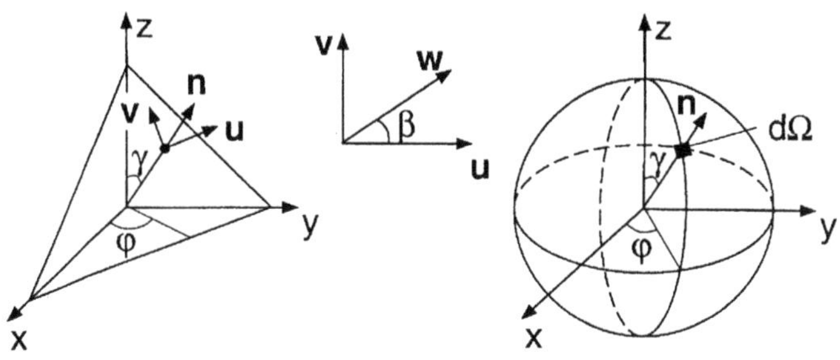

Fig.2 Intersection plane and spherical surface.

For an inhomogeneous stress distribution in the volume, a stress integral is employed for calculating the statistical size effect:

$$I = \int_V \left(\frac{\sigma_V}{\sigma_{V\max}} \right)^{\kappa} \cdot dV \tag{3}$$

The probability of survival can thus be described for the overall system:

$$P_{\ddot{U}} = \exp\left[-I \cdot \left(\frac{\sigma_{V\max}}{\sigma_0} \right)^{\kappa} \right] \tag{4}$$

The statistical size effect is described by means of Eqs. (3) and (4) **(25-27)**. In deriving the equations, it has been assumed that failure originates at the interior of the volume with the same probability as for the surface. As a rule, however, failure occurs at the surface; consequently, the surface area A must be inserted into the preceding equations, instead of the volume V.
The local failure criterion, that is, the equivalent stress $\sigma_{\gamma\phi e}$, must be selected in correspondence with materials. A distinction must be made between ductile and brittle materials. In the case of brittle materials such as ceramics, the defect can be considered to be a crack as a first approximation. The normal stress, which is perpendicular to the crack plane, is decisive for the failure. The normal stress $\sigma_{\gamma\phi}$ is selected as equivalent stress if the crack is not sensitive to shear stress. The equivalent stress is then given by

$$\sigma_V = \left[\frac{1}{4\pi} \int_{\gamma=0}^{\pi} \int_{\varphi=0}^{2\pi} (\sigma_{\gamma\varphi})^{\kappa} \cdot \sin\gamma \cdot d\varphi \cdot d\gamma \right]^{\frac{1}{\kappa}} \tag{5}$$

In Fig.3, the failure limits are plotted with the use of Eq. (5) for various Weibull exponents κ in the range from 2 to ∞ in a σ_1-σ_2 diagram for the planar stress state. For an infinitely large Weibull exponent, $\kappa \to \infty$, the same failure limit is obtained as with the principal normal stress criterion. In accordance with the maximum norm of the algebra **(28)** the equivalent stress given by Eq. (5) is the major principal stress.

$$\sigma_V \xrightarrow{\kappa \to \infty} \sigma_{\max} \tag{6}$$

For a Weibull exponent, κ=2, an elliptical limit curve is obtained.

$$\sigma_V = \frac{1}{2}\left(3\sigma_x^2 + 2\sigma_x \cdot \sigma_y + 3\sigma_y^2 + 4\tau_{xy}^2\right)^{\frac{1}{2}} \tag{7}$$

The failure limits for various Weibull exponents differ especially for the stress states with biaxial tension, $\sigma_2=\sigma_1$ and pure shear with $\sigma_2=\sigma_1=\tau$. For the range $\kappa=10\div30$, the difference from the principle stress criterion is small. At κ=2, the ratio of the tolerable stresses with reference to the tolerable uniaxial stress σ_A is only 0.61 for biaxial tension, $\sigma_2=\sigma_1$, and 0.866 for pure shear, $\sigma_2=-\sigma_1=\tau$.
For ductile materials, such as steel, the beginning of plastic deformation, that is, the beginning of slip motion of the slip system under shear stress, is usually employed as failure limit. The slip system comprises the direction most densely occupied by atoms (slip direction) in the most densely occupied intersection plane (slip plane). The local failure criterion depends on the orientation of the slip plane $\gamma\varphi$ as well as on the

orientation of the slip direction. The shear stress, $\tau_{\gamma\varphi\beta}$, is selected as local equivalent stress. Thus, Eq. (2) must be extended by integration over the angular range β=0 to π. The equivalent stress is then given by

$$\sigma_v = \left[-\frac{1}{4\pi^2} \int_{\gamma=0}^{\pi} \int_{\varphi=0}^{2\pi} \int_{\beta=0}^{\pi} \left(\tau_{\gamma\varphi\beta}\right)^{\kappa} \cdot \sin\gamma \cdot d\varphi \cdot d\gamma \cdot d\beta \right]^{\frac{1}{\kappa}} \tag{8}$$

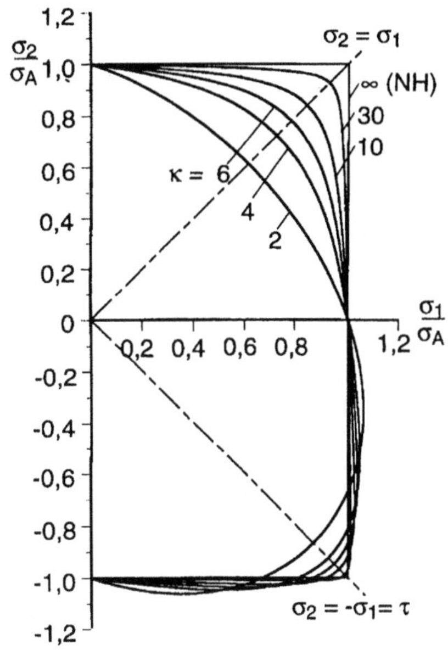

Fig.3 Failure limits with various Weibull exponents by Eq. (5).

In Fig.4, the failure limits given by Eq. (8) for the Weibull exponents from κ=2 to ∞ are plotted in a σ_1-σ_2 diagram. For infinitely large Weibull exponents, $\kappa\rightarrow\infty$, the resulting failure limit, in accordance with the maximum norm of the algebra **(27)**, is the same as that from the maximum shear stress criterion. At κ=2, the same failure limit is obtained as from the von Mises criterion. All failure limits with various Weibull exponents are situated between the two limiting curves, κ=2 and ∞. For the stress state $\sigma_2=\sigma_1$ (biaxial tension), the failure limit is independent of the Weibull exponent. For pure shear, $\sigma_2=-\sigma_1=\tau$, the ratio of the tolerable stress to the uniaxial tolerable stress, σ_A, varies from 0.5 at $\kappa=\infty$ (from the maximum shear stress criterion) to 0.577 at κ=2 (in correspondence with the von Mises criterion).
The von Mises criterion has been interpreted differently in the past:

- Distortion energy (Maxwell 1856, Huber 1904, Hencky 1924)
- Octahedral shear stress (Nadaj 1939)
- Root mean square of the principal shear stresses (Paul 1968)
- Root mean square of the shear stresses for all intersection planes (Novoshilov 1952)

Novoshilov (29) has shown that the root mean square of the shear stresses for all intersection planes is identical with that from the von Mises criterion:

$$\tau_{int} = \left[\frac{1}{4\pi} \int_{\gamma=0}^{\pi} \int_{\varphi=0}^{2\pi} \left(\tau_{\gamma\varphi}\right)^2 \cdot \sin\gamma \cdot d\varphi \cdot d\gamma \right]^{\frac{1}{2}} \cong \left(3I_2'\right)^{\frac{1}{2}} \tag{9}$$

The interpretation given by Novoshilov has led to the development of the hypothesis of effective shear stresses and the shear stress intensity hypothesis (2,3).

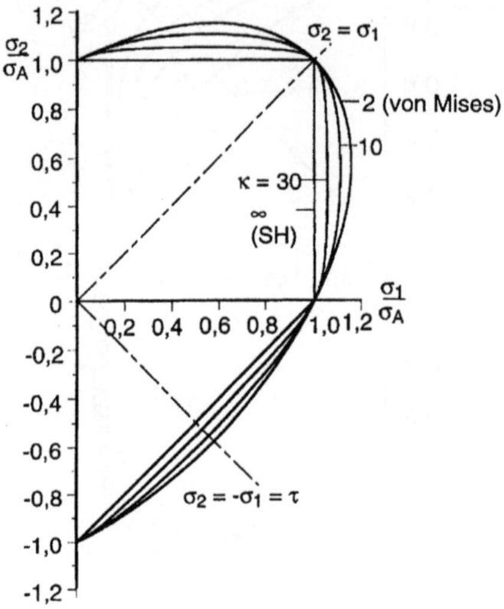

Fig.4. Failure limits with various Weibull exponents by Eq. (8).

It can be proved that the integration over the angle β is proportional to the resultant shear stress, $\tau_{\gamma\varphi}$:

$$\left[\int_{\beta=0}^{\pi} \left(\tau_{\gamma\varphi\beta}\right)^{\kappa} d\beta \right]^{\frac{1}{\kappa}} \cong \tau_{\gamma\varphi} \tag{10}$$

Hence, the integration over the angle β can be omitted for static loads. The interpretation according to Eq. (9) can be regarded as a special case of Eq. (8) with the exponent κ=2.

The classical multiaxial criteria, the principal normal stress criterion as well as the maximum shear stress criterion and the von Mises criterion can thus be considered as special cases of the weakest link theory, Eq. (2). This fact is utilised for formulating a general strength hypothesis in the following section.

GENERAL FATIGUE CRITERION FOR MULTIAXIAL STRESS

A multiaxial fatigue criterion must first satisfy the invariance condition; that is, the calculated equivalent stress must be independent of the selected fixed coordinate system with respect to the body. Moreover, for multiaxial fatigue stresses, the criterion must take into account the variable principal stress direction, see Fig.1. In order to satisfy these conditions, a multiaxial criterion can basically be formulated in two ways:

- as a hypothesis of the integral approach, and
- as a hypothesis of the critical plane approach.

In the case of multiaxial fatigue stresses with a periodically varying stress tensor $\sigma_{ij}(t)$, the stress components can be calculated in an arbitrarily oriented intersection plane at any time. The normal and shear stresses in the intersection plane, which vary with time, are described by mean values and amplitudes. The amplitude and mean value of normal stress, $\sigma_{\gamma\varphi}$, in the intersection plane and of the shear stress, $\tau_{\gamma\varphi\beta}$, in direction **w** can be simply calculated from the maximum and minimum during a period, Fig.2. If the local failure criterion is selected independently of the direction **w** in the intersection plane, the maxima of $\tau_{\gamma\varphi\beta a}$ and $\tau_{\gamma\varphi\beta m.}$ can be employed. Thus, four stress components, $\sigma_{\gamma\varphi a}$, $\tau_{\gamma\varphi a}$, $\sigma_{\gamma\varphi m}$, and $\tau_{\gamma\varphi m.}$, are present in the intersection plane, Fig.5. The calculation of the amplitudes and mean values of the stress components in the intersection plane is described in more detail in **(1, 11)**.

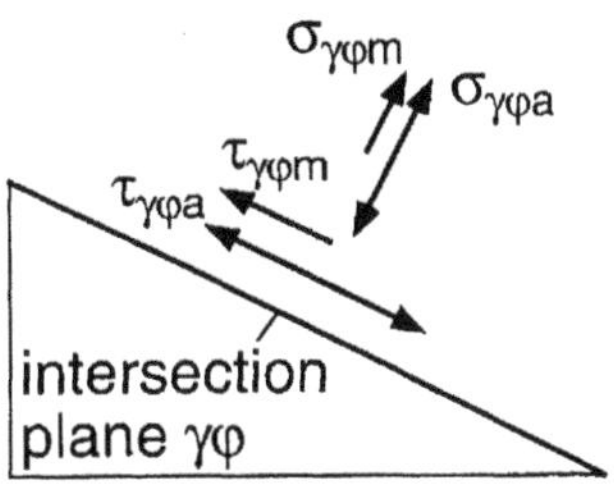

Fig.5. Stress components in an intersection plane

Let $\Sigma_{\gamma\varphi}$ and $T_{\gamma\varphi}$ be two stress components, or two arbitrary combinations of the four stress components in the intersection plane. In the following, the equivalent stress is formulated for $\Sigma_{\gamma\varphi}$ in the sense of the weakest link theory:

$$\Sigma_{\Sigma v} = \left[\int_{\Omega} (\Sigma_{\gamma\varphi})^{\mu} \cdot d\Omega \right]^{\frac{1}{\mu}} ; \qquad \Sigma_{\gamma\varphi} > 0 \tag{11}$$

In this form, the effect of the stress components which are decisive for damage can be described, for instance, the shear amplitude for ductile and "flawless" materials, and the normal stress amplitude for brittle and "defective" materials.

If the exponent μ approaches infinity, the resulting equivalent stress is the maximal stress Σ_{max} in accordance with the maximum norm. In this case, the formulation is applied for the multiaxial criterion of the critical intersection plane approach; accordingly, the stresses in the intersection plane of the maximal stress are decisive for fatigue life failure.

If a defined real number is chosen as the exponent, Eq.(11) corresponds to the formulation for the fatigue hypothesis of integral approach. For the sake of simplicity, the exponent is set equal to 2 for the shear stress intensity hypothesis.

Mean stresses alone cannot cause fatigue failure. In the presence of fatigue stress, however, they decrease or increase the tolerable stress amplitude. The effect of stress components such as the mean normal and mean shear stress can be assessed by means of the following formulation:

$$T_{\Sigma_V} = \left[\frac{\int\limits_{\Omega} (\Sigma_{\gamma\varphi})^{\nu} \cdot (T_{\gamma\varphi})^{\mu} \cdot d\Omega}{\int\limits_{\Omega} (\Sigma_{\gamma\varphi})^{\nu} \cdot d\Omega} \right]^{\frac{1}{\mu}} \quad ; \quad \Sigma_{\gamma\varphi}, T_{\gamma\varphi} > 0 \qquad (12)$$

If μ and ν approach infinity, and ν is much larger than μ, the formulation of the stress component T in the intersection plane corresponds to the maximal stress component Σ.

With the equivalent stresses thus formulated for the stress components, or combinations of same, Eqs. (11) and (12), the failure criterion for multiaxial stress can be established. The fatigue criterion is generally applicable, since all known multiaxial criteria can thus be derived. In the following, this is illustrated for the example of the critical shear stress criterion, as indicated by Nøkleby **(6)**.

In accordance with the criterion of critical shear stress after Nøkleby **(6)**, the critical intersection plane is defined as that with the maximal equivalent stress:

$$\tau_{\gamma\varphi v} = 2\tau_{\gamma\varphi a} + 2\alpha \cdot \sigma_{\gamma\varphi a} + 2\beta \cdot \sigma_{\gamma\varphi m} \qquad (13)$$

The failure criterion is given by

$$\sigma_V = c \cdot \max\{\tau_{\gamma\varphi v}\} = \sigma_W \qquad (14)$$

where c is compensation factor, which depends on the tensile-compressive fatigue strength. In accordance with the general multiaxial fatigue criterion, the failure criterion can be expressed as:

$$\sigma_V = c \cdot \left[\int\limits_{\Omega} (\tau_{\gamma\varphi v})^{\mu} \cdot d\Omega \right]^{\frac{1}{\mu}} = \sigma_W \qquad (15)$$

with $\mu \rightarrow \infty$.

From the general fatigue criterion, arbitrary fatigue criteria can be formulated. For this purpose, the exponents must be defined differently, or only the stress components and combinations of stress components must be selected differently. For the sake of simplicity, the exponents are set equal to 1, 2, or ∞ in Eqs. (11) and (12).

FURTHER DEVELOPMENT OF THE SHEAR STRESS INTENSITY HYPOTHESIS SIH

In the sense of the general fatigue criterion, the shear stress intensity hypothesis, SIH **(1, 3, 12)**, is modified in the following. For the modification, the shear stress amplitude and the normal stress amplitude are evaluated as the integral of the stress over all intersection planes. The mean shear stress is weighted over the shear stress amplitude and the mean normal stress over the normal stress amplitude. Thus, an equivalent stress is formed for each of the four stress components in the intersection plane.

$$\tau_{va} = \left\{ \frac{15}{8\pi} \int_{\gamma=0}^{\pi} \int_{\varphi=0}^{2\pi} \tau_{\gamma\varphi a}^{\mu_1} \cdot \sin\gamma \cdot d\varphi \cdot d\gamma \right\}^{\frac{1}{\mu_1}} \tag{16a}$$

$$\sigma_{va} = \left\{ \frac{15}{8\pi} \int_{\gamma=0}^{\pi} \int_{\varphi=0}^{2\pi} \sigma_{\gamma\varphi a}^{\mu_2} \cdot \sin\gamma \cdot d\varphi \cdot d\gamma \right\}^{\frac{1}{\mu_2}} \tag{16b}$$

$$\tau_{vm} = \left\{ \frac{\int_{\gamma=0}^{\pi} \int_{\varphi=0}^{2\pi} \tau_{\gamma\varphi a}^{\mu_1} \cdot \tau_{\gamma\varphi m}^{\nu_1} \cdot \sin\gamma \cdot d\varphi \cdot d\gamma}{\int_{\gamma=0}^{\pi} \int_{\varphi=0}^{2\pi} \tau_{\gamma\varphi a}^{\mu_1} \cdot \sin\gamma \cdot d\varphi \cdot d\gamma} \right\}^{\frac{1}{\nu_1}} \tag{16c}$$

$$\sigma_{vm} = \left\{ \frac{\int_{\gamma=0}^{\pi} \int_{\varphi=0}^{2\pi} \sigma_{\gamma\varphi a}^{\mu_2} \cdot \sigma_{\gamma\varphi m}^{\nu_2} \cdot \sin\gamma \cdot d\varphi \cdot d\gamma}{\int_{\gamma=0}^{\pi} \int_{\varphi=0}^{2\pi} \sigma_{\gamma\varphi a}^{\mu_2} \cdot \sin\gamma \cdot d\varphi \cdot d\gamma} \right\}^{\frac{1}{\nu_2}} \tag{16d}$$

In the shear stress intensity hypothesis, μ_1 and μ_2 are again set equal to 2. For simplicity, a value of 2 is also selected for ν_1. For evaluating the mean normal stress, the exponent ν_2 is set equal to unity; hence, the difference between a positive and a negative mean stress can be taken into consideration. The failure condition can be formulated by a combination of equivalent stresses.

$$a\tau_{va}^2 + b\sigma_{va}^2 + m \cdot \tau_{vm}^2 + n \cdot \sigma_{vm} = \sigma_W^2 \tag{17}$$

The coefficients a, b, m and n are determined by the requirement that the failure criterion can be satisfied for the uniaxial stress state.

$$a = \frac{1}{5}\left[3\left(\frac{\sigma_W}{\tau_W}\right)^2 - 4\right] \tag{18a}$$

$$b = \frac{1}{5}\left[6 - 2\left(\frac{\sigma_W}{\tau_W}\right)^2\right] \tag{18b}$$

$$m = \frac{\sigma_W^2 - \left(\frac{\sigma_W}{\tau_W}\right)^2 \cdot \left(\frac{\tau_{Sch}}{2}\right)^2}{\frac{4}{7}\left(\frac{\tau_{Sch}}{2}\right)^2} \tag{18c}$$

$$n = \frac{\sigma_W^2 - \left(\frac{\sigma_{Sch}}{2}\right)^2 - \frac{4m}{21} \cdot \left(\frac{\sigma_{Sch}}{2}\right)^2}{\frac{5}{7}\left(\frac{\sigma_{Sch}}{2}\right)} \tag{18d}$$

For this purpose, the characteristic parameters for alternating strength σ_W, pulsating tensile strength σ_{Sch}, alternating torsional strength τ_W, and pulsating torsional τ_{Sch} are required.

Since the coefficients a and b cannot be negative, the shear stress intensity hypothesis is applicable in the following range of fatigue strength ratio with respect to the materials:

$$\sqrt{\frac{4}{3}} < \frac{\sigma_W}{\tau_W} < \sqrt{3} \tag{19}$$

For extending the range of applicability, the exponents, μ_1 and μ_2, can be taken larger than 2.

For calculating the equivalent stress in accordance with the SIH, an integration is necessary for the general case. For synchronous biaxial stresses,

$$\begin{aligned} \sigma_x &= \sigma_{xm} + \sigma_{xa} \cdot \sin \omega t \\ \sigma_y &= \sigma_{ym} + \sigma_{ya} \cdot \sin \omega t \\ \tau_{xy} &= \tau_{xym} + \tau_{xya} \cdot \sin \omega t \end{aligned} \tag{20}$$

the equivalent stresses can be calculated analytically:

$$a\tau_{va}^2 + b\sigma_{va}^2 = \sigma_{xa}^2 + \sigma_{ya}^2 + \left[2 - \left(\frac{\sigma_W}{\tau_W}\right)^2\right] \cdot \sigma_{xa} \cdot \sigma_{ya} + \left(\frac{\sigma_W}{\tau_W}\right)^2 \cdot \tau_{xya}^2 \tag{21a}$$

$$\begin{aligned} \tau_{vm}^2 = \frac{1}{21} \cdot [\, & A_{11} \cdot \sigma_{xm}^2 + A_{12} \cdot \sigma_{ym}^2 + A_{13} \cdot \sigma_{xm} \cdot \sigma_{ym} + \\ & + A_{21} \cdot \tau_{xym}^2 + A_{22} \cdot \sigma_{xm} \cdot \tau_{xym} + A_{23} \cdot \sigma_{ym} \cdot \tau_{xym} \,] \end{aligned} \tag{21b}$$

$$\sigma_{vm} = \frac{3}{7} \cdot \left[A_{31} \cdot \sigma_{xm} + A_{32} \cdot \sigma_{ym} + A_{33} \cdot \tau_{xym}\right] \tag{21c}$$

The coefficients A_{ij} depend only on the mutual ratios of the stress amplitudes, σ_{xa}, σ_{ya}, and τ_{xya}. These coefficients are summarised in table 1.

Table 1. Coefficients A_{ij} for the estimation of the equivalent stresses according to Eq. (21); with $x = \sigma_{xa}$, $y = \sigma_{ya}$ and $z = \tau_{xya}$

i\j	1	2	3
1	$\frac{4x^2+3y^2-4x\cdot y+7z^2}{x^2+y^2-x\cdot y+3z^2}$	$\frac{3x^2+4y^2-4x\cdot y+7z^2}{x^2+y^2-x\cdot y+3z^2}$	$\frac{-4x^2-4y^2+6x\cdot y-6z^2}{x^2+y^2-x\cdot y+3z^2}$
2	$\frac{7x^2+7y^2-6x\cdot y+36z^2}{x^2+y^2-x\cdot y+3z^2}$	$\frac{10x\cdot y-6y\cdot z}{x^2+y^2-x\cdot y+3z^2}$	$\frac{-6x\cdot y+10y\cdot z}{x^2+y^2-x\cdot y+3z^2}$
3	$\frac{5x^2+y^2+2x\cdot y+4z^2}{3x^2+3y^2+2x\cdot y+4z^2}$	$\frac{x^2+5y^2+2x\cdot y+4z^2}{3x^2+3y^2+2x\cdot y+4z^2}$	$\frac{8(x+y)\cdot z}{3x^2+3y^2+2x\cdot y+4z^2}$

ENDURANCE BEHAVIOUR UNDER BIAXIAL STRESS

On the basis of a few load cases, the endurance behaviour is analysed and compared with calculation. More detailed descriptions are presented in **(12)** with the older formulation of the shear stress intensity hypothesis SIH.

The effect of the mean stress is first considered. This is described for the load case of an alternating normal stress with a static normal stress. If the direction of the static normal stress is the same as that of the alternating normal stress, the tolerable stress amplitude can be calculated from Eqs. (20) and (21):

$$\sigma_{xaD} = \sqrt{\sigma_W^2 - \frac{4m}{21}\sigma_{xm}^2 - \frac{5n}{7}\sigma_{xm}} \quad (22)$$

If the static normal stress is perpendicular to the alternating normal stress, the tolerable stress amplitude is given by

$$\sigma_{xaD} = \sqrt{\sigma_W^2 - \frac{m}{7}\sigma_{ym}^2 - \frac{n}{7}\sigma_{ym}} \quad (23)$$

In Fig.6, Eqs. (22) and (23) are compared with the test results in the Haigh diagram. Accordingly, a mean normal stress, which is perpendicular to the alternating normal stress, is less detrimental than one, which is parallel with the alternating normal stress.

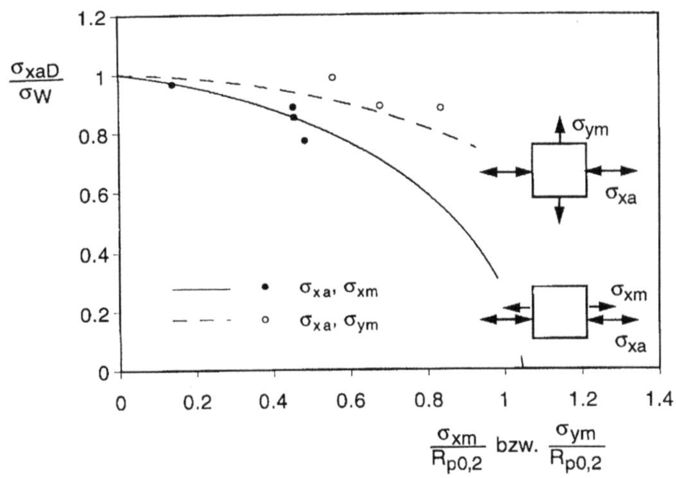

Fig.6 Effect of a static normal stress on the endurance limit of an alternating normal stress.

An interesting load case is that of a normal stress and a shear stress of different vibrational frequencies. For instance, this case occurs with a shaft under flexural and torsional load, if the natural frequencies of the two vibrations are different. For this case, a simple, explicit, approximate equation can be derived with the SIH **(11)**, if the frequencies differ sufficiently.

$$\left(\frac{\sigma_{xa}}{\sigma_W}\right)^2 + \left(\frac{\tau_{xya}}{\tau_W}\right)^2 + \left(\frac{13}{\tau_W^2} - \frac{4}{\sigma_W^2}\right) \cdot \frac{\sigma_{xa} \cdot \tau_{xya}}{5\pi} = 1 \quad (24)$$

In Fig.7, Eq.(24) is compared with corresponding test results. Accordingly, Eq.(24) is applicable for a frequency ratio λ_{xy}>or<2.

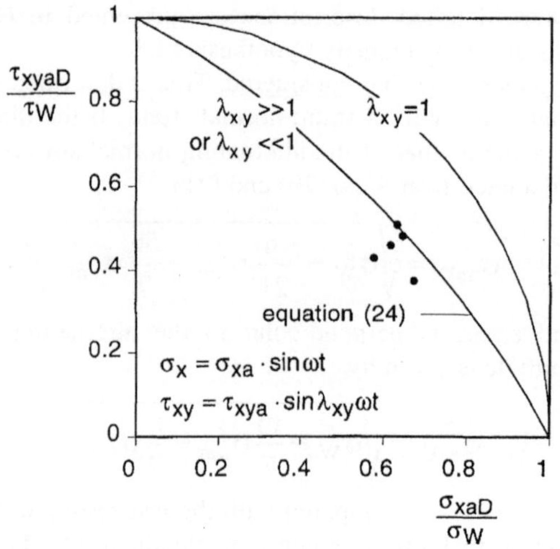

Fig.7 Fatigue limit curves for an alternating normal stress and an alternating torsional stress with different frequencies.

Finally, the improved shear stress intensity hypothesis is verified on the basis of the data compiled in **(11)**. A total of 214 test series for multiaxial load cases with superimposed mean stresses, with phase shifts, with various vibration forms and with frequency difference among the stress components have been recorded in this data base.

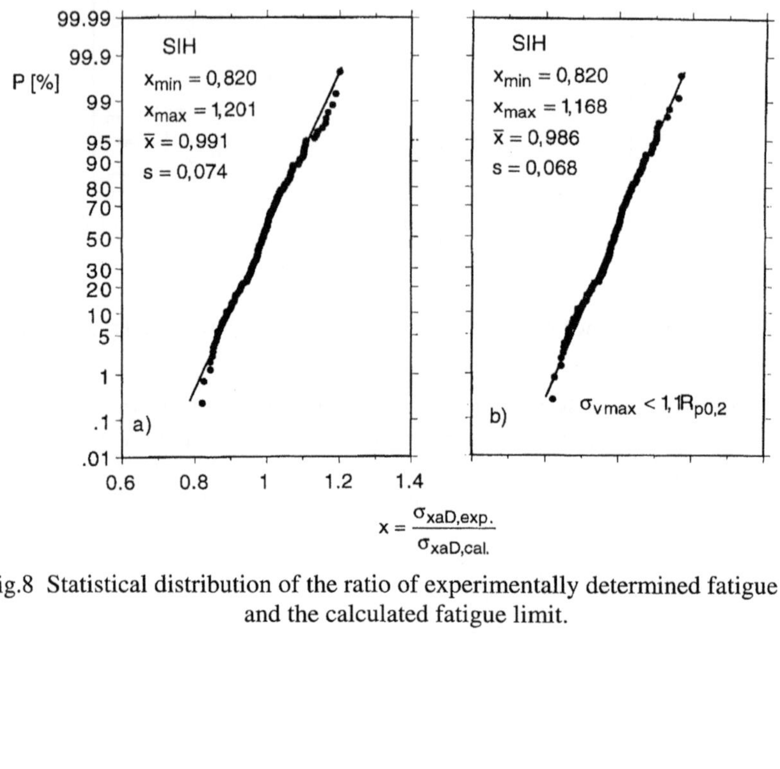

Fig.8 Statistical distribution of the ratio of experimentally determined fatigue limit and the calculated fatigue limit.

The ratio of the experimentally determined endurance limit to the value calculated with the SIH has been computed. The statistical distribution of this ratio is plotted on a Gaussian probability grid, Fig.8. On the average, the ratio is close to unity and exhibits relatively low scatter (standard deviation: s=0.074), Fig.8a. If the static failure limit is taken into account, and the maximal equivalent stress is limited to $\sigma_{vmax}<1.1R_{p0,2}$, the scatter is decreased even further (to s=0.068), Fig.8b.
As further modifications of the SIH, the equivalent stress of the shear stresses, Eqs.(16a) and (16c), can be extended by integration over the angle β, or larger values can be selected for the exponents μ_1, μ_2, ν_1 and ν_2. However, this does not appreciably affect the accuracy of prediction.

CONCLUSIONS

From the weakest link theory, the classical multiaxial criteria, the principal normal stress criterion, the maximum shear stress criterion and the von Mises criterion, have been derived as special cases. On the basis of this analysis, a general fatigue criterion is formulated for multiaxial stresses. The existing multiaxial criteria of integral approach and of the critical plane approach can be derived as special cases from the general fatigue criterion. On this basis, a new modification of the shear stress intensity hypothesis SIH that provides satisfactory agreement between experimental and calculated results is proposed.
An application of the proposed fatigue hypothesis to the service loading is also possible, but has not yet been attempted. For this purpose, an examination is necessary, especially to determine the extent to which the elastoplastic cyclic deformation can and must be taken into consideration in the strength analysis.

REFERENCES

(1) Heidenreich R., Zenner H. and Richter I., (1983), Dauerschwingfestigkeit bei mehrachsiger Beanspruchung. Forschungshefte FKM, Heft 105

(2) Simbürger A., (1975), Festigkeitsverhalten zäher Werkstoffe bei einer mehrachsigen, phasenverschobenen Schwingbeanspruchung mit körperfesten und veränderlichen Hauptspannungsrichtungen. Diss. TH Darmstadt

(3) Zenner H. and Richter I., (1977), Eine Festigkeitshypothese für die Dauerfestigkeit bei beliebigen Beanspruchungskombinationen. Konstruktion 29, pp. 11-18

(4) Troost A. and El-Magd E., (1977), Allgemeine quadratische Versagensbedingung für metallische Werkstoffe bei mehrachsiger schwingender Beanspruchung. Metall 31, pp 759-764

(5) McDiarmid D. L., (1985), The effects of mean stress and stress concentration on fatigue under combined bending and twisting. Fatigue Fract. Engng. Mater. Struct. 8, pp. 1-12

(6) Nøkleby J. O., (1918), Fatigue under Multiaxial Stress Conditions. Report MD-81 001, Div. Masch. Elem., The Norw. Institute of Technology, Trondheim/Norwegen

(7) Bhongbhibhat T., (1986), Festigkeitsverhalten von Stählen unter mehrachsiger phasenverschobener Schwingbeanspruchung mit unterschiedlichen Schwingungsformen und Frequenzen. Diss. Uni. Stuttgart

(8) Troost A., Akin O. and Klubberg F., (1987), Dauerfestigkeitsverhalten metallischer Werkstoffe bei zweiachsiger Beanspruchung durch drei phasenverschoben schwingende Lastspannungen. Konstruktion 39, pp. 479-488

(9) Martens H., (1988), Kerbgrund- und Nennspannungskonzepte zur Dauerfestig-

keitsberechnung- Weiterentwicklung des Konzeptes der Richtlinie VDI 2226. VDI-Berichte Nr 661, pp. 1-25

(10) Lüpfert H. P., (1989), B3BAF–Bewertung dreiachsiger Spannungen. Schmierungstechnik 20, pp. 125-127

(11) Liu J., (1991), Beitrag zur Verbesserung der Dauerfestigkeitsberechnung bei mehrachsiger Beanspruchung. Diss. TU Clausthal

(12) Liu J. And Zenner H., (1993), Berechnung der Dauerschwingfestigkeit bei mehrachsiger Beanspruchung. Mat.-wiss. und Werkstofftech. 24, part1: pp. 240-249, part 2: pp. 296-303 and part 3: pp. 339-347

(13) Martens H. and Hahn M., (1993), Vergleichsspannungshypothese zur Schwingfestigkeit bei zweiachsiger Beanspruchung ohne und mit Phasenverschiebungen. Konstruktion 45, pp. 196-202

(14) Lüpfert H. P., (1994), Beurteilung der statischen Festigkeit und Dauerfestigkeit metallischer Werkstoffe bei mehrachsiger Beanspruchung. Habilitationschrift TU Bergakademie Freiberg

(15) Papadopoulos I. V., (1994), A new criterion of fatigue strength for out-of-phase bending and torsion of hard metals. Int. J. Fatigue 16, pp. 377-384

(16) El-Magd E. and Wahlen V., (1994), Energiedissipationshypothese zur Festigkeitsrechnung bei mehrachsiger Schwingbeanspruchung. Mat.-wiss. Und Werkstofftech. 25, pp. 218-223

(17) Häfele P. and Dietmann H., (1994), Weiterentwicklung der Modifizierten Oktaederschubspannungshypothese (MOSH). Konstruktion 46, pp. 52-58

(18) Weibull W., (1939), A Statistical Theory of Strength of Materials. Ingeniörs Vatenskaps Akademiens Handlingar Nr 151 Generalstabens Litografiska Anstalts Förlag, Stockholm, 1939

(19) Batdorf S. B. and Crose J. G., (1974), Statistical theory for the fracture of brittle structures subjected to nonuniform polyaxial stresses. J.Appl.Mech. 41,pp.459-464

(20) Batdorf S. B., (1977), Some approximate treatments of fracture statistics for polyaxial tension. Int. J. Fract. 13, pp. 5-11

(21) Batdorf S. B. and Heinisch H. L. Jr., (1978), Weakest link theory reformulated for arbitrary fracture criterion. J. Am. Ceram. Soc. 61, pp. 355-358

(22) Evans A. G., (1978), A general approach for the statistical analysis of multiaxial fracture. J. Am. Ceram. Soc. 61, pp. 302-308

(23) Lamon J., (1988), Statistical approaches to failure for ceramic reliability assessment. J. Am. Ceram. Soc. 71, pp. 106-112

(24) Munz D. and Fett T., (1989), Mechanisches Verhalten keramischer Werkstoffe. Berlin: Springer-Verlag

(25) Scholz F., (1988), Untersuchungen zum statistischen Größeneinfluß bei mehrachsiger Schwingbeanspruchung. Forstschritt-Berichte VDI, Reihe 18, Nr 50, Düsseldorf: VDI-Verlag

(26) Ziebart W. and Heckel K., (1978), Ein Verfahren zur Berechnung der Dauerschwingfestigkeit in Abhängigkeit von der Form und der Größe eines Bauteils. VDI-Z. 120, pp. 677-679

(27) Böhm J. And Heckel K., (1982), Die Vorhersage von Dauerschwingfestigkeit unter Berücksichtigung des statistischen Größeneinflusses. Z. Werkstofftech. 13, pp. 120-128

(28) Heuser H., (1986), Funktionalanalysis. Stuttgart: B. G. Teubner

(29) Novoshilov V. V., (1961), Theory of Elasticity (J. J. Sherrkon trans.). Jerusalem: Israel Program for Scientific Translation

THERMOMECHANICAL DEFORMATION BEHAVIOUR OF IN 738 LC AND SC 16

Jürgen MEERSMANN, Josef ZIEBS*, Hellmuth KLINGELHÖFFER and Hans-Joachim KÜHN

Bundesanstalt für Materialforschung und –prüfung (BAM)
Unter den Eichen 87, D – 12205 Berlin, Germany

ABSTRACT

This article describes the study of uniaxial and biaxial thermo-mechanical fatigue (TMF) response of IN 738 LC and the initial experiments of single crystal superalloy SC 16. A life prediction assessment is proposed based on the inelastic work $\Sigma\sigma_{ij}\,\Delta\varepsilon_{ij}^{in}$ at $N_f/2$. It is shown that the J_2-theory is applicable to TMF-loadings. Initial experiments on single crystal superalloy SC 16 prove that there is a non-uniform strain distribution in the plastic region along the circumference of [001] orientated specimens. These findings must be weighed when performing TMF-tests.

KEY WORDS

Multiaxial thermomechanical fatigue, IN 738 LC, SC 16, life prediction, low cycle fatigue

NOTATION

A	coefficient, constant
A_1, A_2, A_3	constants
IP	in phase
OP	out of phase
MF	multiaxiality factor
T	temperature
T_0	room temperature
t	time
N_f	cycles to failure
m	exponent
Γ	torsion angle
α (T)	thermal coefficient of expansion
ε, $\bar{\varepsilon}$	strain, equivalent strain

* Corresponding author

ε_{th}	thermal strain
ε_m, $\bar{\varepsilon}_m$, $\Delta\bar{\varepsilon}_m$	mechanical strain, equivalent mechanical strain, strain range
$\dot{\varepsilon}$, $\dot{\bar{\varepsilon}}_m$	strain rate, equivalent mechanical strain rate
ε_{45}	readings of 45° strain gauge
γ	torsional strain
σ, $\bar{\sigma}$	axial stress, equivalent stress

INTRODUCTION

High temperature energy system components will often experience inelastic deformation resulting from mechanical loading, thermal transient cycles, and thermal gradients. Such forms of multiaxial thermomechanical fatigue (TMF) can involve damage mechanisms that differ significantly from those in isothermal fatigue. To understand these mechanisms, it is imperative that simple and simulated in service thermomechanical cycling should be explored.

The main purpose of the research reported in this paper is to develop a model to predict the TMF life of industrial gas turbine blades for multiaxial loading. TMF data for IN 738 LC under uniaxial loading has been given by Russel (1), Kuwabara et al. (2) and by Bernstein et al (3). The model developed is a semi-empirical model, similar to most engineering models that are actually used to predict low cycle fatigue. A secondary objective of this research is to better understand the multiaxial TMF behaviour of IN 738 LC and SC 16. Concerning the single crystal superalloy SC 16 the main features of the deformation behaviour are the anisotropic (structural and induced by the plastic flow) and inhomogeneous effects. A further objective of this research is to demonstrate how experimental results can give guidance to the development of time and temperature dependent constitutive models.

MATERIALS DETAILS AND EXPERIMENTAL PROCEDURES

The materials studied were the cast nickel base alloys IN 738 LC and SC 16. The chemical composition and the heat treatment of the alloys are presented in Table 1.

Table 1. Chemical composition of IN 738 LC and SC 16

C	Cr	Co	Mo	Ta	Ti	Al	W	Si	Mn	Nb	Fe	Zn
IN 738 LC												
0.105	15.99	8.7	1.77	1.9	3.45	3.4	2.71	0.09	0.03	0.82	0.3	0.037
Solution treated at 1120°C for 2 h, air-cooled, aged at 850°C for 24 h.												
SC 16												
0.01	15.4	0.17	2.8	3.5	3.48	3.45			<0.003	<0.003		
Solution treated at 1260 °C for 2 h, vacuum, high-temperature ageing at 1100 °C for 4 h, final ageing at 850 °C for 24 h, then finally air cooled.												

The test program comprised two groups of experiments: 1. uniaxial and tension-torsion thermomechanical fatigue with linear, diamond and sinusoidal cycling (simple TMF-tests) and 2. complex "bucket" uniaxial and tension-torsion thermo-mechanical tests. As can be seen in Table 2 the simple TMF-tests differ in the ε/γ-ratio, in the ε/T-and ε/γ-paths: tension-compression-tests with ε/T phase angles $\varphi_T = 0°$, linear in phase (IP) and 180° out of phase (180 OP) and diamond ε/T-cyclings, ± 90° (OP); torsion tests with γ/T phase angles $\varphi_T = 0°$ (IP) and $\varphi_T = 90°$ (OP) diamond and circular cycling; proportional tension-torsion tests with ε/T and γ/T phase angles $\varphi_T=0°$ (IP) and ± 90° (OP) sinusoidal and diamond cycling; non-proportional tension-torsion tests with ε/T phase angles $\varphi_T = 180°$ and γ/T phase angles $\varphi_T = -90°$ (diamond, sinusoidal). The complex "bucket" thermo-mechanical fatigue tests followed an assumed strain-temperature history representative of the leading edge of the first stage bucket in service. The tests were performed at equivalent strain ranges $\Delta\bar{\varepsilon}_m$ = 0.6%, 0.93 %, 1.24 %, equivalent strain rate $\dot{\bar{\varepsilon}}_m = 10^{-4}s^{-1}$, $10^{-5}s^{-1}$ the temperature range 450 °C < T < 950 C and temperature rate $\dot{T}$=1.67, 2.08, 2.58 and 4.21 K s^{-1}.

An effective stress range $\Delta\bar{\sigma}$ and an effective plastic strain range $\Delta\bar{\varepsilon}$ based on the von Mises criterion were used for correlating the multiaxial tension/torsion cyclic data:

$$\Delta\bar{\sigma} = \left[(\Delta\sigma)^2 + 3(\Delta\tau)^2\right]^{1/2} \tag{1}$$

and

$$\Delta\bar{\varepsilon} = \left[(\Delta\varepsilon_m)^2 + \frac{1}{3}(\Delta\gamma)^2\right]^{1/2} \tag{2}$$

where $\Delta\sigma$ and $\Delta\tau$ are the axial stress and shear stress ranges, respectively. $\Delta\varepsilon_m$ and $\Delta\gamma$ are the corresponding components of the axial and shear strain ranges.

Total specimen strain ε_{tot} was calculated by adding thermal ε_{th} and mechanical strain ε_m. The thermal strain was determined by

$$\varepsilon_{th} = \alpha\,(T)\,(T - T_o) \tag{3}$$

where T is the instantaneous temperature, T_o = 20 °C and α the coefficient of thermal expansion. Based on experimental data α (T) could be approximated by

$$\alpha(T) = A_1 + A_2(T - T_o) + A_3(T-T_o)^2 \tag{4}$$

where A_1, A_2, A_3 are constants.

Details of the corresponding isothermal low cycle fatigue data along with the cyclic hardening and softening behaviour can be found in (4).

Monotonous and sequential tension, torsion and tension-torsion loadings have been performed along <001> and within the standard triangle type orientations on hollow specimens of SC 16 to demonstrate the various aspects of non-homogeneous deformation. Seven or eight strain gauge rosettes have been attached on the circumference of the cross section with respect to the orientation of the axis at RT. Each strain gauge had dimensions of 0.76 x 0.76 mm. The tests were conducted under axial and torsional strain or displacement (stroke) and angle position control.

Table 2. Details of simple and complex thermomechanical strain paths

Temperature °C			Temperature rate K s^{-1}		Phase angle °		Strain versus temperature
T_1	T_2	T_s	$\dot{\varepsilon}=10^{-4}s^{-1}$	$\dot{\varepsilon}=10^{-5}s^{-1}$	ε - T	γ - T	
TMF Tensile-compressive strain ε_m = 0,6 %, 0,5 %, 0,4 % (CCD, CWD ε_m=0,6 %)							
950	450	700	4,17	0,42	0, 180		
850	600	725	2,08	0,21	0, 180		
750	550	650	1,67	0,17	0, 180		
950	450	450	4,17		-90		
850	600	600	2,08		-90		
450	950	950	4,17		90		
600	850	850	2,08		90		
TMF Shear strain $\gamma/\sqrt{3}$ = 0,6 %, $\dot{\gamma}/\sqrt{3}$ = $\dot{\varepsilon}$							
950	450	700	4,17	0,42		0, 90	
760	450	605	2,58	0,26		0, 90	
450	950	950	4,17	0,42		90	
450	760	760	2,58	0,26		90	
TMF Proportional axial and shear strain $\bar{\varepsilon}_m$ = 0,6 %							
950	450	700	4,17	0,42	0	0	
760	450	605	2,58	0,26	0	0	
450	950	700	4,17		-90	-90	
450	950	950	4,17	0,42	90	90	
450	760	760	2,58	0,26	90	90	
TMF Nonproportional axial and shear strain $\bar{\varepsilon}_m$ = 0,4 %, 0,5 %, 0,6 %							
450	950	700	4,17		180	-90	
600	850	725	2,08		180	-90	

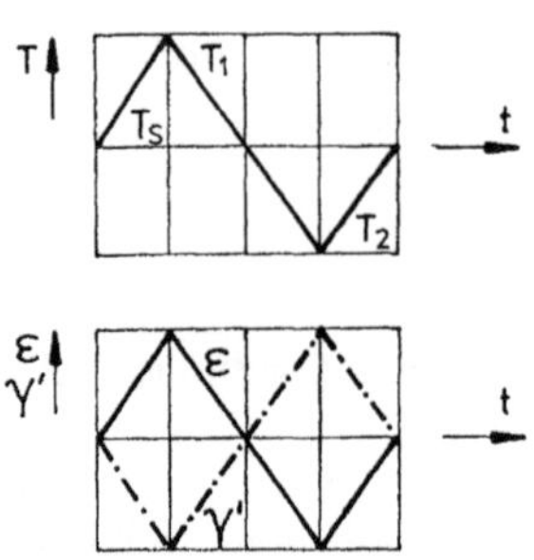

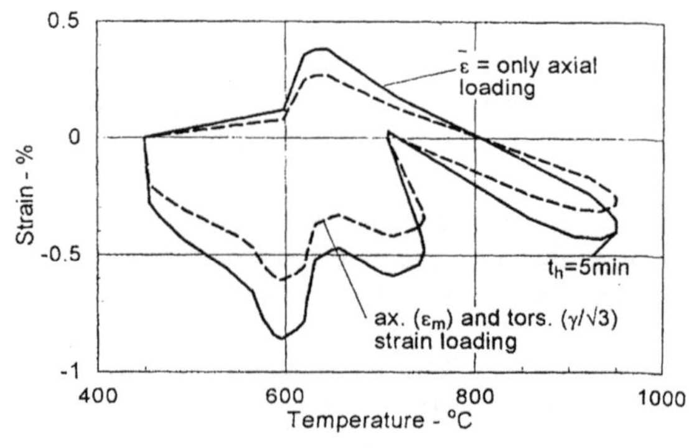

LIFE PREDICTION RESULTS

The simple TMF experiments were designed to investigate three points. Firstly, strain-temperature phase effects were studied by using in-phase, out-of-phase and diamond TMF cycles, where the axial strain, shear strain or the axial and shear strain amplitude were held constant, $\varepsilon_m = \bar{\varepsilon} = \pm$ 0.4, 0.5 and 0.6 %. Does this result in different fatigue lives? Secondly, how do different temperature ranges 450°C - 950°C, 450°C –760°C and 600°C – 850°C affect fatigue lives? Thirdly, can the isothermal data and the J_2-theory be used to describe the thermo-mechanical response of IN 738 LC?

The longest lives in air were exhibited by those samples which had undergone diamond-type loading (5). For those tests, the strain at the maximum temperature was zero. The shortest lives were seen in those tested under sinusoidal non-proportional strain paths.

Different temperature mechanical strain phasings, that is linear out-of-phase or in phase cycling, had a significant effect in fatigue lives in the temperature range 450°C to 950°C. A crossover of the linear TMF lines with temperature phasing of 0° or 180° is seen at about 10 cycles to failure, Fig.1. The linear out-of-phase TMF cycle best simulates the complex blade TMF cycle for uniaxial loading (axial strain) because the highest temperatures occur while the surface is in compression. Nevertheless more damaging thermal mechanical loading was found in sinusoidal tension-torsion cycles with or without hold times. On the other hand complex TMF tests gave longer lives at low strain ranges when compared with simple linear TMF tests. Since the high temperature deformation for complex TMF tests occurs at larger compressive inelastic strain, a shift in mean stress is seen in all tests.

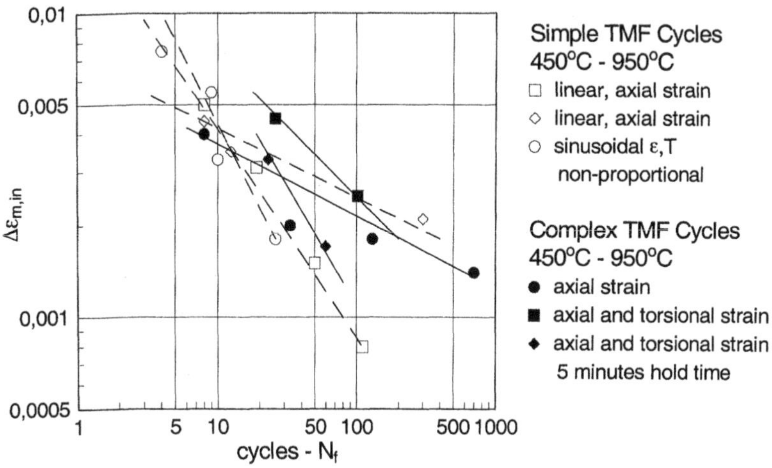

Fig.1 Inelastic strain range, $\Delta\varepsilon_{m,\,in}$, versus cycles to failure for simple and complex TMF tests of IN 738 LC, $\dot{\varepsilon} = 10^{-4}\,s^{-1}$.

The fatigue life was found to be mainly spent in the propagation of microcracks as for isothermal tests. Fatigue lives for the different TMF cycles are significantly less than for isothermal cycles at the same maximum temperature (4). This suggests that the thermal cycling introduce additional damage associated with thermal inhomogenities, either microscopically or macroscopically. Linear TMF tests conducted in the temperature range 450°C to 750°C gave longer lives, twice as long as the tests with the temperature range 450°C to 950°C (4). However, the environmental attack should also be considered because it is greater at higher temperatures.

The applied axial-torsional strain temperature history for proportional tension-torsion, non-proportional diamond and complex "bucket" loadings and the resulting axial and shear stress temperature/cycles to failure responses are presented in Fig.2.

The applied in-phase (φ_T = 0°) proportional strain-temperature path results in an asymmetric stress response (Fig.2a). In the range of high temperatures, 750°C to 950°C, a significant stress relaxation is observed due to dynamic recovery. The stress response for tension and torsion is similar.

The non-proportional diamond (φ_T = 90°) strain-temperature path achieves extremes of strain at T = 700°C, Fig.2b. The TMF stress response meets at comparable mechanical strains the response of isothermal tests in the temperature range 700°C to 450°C to 700°C. In the 850°C to 950°C to 850°C range softening is observed.

As shown in Fig.2c compressive stress peaks occur after warm-up, during acceleration and at base load for the blade strain temperature cycles. Tensile peaks occur before 'load' and during 'unload'. The largest inelastic strain is achieved after warm up. It is mainly reduced during the unload phase by tensile stress. The effects of different strain ranges ($\Delta\varepsilon$ = 1.24 %, 0.93 %, 0.60 %) upon the fatigue lives is shown in Fig.3 for the complex bucket cycle. Peak stresses ($\overline{\sigma}$, σ_{max}, σ_{min}, τ_{max}, τ_{min}) are plotted versus the number of cycles. The hardening/softening behaviour is similar to the behaviour of the isothermal counterparts. However, the different strain ranges generate increasing tensile mean stresses. The effect of these tensile mean stresses on fatigue life typically increases as the total strain range increases. Note that the simple TMF tests generate compressive mean stresses in the hysteresis loops (Fig.2). However, the magnitudes of these compressive mean stresses are relatively small in comparison to the magnitudes of those generated in the 'bucket' cycles in the tension mode. The detrimental effect on the fatigue lives was already mentioned above.

TEMPERATURE HISTORY EFFECTS

A comparison between the isothermal and the sinusoidal non-proportional tension-torsion TMF responses at temperatures 450°C to 950°C with an effective mechanical strain of $\overline{\varepsilon}$ =0.6% is shown in Fig.4. In this figure the thick solid line represents the stress loci of the TMF tests in the $\sqrt{3}\tau/\sigma$ - stress space and the circles the stress loci of isothermal tests under the same effective strain but at different temperatures. The experimental stress loci of the TMF and isothermal tests coincide only at the highest and lowest temperatures (600°C to 450°C), but discrepancies can be observed at other temperatures. The study of thermal history effects has brought out an important point about thermo-mechanical cycling. If the strain range and temperature levels activate similar deformation mechanisms, the isothermal values can be expected to model the TMF-response satisfactorily by interpolation. However, if different deformation mechanisms are activated, temperature history effects may be very important.

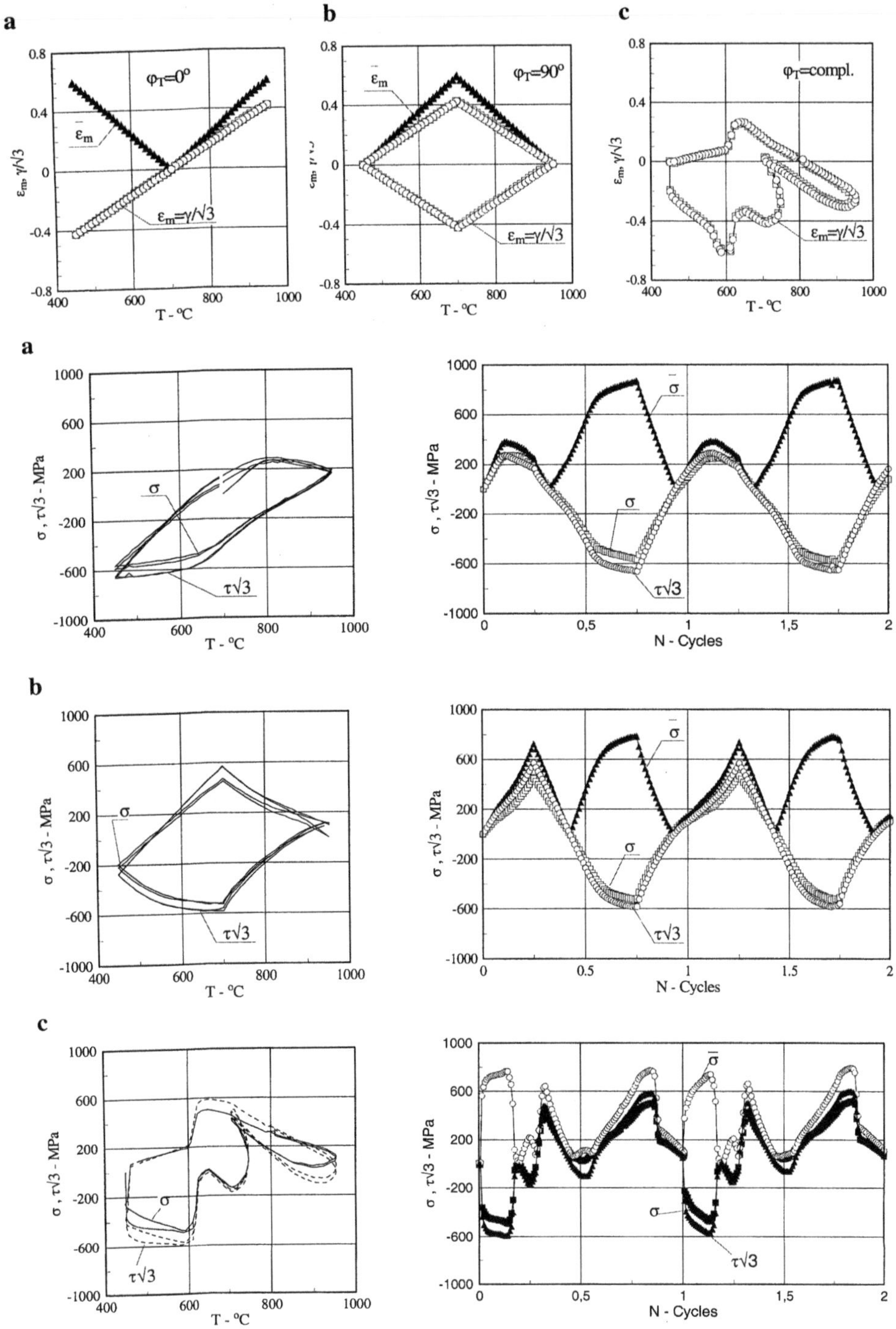

Fig.2 Applied strain and temperature phasing and stress responses for simple and complex TMF tests.

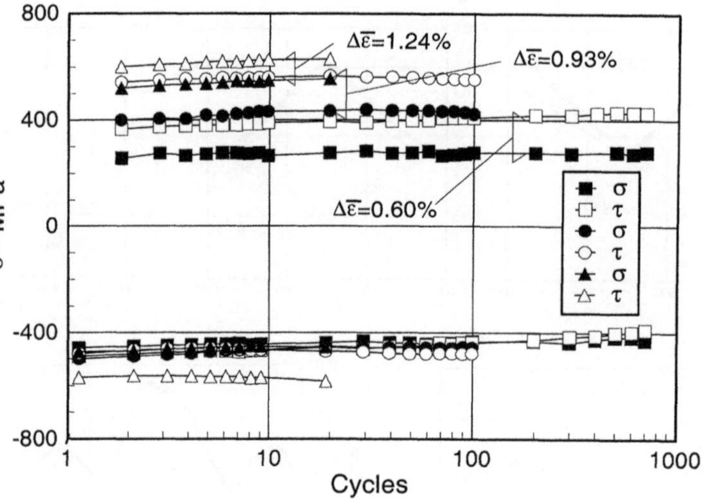

Fig.3 Axial and torsional stress versus number of cycles for 'bucket' cycles, $\Delta\bar{\varepsilon}$=0.6%, 0.93%, 1.24%.

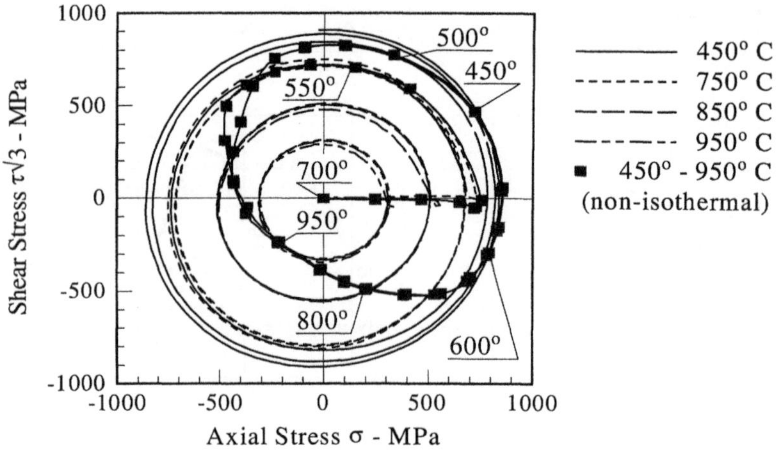

Fig.4 Comparison of TMF and isothermal stress responses for 90° out-of-phase strain cycling, $\bar{\varepsilon}$=0.6%, of IN 738 LC.

EXPERIMENTAL VERIFICATION OF THE J_2-THEORY FOR TMF LOADING

The equivalent stresses and strains of simple and complex TMF loadings were calculated by the von Mises relation, Fig.5. The curves of equivalent stresses versus equivalent strains lead to very complex shapes because all values are positive. Therefore the curves were changed by signs to get the usual hysteresis loops. Pure tension-compression- and torsion-tests as well as tension-compression-torsion-tests with $\varphi_T = 0$ (IP) are compared in Fig.5a. Fig.5b represents the OP-path with $\varphi_T = 90°$ and Fig.5c the complex path. In this diagram the pure torsional load is left. The equivalent stress values of pure tension-compression-tests are about 100 MPa lower than the other. The deviations are inside the scatterband of this material. These diagrams verify that the von Mises-hypothesis is altogether applicable to the deformation behaviour of IN 738 LC at TMF loading.

LIFE RELATIONS FOR THERMOMECHANICAL FATIGUE

Although numerous life prediction methods have been forwarded for simple TMF-tests (**6** – **9**), few studies were directly concerned with the fatigue life of 'bucket' TMF paths (**1**). At elevated temperatures, under cyclic loading, the inelastic deformation of metals is comprised of both plastic and viscous components. Since both plastic and viscous deformation mechanisms can serve as driving forces for microcrack initiation and propagation, cumulative fatigue and creep damage mechanisms are intimately coupled under TMF loading. The driving force for the propagation of the nucleated cracks in metals is a function of the shear and normal stresses acting on the crack plane (**10**). In multiaxial loading conditions the magnitude of the stresses depends on the strain paths. A global measure which includes all of the above mentioned parameters is the inelastic work. Therefore the general form of the criterion is:

$$\sum \sigma_{ij} \Delta\varepsilon_{ij}^{in} = A N_f^m \tag{5}$$

where the coefficient A and the exponent m is calculated by linear regression for each TMF history. Fig.6 shows the data of five TMF histories with their straight-lines of regression. The most outside data points limit a scatterband with a N_f – width factor 2.25 with respect to the scatterband midlife. N_f was defined as the cycle at which the maximum stress dropped off to 1 % of a steady-state value including also points after saturation. As is shown in Fig.6 the results of the uniaxial and biaxial TMF tests fall in a narrow band.

The quality of the lifetime prediction is also presented in Fig.6. As can be seen the correlation is generally within a factor of ± 2 of the medium between predicted and actual life.

Manson and Halford (11) proposed a stress based multiaxiality factor MF that modifies the inelastic work. The factor accounts for the change in ductility of a material as the state of stress changes

$$\sum MF\, \sigma_{ij}\, \Delta\varepsilon_{ij}^{in} = A N_f^m \tag{6}$$

where

$$MF = \frac{1}{2 - TF}, \qquad TF \le 1 \tag{7}$$

$$MF = TF, \qquad TF \le 1 \tag{8}$$

and

$$TF = \frac{\sigma_1 + \sigma_2 + \sigma_3}{\frac{1}{\sqrt{2}}\sqrt{(\sigma_1 - \sigma_2)^2 + (\sigma_2 - \sigma_3)^2 + (\sigma_3 - \sigma_1)^2}} \tag{9}$$

However, the Eq. (6) estimates the fatigue lives with nearly the same degree of accuracy as without the multiaxiality factor.

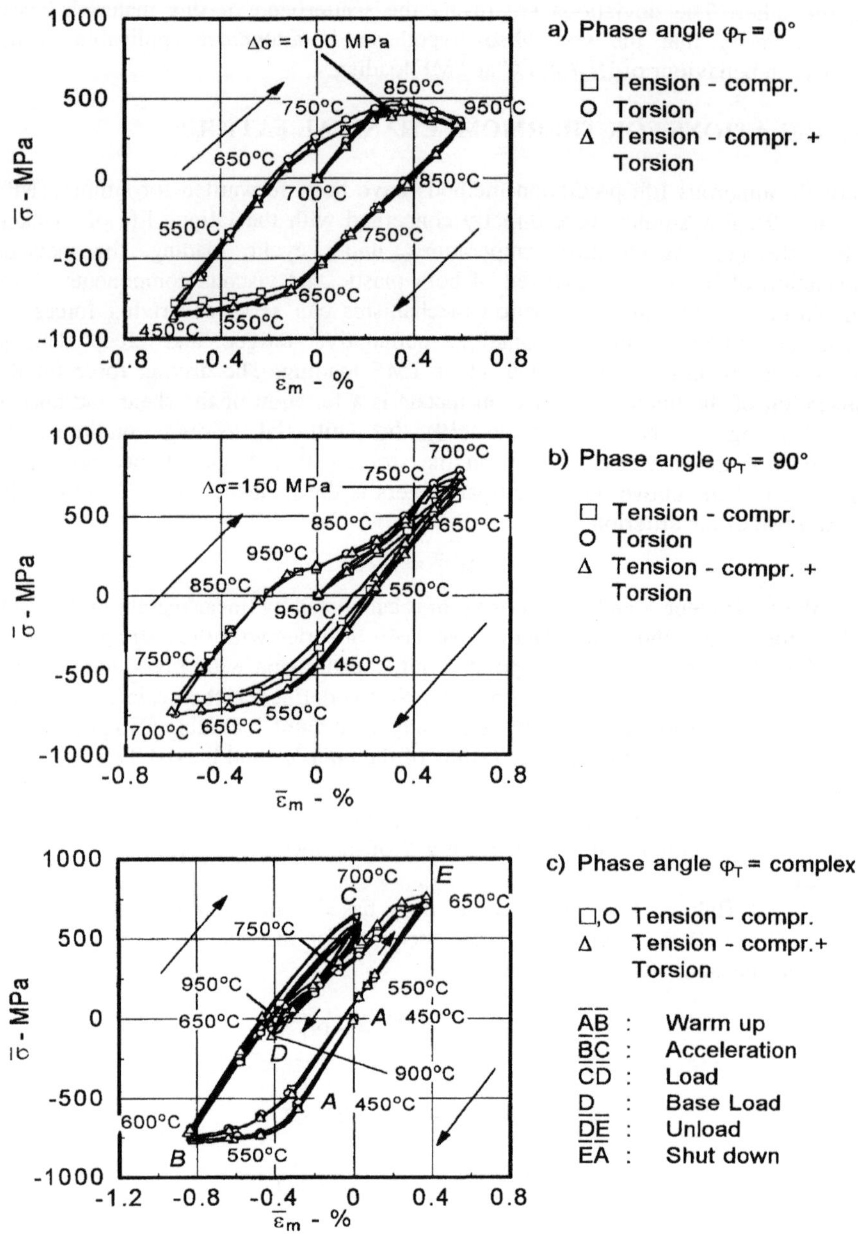

Fig.5 Comparison of uni-and multiaxial TMF-tests (von Mises relation) 450 °C >T > 950 °C, $\dot{\bar{\varepsilon}}=10^{-4}\ s^{-1}$, $\Delta\bar{\varepsilon}_m=1.24\%$.

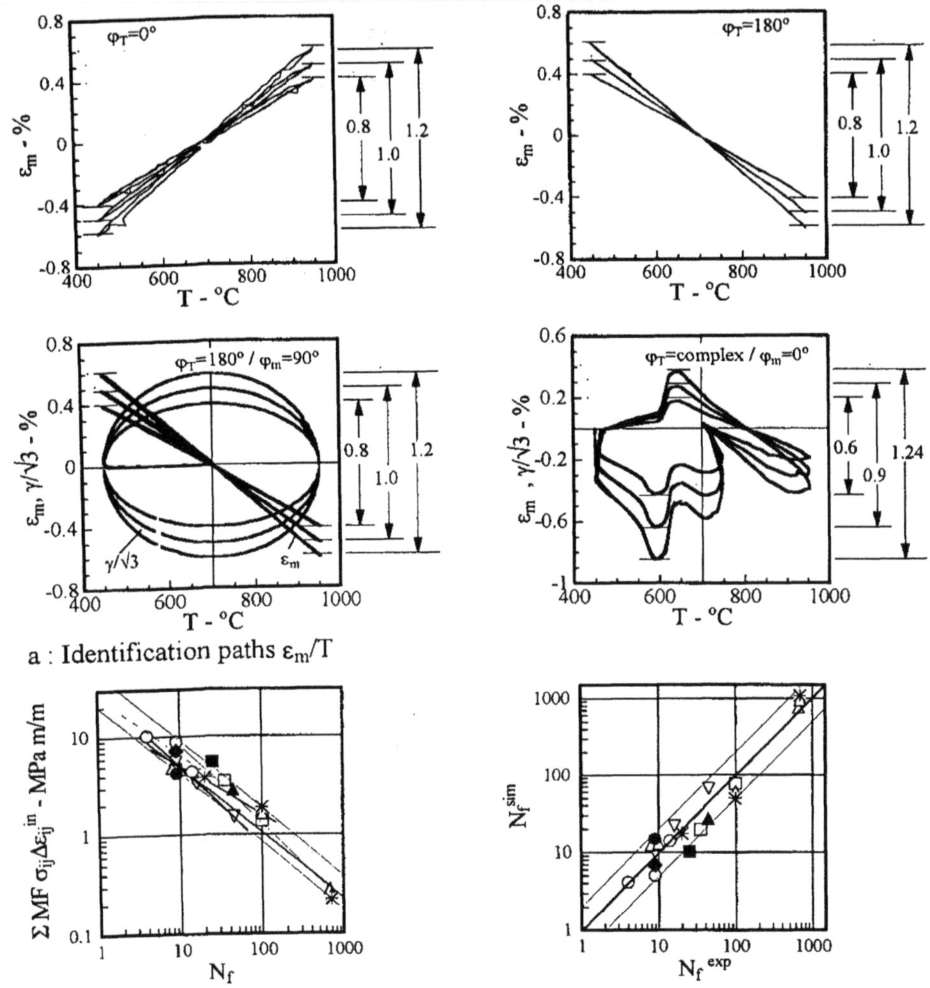

b : Determination of A, m by ID-tests.

c : Verification of A, m.

Identification (ID-) - tests

tension-compr.:
- ▽ Ψ_T = 0°
- □ Ψ_T = 180°
- △ Ψ_T = compl.

tension-compr.+ torsion:
- ✱ Ψ_T = compl.
- ○ Ψ_T = 180°, sinus.

Verification-tests

....

- ▲ Ψ_T = 90°, Ψ_m = - , ΔT = 850 - 600°C
- ◆ Ψ_T = 180°, Ψ_m = 90°, ΔT = 950 - 450°C
- ■ Ψ_T = 180°, Ψ_m = 90°, ΔT = 850 - 600°C
- ● Ψ_T = 0°, Ψ_m = 0°, ΔT = 950 - 450°C

Fig.6 $\sum \sigma_{ij}\Delta\varepsilon_{ij}^{in}$ versus number of cycles for different tests including applied strain paths (at the top) and simulated versus experimental life.

EXPERIMENTAL STUDY OF THE ANISOTROPIC AND NON-HOMOGENEOUS DEFORMATION BEHAVIOUR OF SC 16 SINGLE CRYSTALS

In order to take full advantage of Nickel based single crystal superalloys it is necessary to understand the anisotropic and non-homogeneous deformation behaviour. Therefore the local deformation characteristics of the SC 16 alloy were investigated via detailed studies of the local deformation with strain gauges at RT. The observations were analysed and interpreted in terms of crystallographic slip. In discussing TMF testing on single crystal superalloys it will be necessary to deal with non-uniform strain distribution along the circumference of different oriented tubular specimens under tension, torsion and tension-torsion loading.

EXPERIMENTAL EVIDENCE OF THE TENSILE BEHAVIOUR

The mechanical response of SC 16 alloy in a tensile test depends on material orientation, temperature, strain and specimen shape and is quite different from that of a polycrystalline alloy. Initially round tubular specimens of different diameters (26.5 mm outside diameter, 1.5 mm thickness; 20 mm outside diameter, 1.5 mm thickness) in the near [001] or within the standard triangle type orientations deform into cross sections with a complex shape, Fig.7. In contrast the post yield response of solid specimens shows a cross section with a circular or elliptical shape. There is not a clear explanation of this fact up to now. The reason is mainly to be found in the complexities of the deformation modes present in two- phase materials. The active deformation modes and their critical stresses depend on the composition, the temperature, the strain rate, the stress state and the previous deformation history. The work of Chin and Mammel (12) was the first attempt to do a systematic analysis of activated slip systems in the standard stereographic triangle. For {111} <101> slip systems, loading of fcc materials in or near the [001] direction will activate eight slip systems simultaneously if the stress is equivalent to the critical resolved shear stress (CRSS). Theoretical predictions of the combinations of active slip systems in a single crystal of a particular orientation have usually been based on one or two extreme sets of assumed boundary conditions. In the case of hollow specimens there could be particular boundary conditions on strain. The geometrical part of single-crystal plasticity is then the yield condition. This provides the basis for deriving which combination of slips will be activated among the many that are kinematically possible to achieve a given strain increment, and what particular stress state is necessary to activate this combination.

TORSION TESTS ON <001> SPECIMENS AT RT

Local strain measurements by means of eight strain gauge rosettes at room temperature prove that four 'soft' zones are present in the specimens near <110> and four hard zones in <100> orientations, Fig.8. This fact has already been mentioned by Nouailhas et al (13). The simulation of such non-homogeneous test requires octahedral and cube slip.

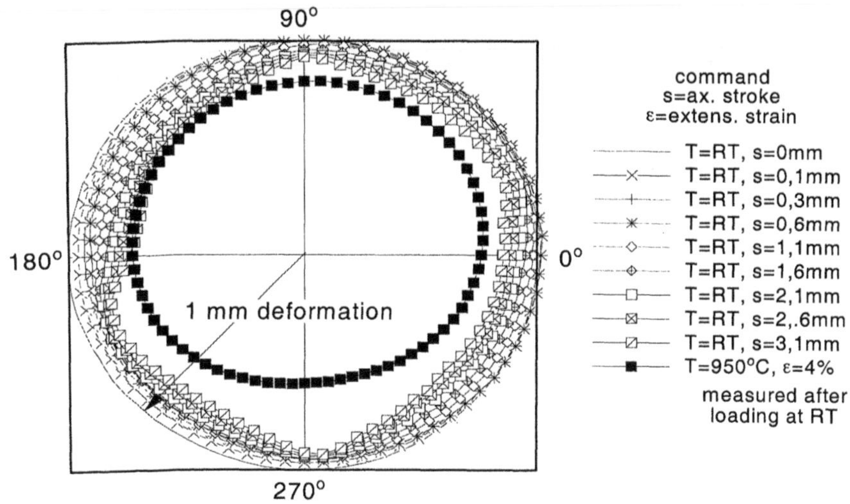

Fig.7 Cross sections (contours) of hollow specimens at RT and T = 950°C.

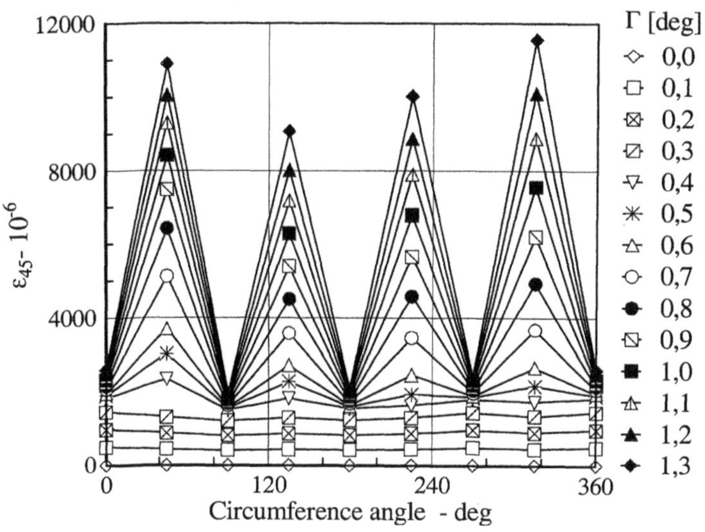

Fig.8a ε_{45}-curves versus the circumference angle.

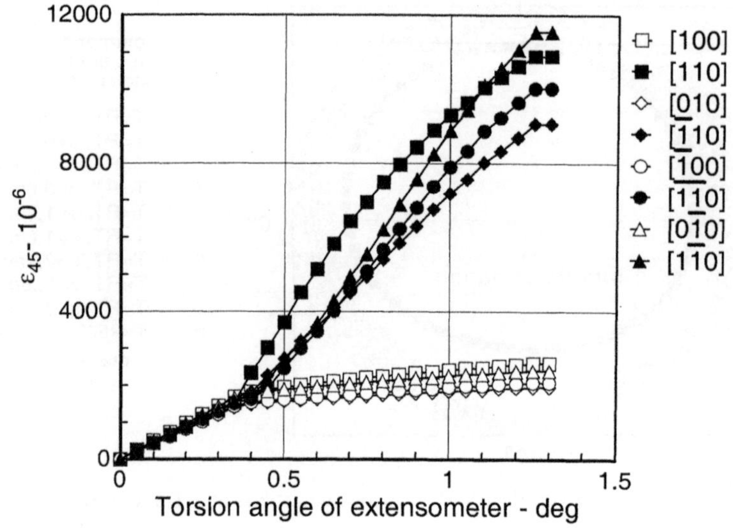

Fig.8b ε_{45} – curves versus the torsion angle of the extensometer.

TENSION-TORSION TESTS ON <001> SPECIMENS AT RT

A completely non-symmetrical non-homogeneous deformation behaviour is observed under tension-torsion testing on [001] oriented specimens at room temperature. Eight strain gauge rosettes were attached along the circumference as above. One of them is located near a <100> direction; the others were regularly disposed. Axial and torsional strains were also measured by means of an axial-torsional extensometer attached to the specimen near [010]. Fig.9 shows the results of the strain gauges in the axial direction and at 45°. Different tension-torsion loading paths were measured. As can be seen in this diagram the deformation is found to be non-uniform and non-symmetrical. There are two 'soft' regions. When tension becomes predominant, quasi – uniform straining is obtained only in the axial directions.

These findings were confirmed by yield surface tests, Fig.10. The surface was determined by following radial (proportional) stress paths from the origin with an axial-torsional extensometer attached at different positions along the circumference. A small von Mises equivalent strain offset of yielding, ~ 10^{-5}mm/mm was selected.

Octahedral slip (oblique segments) and cube slip (horizontal segments) are both involved in this diagram. The experiments prove, Fig.10 a, that there is a non-uniform inelastic strain distribution along the circumference of <001> orientated specimens due to different slip systems (octahedral and cube slip). However, the assumption of homogeneous stresses in the specimen lead to same experimental yield surfaces, Fig.10 b. This diagram also confirms that a quadratic expression (14, 15) of a macroscopic yield criterion is not acceptable for a single crystal.

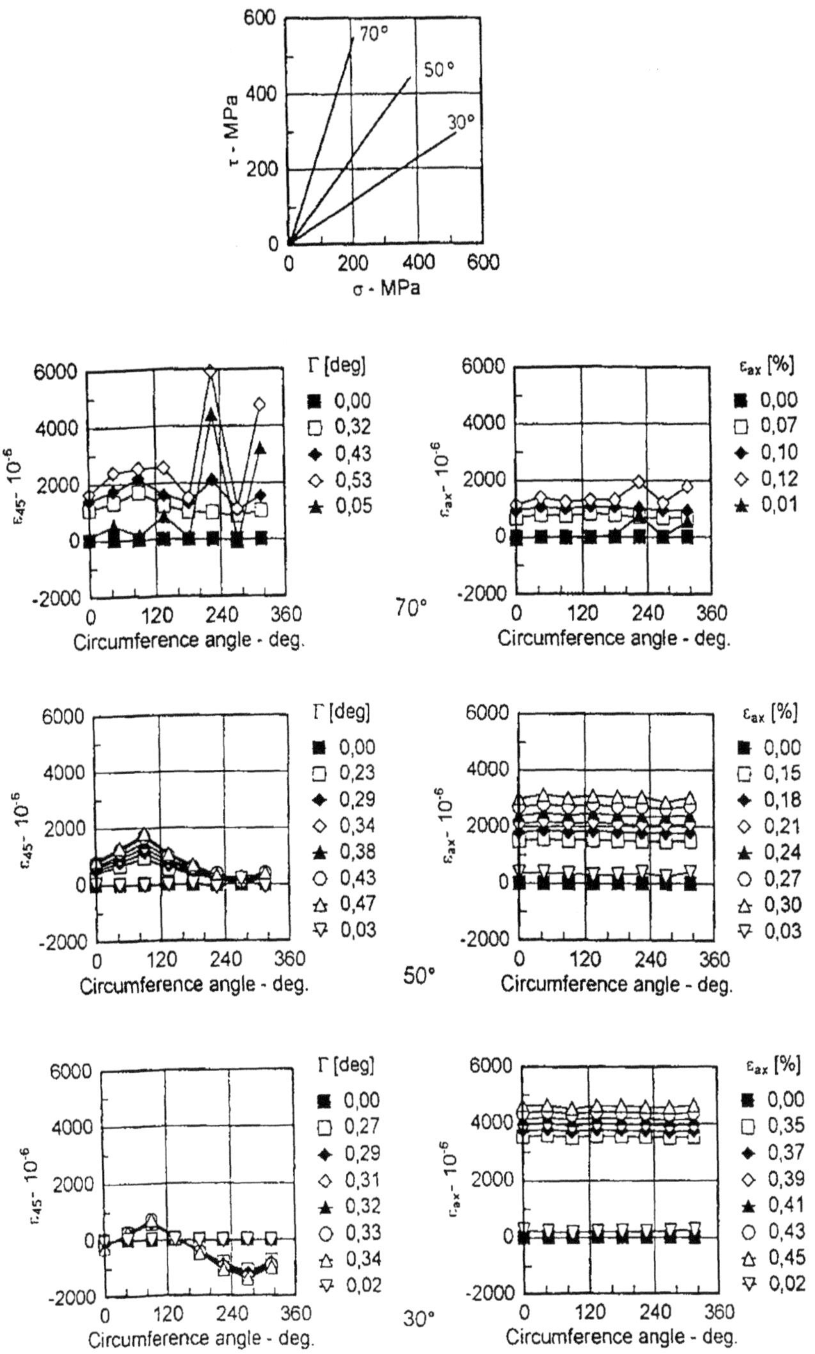

Fig.9 Readings of eight strain gauge rosettes in the axial direction and at 45° versus the circumference angle in dependence of the loading paths 30°, 50°, 70° (at the top).

a

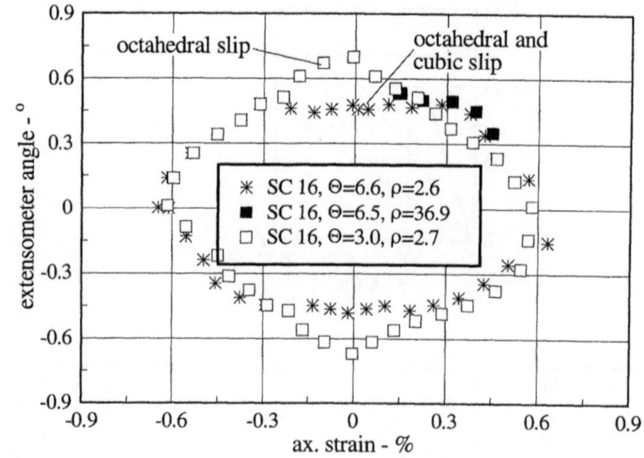

b

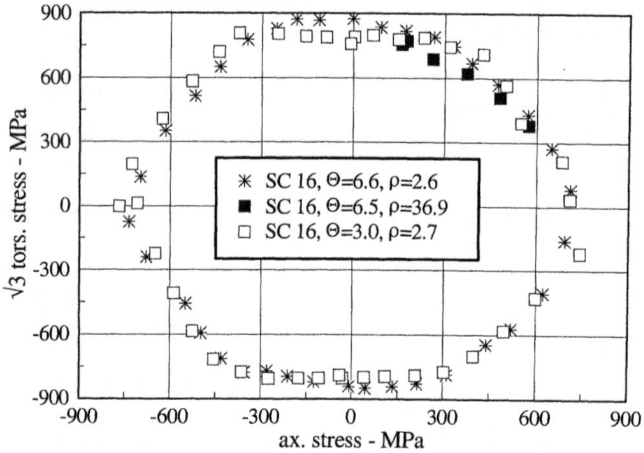

Fig.10 Yield surfaces of a [001] oriented specimens,

a) applied axial-torsional strain paths, extensometer at different positions along the circumference
b) shear stress-axial stress plane(Θ, ρ: orientation of the crystal axes in terms of the angles Θ and ρ in the stereographic triangle).

CONCLUSIONS

This study has demonstrated the versatility of the life prediction assessment. It can be applied to any arbitrary temperature-strain phasing. The sinusoidal TMF-time histories show the greatest inelastic works and result in the fewest cycles to failure. The scatterband with a N_f – width-factor is nearly the usual of 2.0.

It was verified by experiments that the v. Mises-hypothesis is altogether applicable to the deformation behaviour of IN 738 LC at TMF loading.

Initial experiments on single crystal superalloy SC 16 prove that there is a non-uniform strain distribution in the plastic region along the circumference of [001] orientated specimens under torsion or tension-torsion loading. This fact must be weighed when exact TMF tests are performed. The behaviour reported can be explained in terms of slip on a finite number of slip systems.

REFERENCES

(1) Russel E. S., (1986), Practical life prediction methods for thermal-mechanical fatigue of gas turbine buckets, Proc. Conference on Life Prediction for High-Temperature Gas Turbine Materials, (Edited by V. Weiss and W.T. Bakker, EPRI AP-4477, Electric Power Research Institute, Palo, Alta, CA), pp 3-1 – 3-39

(2) Kuwabara K., Nitta A., and Kitamura T., (1983), Thermal-mechanical fatigue life prediction in high-temperature component. Materials for power plant, Proc. of the ASME International Conference on Advances in Life Prediction Methods, Albany NY, pp 131 – 141

(3) Bernstein H.L., Grant T.S., McClung R.C. and Allen J.M., (1993), Prediction of thermal-mechanical fatigue life for gas turbine blades in electric power generation, Thermo-mechanical Fatigue Behaviour of Materials, ASTM STP 1186 (Edited by H. Sehitoglu), ASTM, Philadelphia, pp. 212 – 238

(4) Ziebs J., Meersmann J., Kühn H.-J. and Ledworuski, S., (1992), High temperature inelastic deformation of IN 738 LC under uniaxial and multiaxial loading, Low Cycle Fatigue and Elasto-Plastic Behaviour of Materials - LCF 3, (Edited by K.T. Rie), Elsevier Applied Science, pp. 248 – 255

(5) Meersmann J., Ziebs J. and Kühn, H.-J., (1994) The thermo-mechanical behaviour of IN 738 LC, Materials for Advanced Power Engineering, Part I, (D. Coutsouradis et al, Eds.), Kluwer Academic Publishers, pp 841 - 852

(6) Swanson G.A., Linask,J. Nissley D.M.,Norris P.P., Mayer T.G. and Walker K.P., (1987), Life prediction and constitutive models for engine hot section anisotropic materials program, NASA Report 179594, Washington, D.C.

(7) Remy L., Bernard H., Malpertu I.L. and Rezai-Aria F., (1993), Fatigue life prediction under thermal-mechanical loading in a Nickel-base superalloy, Thermomechanical Fatigue Behaviour of Materials for Testing and Materials, ASTM STP 1189, (H. Sehitoglu, Eds), ASTM, Philadelphia, pp 3 – 16

(8) Neu R.W. and Sehitoglu H., (1989), Thermomechanical fatigue, oxidation and creep: part I. Damage mechanisms, and part II. Life prediction, Metallurg. Transactions A., vol. 20 A, pp 1769 – 1783

(9) Bakis C.E., Castelli M.G. and Ellis J.R., (1993), Thermo-mechanical behaviour, Advances in Multiaxial Fatigue, ASTM STP 1191, (D.L. Mc Dowell and R. Ellis, Eds.), ASTM, Philadelphia, pp 223 – 243

(10) Socie D., (1993), Critical plane approaches for multiaxial fatigue damage assessment, Advances in Multiaxial Fatigue, ASTM STP 1191, (D.L. Mc Dowell and R. Ellis, Eds.), ASTM, Philadelphia, pp 7 – 35

(11) Manson S.S. and Halford. G.R., (1971), Discussion, multiaxial low cycle fatigue of type 304 stainless steel, J. Eng. Mat. Techn., ASME, vol 99, pp 283 – 286

(12) Chin G.Y. and Mammel W.L., (1967), Computer solution of the Taylor analysis for axisymmetric flow, Transactions Metallurg. Soc. AIME, vol. 239, pp 1400 – 1405

(13) Nouailhas D. and Cailletaud G., (1995), Tension-torsion of single crystals superalloys: experiment, Int. J. Plasticity, vol. 11, No. 4, pp 451 – 470

(14) Hill R., (1948), A theory of the yielding and plastic flow of anisotropic metals, Proc. of the Royal Society of London, Series A, pp 281 – 297

(15) Lee D. and Zaverl F., (1978), A generalized strain rate dependent constitutive equation for anisotropic metals, Acta Metallurgica, pp 1771 - 1780

Acknowledgements

The study presented here is part of an extensive investigation into IN 738 LC and SC 16 under multiaxial states of stress and temperature history. The financial support for this research, provided by Deutsche Forschungsmeinschaft (DFG), is gratefully acknowledged.

A MESOSCOPIC APPROACH FOR FATIGUE LIFE PREDICTION UNDER MULTIAXIAL LOADING

Franck MOREL*, Narayanaswami RANGANATHAN*, Jean PETIT*
and André BIGNONNET**

* Laboratoire de Mécanique et de Physique des Matériaux, ENSMA - Futuroscope
** Direction des Recherches et Affaires Scientifiques, PSA Peugeot Citroën –Bièvres

ABSTRACT

This paper deals with the presentation of a high cycle multiaxial fatigue life prediction method for metallic materials. By means of the mesoscopic approach introduced by Dang Van and developed by Papadopoulos, accumulated plastic strain due to external loading is estimated at a scale on the order of a grain or a few grains. Its evaluation requires the use of a critical plane type fatigue criterion. As soon as the accumulated plastic mesostrain, considered as the damage variable, reaches a critical value, a crack is considered to be initiated. The complex and combined cases of loading (multiaxial and variable amplitude) can be analysed with this new method. Particular attention is given to a description of the detrimental effect of out-of-phase loadings. A good agreement has been found between the predicted and experimental results for in-phase and out-of-phase sinusoidal constant amplitude loadings by examining a large amount of experimental data.

KEY WORDS

High cycle fatigue, out-of-phase loading, lifetime prediction, variable amplitude loading, critical plane

NOTATION

Macroscopic quantities:

$\underline{\underline{\Sigma}}$ macroscopic stress tensor
$\underline{\underline{E}}$ macroscopic strain tensor
$\underline{C}$ macroscopic shear stress vector
$\underline{T}$ macroscopic resolved shear stress vector acting on an easy glide direction
T_a amplitude of macroscopic resolved shear stress
P macroscopic hydrostatic stress

Mesoscopic quantities :

$\underline{\underline{\sigma}}$ mesoscopic stress tensor
$\underline{\underline{\varepsilon}}$ mesoscopic strain tensor

$\underline{\tau}$ mesoscopic resolved shear stress vector acting on an easy glide direction
γ^p mesoscopic shear plastic strain
τ_y shear yield limit of a crystal
Γ accumulated plastic mesostrain
T_σ measure proportional to an upper bound of the plastic mesostrain accumulated on an elementary material plane Δ, also average value of T_a
T_Σ maximum value of T_σ
H phase-difference coefficient

INTRODUCTION

The mesoscopic approach introduced by Dang Van **(1)** and developed by Papadopoulos **(2-3)** forms the basis of this study. Multiaxial endurance criteria built according to this theory has been successfully **(4)** used to predict fatigue behaviour of mechanical components. Nevertheless, these methods only differentiate a damaging cyclic loading from a non damaging one. When a failure event is to be predicted, it is important to know how many cycles must be applied to reach it. Most of the fatigue life prediction methods proposed are built by extending, to the limited fatigue life regimes, endurance criteria expressed in terms of macroscopic mechanical parameters **(5)**. In the high cycle fatigue field, crack initiation is a phenomenon taking place at the scale of a few grains. Consequently, it seems natural to introduce a damage variable computed at this scale. Papadopoulos **(2)** used the accumulated mesoscopic plastic strain. We will make the same choice to propose a method producing fatigue life prediction for multiaxial constant or variable amplitude loading.

OVERVIEW OF THE MESOSCOPIC APPROACH

To depict fatigue crack initiation phenomenon in polycrystalline metallic materials, two scales of description of a material are distinguished: the usual macroscopic scale and a mesoscopic one. The macroscopic scale is defined with the help of an elementary volume V determined at any point O of a body as the smallest sample of the material surrounding O that can be considered to be homogeneous. Usually, engineers use stresses and strains measured or estimated at this scale. V contains a large number of grains (crystals) and the mesoscopic scale is defined as a small portion of this volume. In the high cycle fatigue regime, some grains undergo local plastic strain while the rest of the matrix behaves elastically (the overall plastic strain is negligible). It seems, therefore, legitimate to use the scheme of an elastoplastic inclusion submitted to uniform plastic strain $\underline{\underline{\varepsilon}}^p$ and embedded in an elastic matrix, both having the same elastic coefficients. If the total strains of the matrix $\underline{\underline{E}}^e$ and of the inclusion $\underline{\underline{\varepsilon}}^e + \underline{\underline{\varepsilon}}^p$ are supposed to be the same (Lin-Taylor hypothesis), it follows that **(1)**:

$$\underline{\underline{\sigma}} = \underline{\underline{\Sigma}} - 2\mu\underline{\underline{\varepsilon}}^p \tag{1}$$

where $\underline{\underline{\Sigma}}$ and $\underline{\underline{\sigma}}$ are the macroscopic and the mesoscopic stress fields, $\underline{\underline{\varepsilon}}^p$ is the plastic mesostrain and μ is the shear modulus.

By assuming that only one glide system (defined by a normal vector $\underline{n}$ to a plane and a direction $\underline{m}$ on this plane) is active per every plastically deforming grain of the metal, Papadopoulos **(3)** established from the last relation a macro-meso passage for a glide system activated in a flowing crystal:

$$\underline{\tau} = \underline{T} - \mu\gamma^{p}\underline{m} \tag{2}$$

where $\underline{\tau}$ and $\underline{T}$ are the mesoscopic and macroscopic resolved shear stresses acting along the slip direction $\underline{m}$:

$$\underline{\tau} = \left(\underline{m}.\underline{\underline{\sigma}}.\underline{n}\right)\underline{m} \tag{3}$$

$$\underline{T} = \left(\underline{m}.\underline{\underline{\Sigma}}.\underline{n}\right)\underline{m} \tag{4}$$

γ^{p} is the magnitude of the plastic mesoscopic shear strain.

A CRITICAL PLANE TYPE FATIGUE CRITERION

The ability of a loading to create a macrocrack will be checked here through an endurance criterion based on a critical plane approach and presented by Papadopoulos in reference **(3)**. We have observed that, during a high cycle fatigue test, some less favourably oriented grains (mesoscopic scale) of V are subjected to plastic glide. The fatigue limit can therefore be related to some characteristic quantities of an elastic shakedown state reached by these plastically deforming crystals. A parameter, T_σ, proportional to an upper bound of the plastic mesostrain accumulated in some crystals of V, has been introduced **(3)**. It has been shown that the limit to apply on this parameter depends on the maximum value P_{max} that reaches the mesoscopic (equal to the macroscopic) hydrostatic stress during a loading cycle. The criterion is written as:

$$\max_{\theta,\varphi}\left(T_\sigma(\theta,\varphi)\right) + \alpha P_{max} \leq \beta \tag{5}$$

T_σ is a function of the orientation of a material plane Δ through the angles θ and φ, spherical co-ordinates of the unit normal $\underline{n}$ to the plane Δ :

$$\underline{n} = \begin{pmatrix} \sin\theta\cos\varphi \\ \sin\theta\sin\varphi \\ \cos\theta \end{pmatrix} \tag{6}$$

$T_\sigma(\theta,\varphi)$ is estimated by an integration carried out through the whole area of the plane Δ :

$$T_\sigma(\theta,\varphi) = \sqrt{\int_{\psi=0}^{2\pi} T_a^{\,2}(\theta,\varphi,\psi)\,d\psi} \tag{7}$$

Where T_a is the amplitude of the macroscopic resolved shear stress acting on a line of the plane Δ directed by $\underline{m}$ (Figure 1). This line is located by the angle ψ that makes with an arbitrary but fixed axis in Δ.

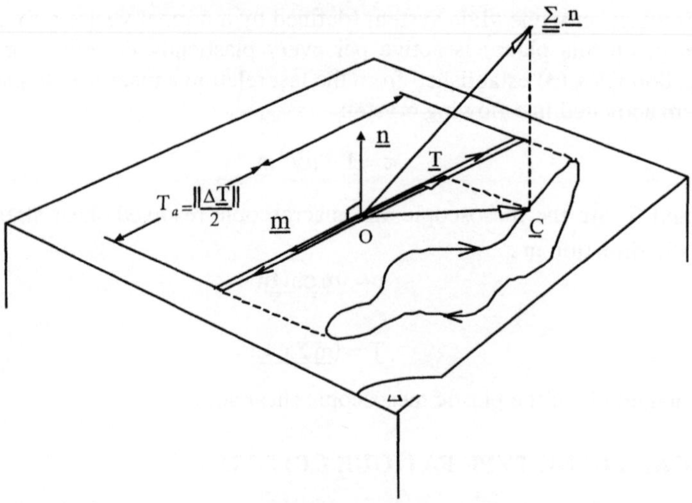

Fig.1 Path of the macroscopic shear stress $\underline{C}$ acting on a material plane Δ and corresponding path of the macroscopic resolved shear stress $\underline{T}$ acting on an easy glide direction.

The material parameters α and β can be related to the fatigue limits of two standard fatigue tests, for example fully reversed tension-compression, s, and fully reversed torsion, t:

$$\alpha = \sqrt{\pi}\frac{t - \frac{s}{2}}{\frac{s}{3}} \qquad \beta = \sqrt{\pi}\, t \tag{8}$$

Hereafter, in order to make relations less cumbersome, maximum value of T_σ will be denoted as T_Σ:

$$T_\Sigma = \max_{\theta,\varphi}\left(T_\sigma(\theta,\varphi)\right) \tag{9}$$

LIFE PREDICTION ASSESSMENT PROCEDURE

Sinusoidal constant amplitude loading case

We first consider the synchronous sinusoidal loadings defined by:

$$\Sigma_{ij}(t) = \Sigma_{ijm} + \Sigma_{ija}\sin(\omega t - \beta_{ij}) \qquad i,j = x,y,z \tag{10}$$

where Σ_{ija} and Σ_{ijm} are amplitudes and mean values of the (i,j) stress components and β_{ij} represents the phase differences between (i,j) stress component and a reference stress Σ_{xx} (β_{xx}=0). On the critical material plane Δ_c related to the maximum measure T_Σ, the loading path described by the shear stress vector is elliptic and the corresponding amplitude of the shear stress, defined as half of the longest chord of the closed curve, is denoted as C_A.

Definition of a multiaxial limit loading

A particular multiaxial loading can be defined according to the endurance limit concept applied to the criterion (T_Σ, P_{max}). If sinusoidal constant amplitude loadings are defined by the same mean values Σ_{ijm} and the same phase angles β_{ij} between the stress components and a simple multiplicative coefficient is applied to all the amplitudes then, in the plane (T_Σ, P_{max}), they are displayed by points lying on a same line (Fig.2). These loadings are said to be "similar". The multiaxial limit loading is defined as one of these loadings. It is displayed by a point, which belongs to the threshold endurance line delimiting the domain of safe operation against fatigue. The corresponding mechanical parameters are denoted as $T_{\Sigma lim}$ and $(P_{m.} + P_a)_{lim}$ where $T_{\Sigma lim}$ is function of the ratio $\frac{T_\Sigma}{P_a}$ and of the mean of the hydrostatic pressure P_m:

$$T_{\Sigma lim} = \frac{-\alpha P_m + \beta}{\alpha + \frac{T_\Sigma}{P_a}} \frac{T_\Sigma}{P_a} \tag{11}$$

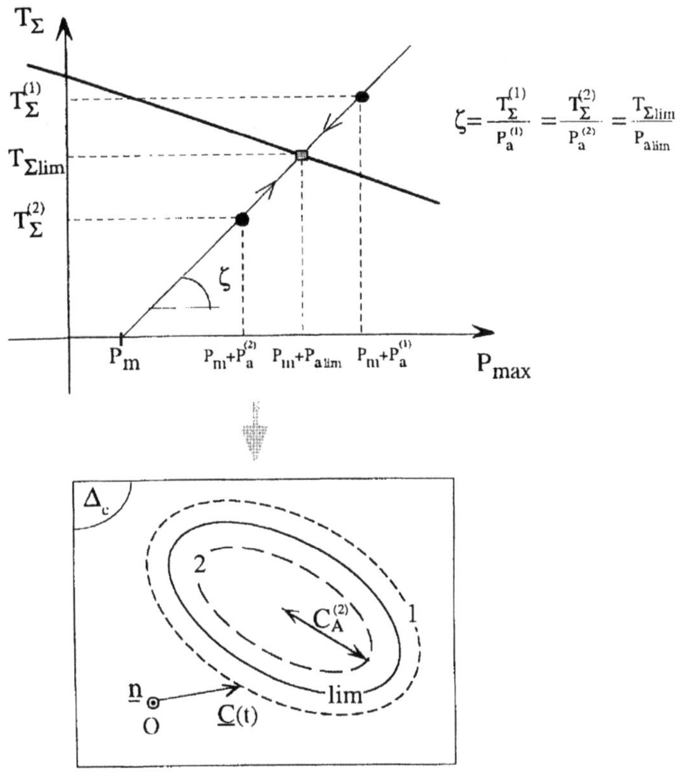

Fig.2 Determination of the limit loading characteristics in the plane of the endurance criterion (T_Σ, P_{max}) from two similar loadings and corresponding elliptic paths on the critical material plane Δ_c.

A newly defined phase-difference coefficient

For a similar loadings group described above, one can show **(6)** that the ratio $\frac{T_\Sigma}{C_A}$ remains constant. It will constitute a phase-difference coefficient denoted as H:

$$H = \frac{T_\Sigma}{C_A} \tag{12}$$

The more the elliptic path is open, the higher is the coefficient H (Fig.3). For a proportional loading, H is equal to $\sqrt{\pi}$. In the case of a particular circular path, H reaches the maximum value $\sqrt{2\pi}$. The linear path and the circular one lead to two bounds of the coefficient H. Since H is the same for these two similar loadings, it follows that:

$$\frac{T_\Sigma}{C_A} = \frac{T_{\Sigma \lim}}{\tau_{\lim}} \tag{13}$$

where $\tau_{\lim}$ is the amplitude (on the critical plane) of the macroscopic shear stress for the limit loading. From the last relation, expression of $\tau_{\lim}$ can be deduced:

$$\tau_{\lim} = \frac{T_{\Sigma \lim}}{H} \tag{14}$$

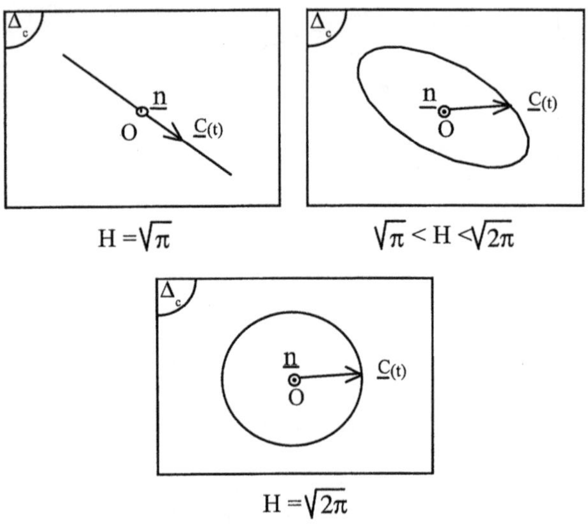

Fig.3 Different paths and corresponding phase-difference coefficient H values.

For a constant value $T_{\Sigma \lim}$, $\tau_{\lim}$ decreases while H increases. The coefficient H is able to reflect the influence of the path shape on the parameter $\tau_{\lim}$ related to the endurance limit. If we imagine that each direction on the critical plane Δ_c is related to an easy glide direction of a crystal and that only one glide system is active per every plastically deforming crystal, then the use of the parameter T_σ (considering all the directions of a material plane) in $\tau_{\lim}$ estimation can be understood as a precise description of the contribution of many grains to damage mechanism. When the crystals are equally stressed (circular path), τ_{lim} reaches its minimum value because H is maximum.

Damage estimation and initiation criterion

Initiation of fatigue cracks in metals is known to be a consequence of cyclic plastic strain localisation **(7).** The cumulative plastic mesostrain will then be considered as the principal cause of damage accumulation. In the same way as in Papadopoulos' work **(3),** the crystal is assumed to follow a combined isotropic and kinematical rule when flowing plastically and the initiation of slip in the crystal is determined by Schmid's law. A crystal starts to deform plastically when the shear stress acting on the slip plane in the slip direction, reaches a critical value denoted as τ_y. Three successive linear isotropic hardening rules are adopted to describe the crystal behaviour from initial yield to failure. The yield limit starts to increase in the initial hardening phase, remains constant in the saturation phase (represented by τ_s) and then decreases in the softening phase (Fig.4b)). The crystal is said to be broken as soon as the yield limit becomes negligible. By using the macro-meso passage of the relation (2) and linear isotropic and kinematical hardening rules, damage (accumulated plastic mesostrain) evolutions in the three phases can be drawn **(6)** (Fig.4a)).

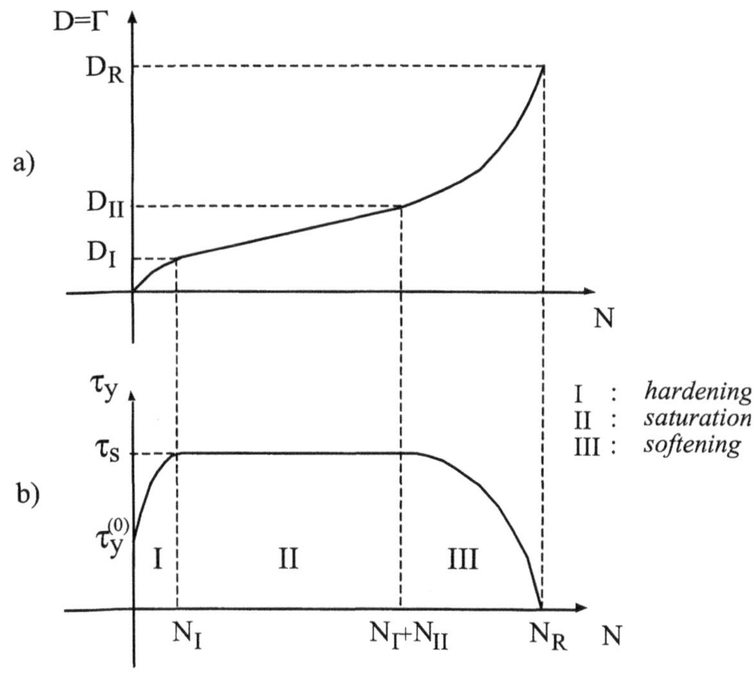

Fig.4 a) Damage and b) yield limit evolutions (mesoscopic scale) in the three behaviour phases (hardening, saturation and softening) when a cyclic loading is applied.

It is assumed that crack initiation occurs by the breaking of the most stressed grains along the plane experiencing T_Σ (maximum value of T_σ) and that only one glide system operates in them **(8).** Consequently, it seems natural (on this critical plane) to be interested in these plastically less resistant grains whose easy glide directions coincide with the direction leading to the maximum value of the macroscopic resolved shear stress C_A. Once the accumulated plastic mesostrain Γ along this particular gliding system reaches a critical value Γ_R, these grains are said to be broken and an analytical expression of the number of cycles to initiation can be achieved **(6)**:

$$\Gamma = \Gamma_R \Rightarrow N_i = p\ln\left(\frac{C_A}{C_A - \tau_{lim}}\right) + q\frac{\tau_{lim}}{C_A - \tau_{lim}} - \frac{r}{C_A} \qquad (15)$$

where p, q and r are functions of the hardening parameters of the three phases defined above.

In the last relation, the detrimental effects of out-of-phase loadings are introduced through τ_{lim}. As the coefficient H increases, τ_{lim} as well as N_i decrease so more damage is accumulated. The identification of the model parameters requires two endurance limits (parameters α and β of the endurance criterion) and a single S-N curve (parameters p, q and r). The location of the critical plane by means of the measure T_σ is the first step of the life prediction procedure. C_A, T_Σ and the phase-difference coefficient H are computed on this plane. Afterwards, τ_{lim} and $T_{\Sigma lim}$ estimations are carried out according to the endurance criterion (T_Σ, P_{max}). Finally, N_i is readily deduced from (15).

APPLICATIONS

A number of data concerning multiaxial (out-of-phase or in-phase) constant amplitude fatigue loading has been found in the scientific literature **(9-12).** The data are from bending-torsion tests, tension-torsion tests and from tests on thin-walled tubes under internal pressure and axial load. Fig.5 shows the comparison between estimated and measured lives.

The diagonal represents perfect agreement between the model and experimental results. The dashed line determines an interval of agreement by a factor of three. It should be noted that the eight points related to Dubar's data have been evaluated through a statistical analysis on 200 tests. Every other point on the graph represents a failed specimen. The theoretically predicted lives using the proposed model and the experimentally measured ones are in fairly good agreement except for Sonsino data. It appears through this case (where only a few S-N curve data are available) that accurate statistical evaluation of the fatigue limits used in the model is an essential condition to good life predictions. In conclusion, more experimental work is needed to fully assess the model's behaviour and accuracy.

Variable amplitude loading case

In general, service loadings, which are applied to mechanical components, vary irregularly with time. For instance, suspension arms of a car are exposed, during their service life, to a large number of cycles of variable amplitudes, caused by external forces and resulting in the possibility of fatigue cracking. The prevention of such fatigue cracks and failures therefore requires relevant fatigue life prediction method. An extension of the previously described method to variable amplitude multiaxial stress history can be proposed by using statistical parameters.

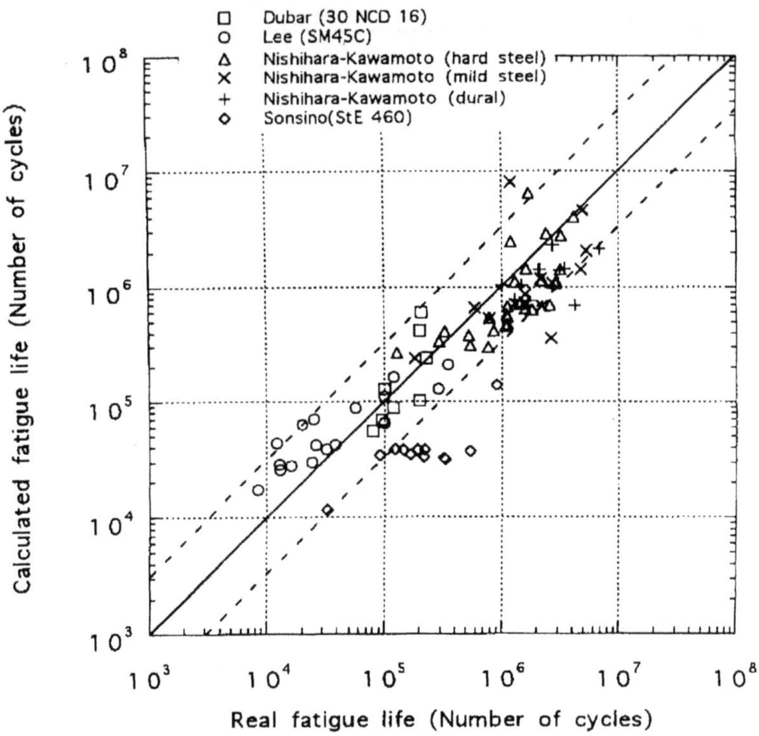

Fig.5 Calculated fatigue life versus fatigue life from multiaxial constant amplitude fatigue tests.

Damage accumulation

With the present method, damage accumulation is still carried out by adopting three successive linear isotropic hardening rules of crystal behaviour (Fig.4). Besides, the mechanical parameters representative of the loading and used for damage accumulation are the macroscopic resolved shear stress on a particular gliding system and the hydrostatic stress. The yield limit acts as a filter that defines the part of a transition leading to damage. Fig.6 shows, for a complex loading, the evolution of the macroscopic resolved shear stress $\mathcal{T}(t)$ on a particular gliding system and the yield limit $\tau_y^{(i)}$ reached at the ith extremum. The segment length denoted as $\Omega_{i\to i+1}$ is proportional to the plastic mesostrain $\Gamma_{i\to i+1}$ accumulated during the transition from i to i+1 :

$$\Gamma_{i\to i+1} \propto \Omega_{i\to i+1} \tag{16}$$

and

$$\Omega_{i\to i+1} = |T_{i+1} - T_i| - 2\tau_y^{(i)} \tag{17}$$

where T_i and T_{i+1} are values of the extrema i and i+1.

If this sequence is applied successively until failure, the yield limit will first increase in the hardening phase, remains constant during saturation and decrease in the softening phase. With the present way of accumulating damage, no counting method is required. Indeed, damage is deduced step by step from the hardening rules. This fact is quite new because most of the fatigue life prediction methods in the literature apply successively a counting method and a damage law without any links between them.

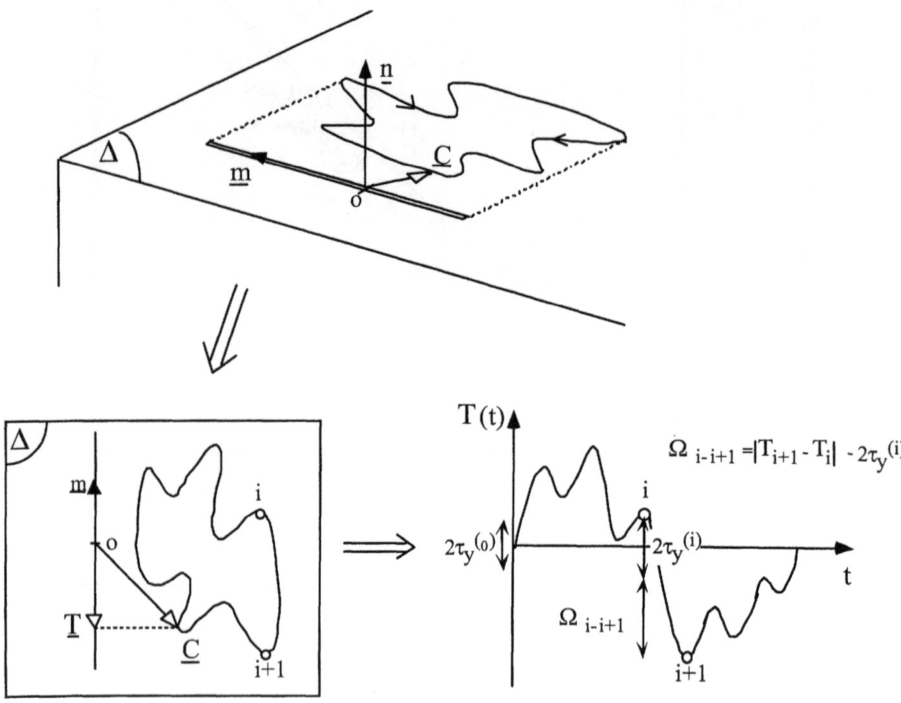

Fig.6 Accumulation of plastic mesostrain when a complex loading is applied.

Statistical analysis

Complete knowledge of hardening rules is achieved when the saturation phase yield limit is known. Its estimation requires the introduction of statistical parameters. The mechanical factors T_σ and C_A are no more relevant when loadings vary irregularly with time. At a first approximation, the root mean square value of the macroscopic resolved shear stress denoted as T_{rms} seems more convenient in this case:

$$T_{rms} = \sqrt{\frac{1}{N}\sum_i (T_i - T_{mean})^2} \tag{18}$$

where T_i are the peak and valley values of T(t) evolution, N their number and T_{mean} the mean value defined by:

$$T_{mean} = \frac{1}{N}\sum_i T_i \tag{19}$$

In the expression of T_σ previously mentioned, T_a is now replaced by T_{rms} and, by this way, a new parameter denoted as $T_{\sigma rms}$ can be defined:

$$T_{\sigma rms}(\theta,\varphi)=\sqrt{\int_{\psi=0}^{2\pi} T_{rms}^2(\theta,\varphi,\psi)d\psi} \tag{20}$$

Like T_σ, $T_{\sigma rms}$ is a function of the orientation of the material plane Δ through the angles θ and φ, spherical co-ordinates of the unit normal $\underline{n}$ to a plane Δ. The maximum value of $T_{\sigma rms}$ will be denoted as $T_{\Sigma rms}$:

$$T_{\Sigma rms}=\max_{\theta,\varphi}\left(T_{\sigma rms}(\theta,\varphi)\right) \tag{21}$$

The critical material plane on which damage estimation will be carried out corresponds to the measure $T_{\Sigma rms}$. On this plane denoted as Δ_c, a variable amplitude loading generates a macroscopic shear stress vector path of complex shape. A "global" phase-difference coefficient, denoted as H, can be introduced to take into account the "out-of-phase content" of this complex loading sequence. H is equal to the ratio between $T_{\Sigma rms}$ and C_{rms} where C_{rms} is the maximum value of T_{rms} on the critical plane:

$$H=\frac{T_{\Sigma rms}}{C_{rms}} \tag{22}$$

As in the previous case of sinusoidal constant loading, H is bounded by $\sqrt{\pi}$ and $\sqrt{2\pi}$.

From this point, $T_{\Sigma lim}$ and τ_{lim} computations follow the same procedure as previously, T_Σ and P_a being replaced respectively by $T_{\Sigma rms}$ and P_{rms} (P_{rms} is the root mean square value of the hydrostatic pressure).

Fatigue life prediction steps (Fig.7)

Required material fatigue characteristics are two endurance limits, an S-N curve and a particular two constant amplitude blocks loading test. The first step of the fatigue life prediction procedure is once again the location of the critical material plane through a maximisation of the parameter $T_{\sigma rms}$. After $T_{\Sigma rms}$, P_{rms}, $T_{\Sigma lim}$ and τ_{lim} computations, damage accumulation can be estimated from the evolution of the macroscopic resolved shear stress $\mathcal{T}(t)$ on a direction of the critical plane. The number of sequences to crack initiation is deduced from a calculation on the direction leading to the highest accumulated plastic mesostrain.

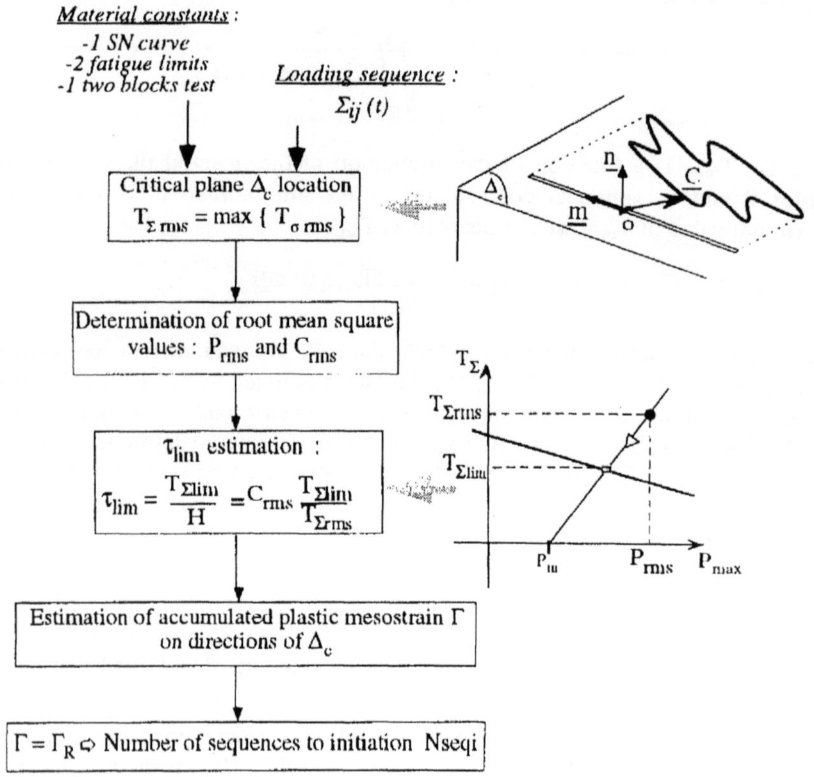

Fig.7 Diagram of algorithm for evaluation of fatigue life of metals under(uniaxial or multiaxial) variable amplitude loading

DISCUSSION

Difficulty in the application of a linear damage rule (Miner rule) results from improper characterisation of damaging events and from not taking the interaction effect into consideration. With our model, some sequence effects are considered. In fact, yield limit evolution in the hardening and softening phases leads to a non-linear damage accumulation. Sequence effects depend on the material behaviour phase reached in the plastically deforming crystals. On the contrary, in the saturation phase, no sequence effect occurs and damage accumulation is linear.
The rainflow counting method first proposed by Endo **(13)** was initially applied to the case of low cycles fatigue. The author considered that a successive application of plastic strain was the principal cause of fatigue crack initiation. Load cycles directly linked to the apparition of closed hysteresis loops in the stress-plastic strain plane was then extracted. The proposed method is similar to the work of Endo. Although the macroscopic strain is purely elastic in the high cycle fatigue field, a loading range responsible for plastic mesostrain can be defined with the help of the macro-meso passage.

CONCLUSIONS

A new fatigue life prediction method that can be applied to any kind of loading (multiaxial and variable amplitude) is presented in this paper. A crack is supposed to initiate by failure of some plastically deforming grains following three successive phases of behaviour: hardening, saturation and softening. The damage variable chosen is the plastic strain accumulated at a mesoscopic scale and its estimation requires the location of the plane subjected to maximum damage. A newly defined coefficient is introduced to represent the effect of phase difference on damage accumulation. Besides, sequence effects are reflected by the non-linear damage accumulation in two of the three behaviour phases adopted.

The predicted results for in-phase and out-of-phase cyclic loading are in good agreement with the experimental results.

A critical plane type fatigue criterion has been used in the paper but an extension to a volume type approach is in progress.

REFERENCES

(1) Dang Van K., Griveau B. and Message O., (1982), On a new multiaxial fatigue limit criterion: theory and application, Biaxial and Multiaxial Fatigue, EGF Publication 3, (Edited by M. W. Brown and K. J. Miller), pp. 479-496

(2) Papadopoulos Y.V., (1987), Fatigue polycyclique des métaux : une nouvelle approche, P.H.D. Thesis, Ecole Nationale des Ponts et Chaussées, Paris

(3) Papadopoulos Y.V., (1993), Fatigue limit of metals under multiaxial stress conditions: the microscopic approach, Technical Note N° I.93.101, Commission of the European Communities, Joint Research Centre, ISEI/IE 2495/93

(4) Ballard P., Dang Van K., Deperrois A. and Papadopoulos Y.V., (1995), High cycle fatigue and finite element analysis, Fatigue and Fracture of Engineering Materials and Structures, Vol. 18, N° 3, pp.397-411

(5) Vidal E., Kenmeugne B., Robert J.L. and Bahuaud J., (1994), Fatigue life prediction of components using multiaxial criteria, Fourth internationnal conference on biaxial/multiaxial fatigue, St Germain-en-Laye(France), pp. 353-366

(6) Morel F, (1997), A fatigue life prediction method based on a mesoscopic approach in constant amplitude multiaxial loading, Fatigue and Fracture of Engineering Materials and Structures, vol. 21, pp.241-256

(7) Lukas P. and Kunz L., (1992), What is the nucleation stage in fatigue, Theoretical Concepts and Numerical Analysis of fatigue, (Ed. A.F. BLOM and C.J. BEEVERS), Birmingham, pp.3-22

(8) Basinski Z.S. and Basinski S.J., (1992), Fundamental aspects of low amplitude cyclic deformation in face-centred cubic crystals, Progress in Materials Science, Vol. 36, pp. 89-148

(9) Dubar L., (1992), Fatigue multiaxiale des aciers - Passage de l'endurance à l'endurance limitée - Prise en compte des accidents géométriques, P.H.D. Thesis, Ecole Nationale Supérieure d'Arts et Metiers, Talence.

(10) Lee S.B., (1985), A criterion for fully reversed out-of-phase torsion and bending, Multiaxial Fatigue, ASTM STP 853, (K.J. Miller and M.W. Brown Eds.), American Society for Testing and Materials, Philadelphia, pp. 553-568

(11) Nishihara T. and Kawamoto M., (1954), The strength of metals under combined alternating bending and torsion with phase difference, Memoirs of the College of Engineering, Kyoto Imperial University, Vol. 11, N° 5, pp.85-115

(12) Sonsino S.M., (1994), Schwingfestigkeit von geschweiβten Komponenten unter komplexen elasto-plastischen, mehrachsigen Verformungen, LBF, Darmstadt, Report N° 6078.

(13) Endo T., (1974), Damage evaluation of metals for random or varying loading - Three aspects of the Rainflow method, Proceedings of the 1974 Symposium on Mechanical Behaviour of Materials, Society of Material Science Japan, pp. 372-380

Acknowledgements

We acknowledge the financial and technical support of this work by PSA.

DEVELOPMENT OF A HIGH-TEMPERATURE BIAXIAL FATIGUE TESTING MACHINE USING A CRUCIFORM SPECIMEN

Takashi OGATA* and Yukio TAKAHASHI*

Central Research Institute of Electric Power Industry

ABSTRACT

In order to perform high-temperature fatigue tests under a wide range of biaxial stress state, a high-temperature biaxial fatigue testing machine, which can apply equibiaxial tension and compression loading to a cruciform specimen, was developed. Strain controlled biaxial fatigue tests on the 316FR stainless steel were performed under proportional and nonproportional loading, in the latter of which phase difference existed between x and y directional strain, at 550°C. Through comparison of the load-strain data between x and y direction obtained from the equibiaxial tests, it was confirmed that the biaxial fatigue tests were successfully performed by using the newly developed machine. Fatigue failure lives did not correlate well with Mises equivalent strain range, giving the shortest life for the nonproportional loading. Based on the experimental results represented on the ε_x-ε_y diagram, a new biaxial fatigue criterion, equivalent normal strain range $\Delta\varepsilon_u$, was proposed. Biaxial fatigue lives correlated well with $\Delta\varepsilon_u$ regardless of the strain ratio and the loading mode.

KEY WORDS

Biaxial fatigue criterion, cruciform specimen, high-temperature, fractography, life prediction

INTRODUCTION

The majority of high-temperature structures and components in power plants are subjected to biaxial/multiaxial fatigue loading depending on their configurations and operating conditions. Therefore establishment of a biaxial/multiaxial fatigue life criterion is important for design and remaining life assessment of actual components to maintain reliable operation. The authors have been performing tension/compression and torsion tests under proportional and nonproportional loading conditions at high-temperature and elucidated fatigue failure mechanism and life property under biaxial loading **(1 - 3)**. However a whole range of biaxial strain state is not covered by the tension-torsion loading. A biaxial fatigue test using a cruciform specimen is able to cover whole strain state ranging from torsion to equibiaxial. Although such a testing machine can be thus effectively used for the study on biaxial fatigue, very limited data is available **(4, 5)** due to technological difficulty, especially at high-temperature. In this study, a new high-temperature biaxial fatigue testing machine using a cruciform specimen is developed and biaxial fatigue tests on 316 stainless steel are performed under proportional and nonproportional loading conditions at 550°C. Cyclic deformation and failure characteristics under biaxial fatigue loading are described, and a new biaxial fatigue criterion is proposed based on the experimental results.

DEVELOPMENT OF A HIGH-TEMPERATURE BIAXIAL FATIGUE TESTING MACHINE

Based on detailed specification of a high-temperature in-plane biaxial fatigue testing machine (HTBFM) provided by CRIEPI, the machine was manufactured and installed by MTS Corporation. An outline of the newly developed HTBFM is addressed in this section.

Loading and heating equipment

Appearance of the main frame of the HTBFM with a heating device and a controller is shown in Fig.1.

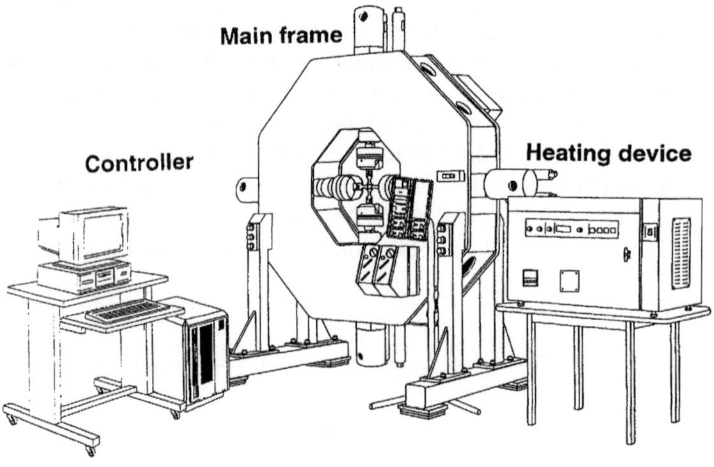

Fig.1 Appearance of biaxial fatigue testing machine.

The HTBFM consists of four hydraulic servo driven actuators and wedge grips mounted in a rigid load frame, which can maintain high stiffness in the system. Tensile and compressive loads can be applied by two pairs of actuators independently and controlled by a digital control system. The maximum applied load is 100kN. Load ratio and phase correlation can be arbitrarily chosen. To mount a cruciform specimen, two actuators in x direction and one actuator in y direction have L shaped plates to determine the specimen position on the grips, which allow the specimen to be mounted easily. In order to maintain specimen centre at a fixed position, the controller adopted the Control Matrix concept, which allows each control loop to be stabilised, and optimised. Specimen centre can be maintained by minimising the differentiation of the LVDT signal between two actuators in the same axis, independently from another axis. Concept of centre control is shown in Fig.2. Movement of the point of specimen centre is less than 2μm during cyclic test with a strain range of 1%. The specimen can be heated up to 1000°C by the induction-heating device with a heating coil. Temperature distribution at the control temperature of 550°C is ranging from 548°C to 553°C within the gage area (15mm diameter in the centre of the specimen).

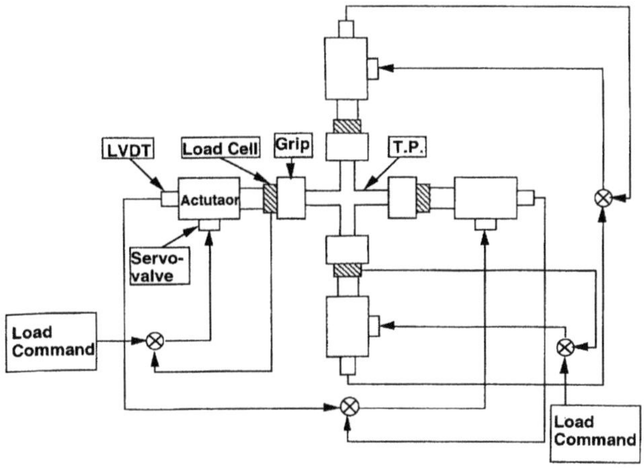

Fig.2 Concept of loading control testing.

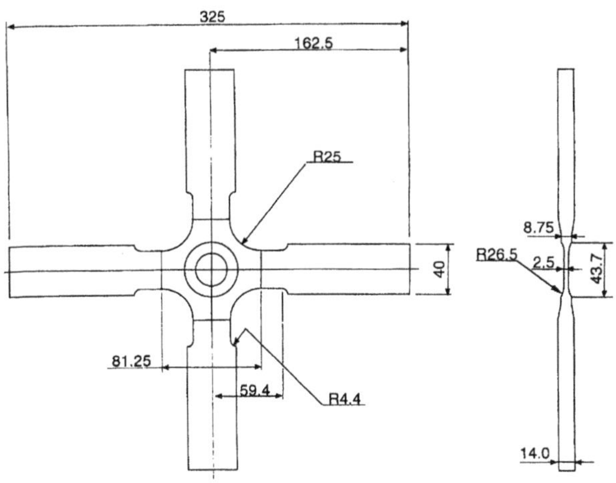

Fig.3 Specimen geometry.

Specimen design and strain measurement

Important things to consider in the design of a cruciform specimen are to provide uniform stress strain field in the centre of the specimen and to avoid crack initiation outside of the gage area due to unpreferable stress concentration. The basic geometry was determined referring to the study by Sakane and Ohnami (**4**). The specimen geometry was designed by a 3-D elastic-plastic finite element analysis using the Marc

K6 with a 3D 8 nodes-cubic element. Some improvements were made from pre-determined geometry to increase specimen stiffness and decrease stress concentration at specimen shoulders. Final specimen geometry is shown in Fig.3. The specimen has a 2.5 mm thickness, 15 mm diameter gage area in the centre and shoulders with smooth curvature of 25 mm radius. Elastic-plastic stress analysis results under equibiaxial loading condition represented by von Mises stress are shown in Fig.4. It is seen that an almost uniform stress field could be obtained in the gage area and unpreferable stress concentration outside of the gage area did not occur.

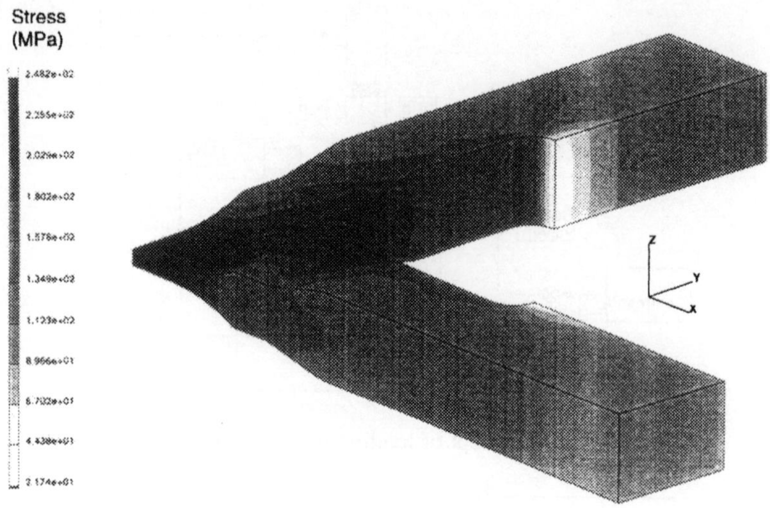

(a) Mises equivalent stress distribution

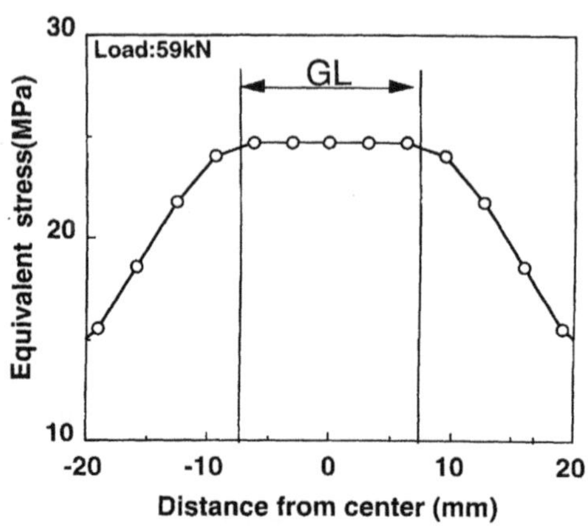

(b) Stress distribution around gage regime

Fig.4 Elastic-plastic stress analysis results of the cruciform specimen

A high-temperature biaxial extensometer (HBE) applied for the cruciform specimen to control x and y directional strains was also developed. The HBE was manufactured by combining two separate uniaxial extensometers into one structure. The x and y directional strains can be controlled independently. Measured noise level at a cyclic condition was lower than 30 mV and interference level by movement of other axis was lower than 0.003%.

EXPERIMENTAL PROCEDURE

A material used in this study was 316 stainless steel specially improved as a fast breeder component material (316FR). Chemical composition is shown in the table1. The feature of chemical composition of this material is medium nitrogen and low carbon contents which suppress grain boundary degradation caused by precipitation of chromium carbides. The cruciform specimen shown in Fig.3 was machined from a hot rolled plate with 50mm thickness.

Strain controlled fatigue tests were performed at 550°C under the test conditions shown in Fig.5. In the proportional loading tests, surface principal strain ratio, ϕ was defined as the ratio of x directional strain, ε_x to y directional strain, ε_y applied to the specimen. ϕ = -1 (pure torsion), -0.5 (uniaxial tension), 0 (plane strain) and 1 (equibiaxial tension) with von Mises strain range of 0.5% and 1.0% were employed. Nonproportional loading tests, in which phase difference, θ of 22.5°, 45°, 90° and 135° exists between x and y directional strains were also performed. Strain path of the 90° nonproportional loading test is shown by a broken line in Fig.5. Strain rate of all tests was 0.1%/sec for von Mises strain. Fatigue failure life was defined as the number of cycles when either x or y directional load reduced 5% from its maximum value. Test conditions and results are summarised in Table 2.

Table 1 Chemical composition.

wt(%)

C	Si	Mn	P	S	Ni	Cr	Co	N
0.008	0.53	0.85	0.026	0.004	11.16	16.88	0.07	0.0754

Table 2 Biaxial fatigue test conditions and results

ϕ	θ	$\Delta\varepsilon_m$	strain range		load range		failure life
			y	x	y	x	N_f
	(deg)	(%)	(%)		(%)		(cycles)
1	0	1.0	0.5	0.5	112.8	114.4	4572
		0.5	0.25	0.25	83.6	84.8	123806
0	0	1.0	0.87	0	112.8	79.8	1420
		0.5	0.44	0	101.5	65.8	49227
-0.5	0	1.0	1.0	0.52	94.4	0	3046
-1	0	1.0	0.87	0.87	52.3	57.2	10256
		0.5	0.44	0.44	42.6	45.2	>170000
1	90	1.0	0.87	0.87	125.7	149.6	816
		0.5	0.44	0.44	104.2	110.9	8630
	22.5	0.5	0.44	0.44	106.4	108.1	7680
	45	0.5	0.44	0.44	107.5	111.4	5135
	135	0.5	0.44	0.44	81.5	86.1	23390

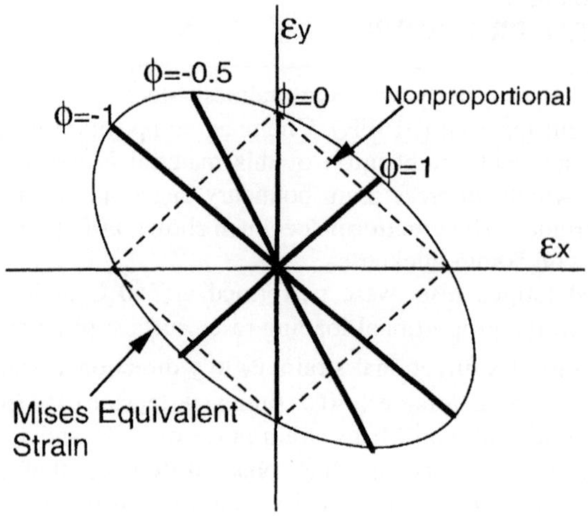

Fig.5 Strain path of proportional and nonproportional tests.

TEST RESULTS AND DISCUSSION

Cyclic deformation property

The 316FR steel shows cyclic hardening behaviour at initial stage and then maintains constant load amplitude until rapid decreasing by cracking. Load-strain hysteresis loops obtained at near mid-life are shown in Fig.6. The coincidence of loops in x-and y-direction in $\phi = 1$ and -1 confirms that the biaxial tests using the newly developed machine are reliably performed. In the proportional loading tests for the same Mises strain range, y directional peak load is the largest in $\phi = 0$ and smallest in $\phi = -1$. In the 90° phase difference nonproportional loading tests, although the y-directional peak load is almost the same as that in $\phi = 0$, the peak load in x-direction is larger than that in y-direction in spite of the same strain range. In the y directional loops in proportional loading tests, it can be seen that slope ratio of load to strain both adjust after passing minimum strain and before reaching maximum strain, which are equivalent to the elastic modules and hardening coefficient respectively, becoming larger with increasing principal strain ratio. In Fig.6(e), numbers designated in the strain waveforms of nonproportional loading, where x directional strain is 90° ahead of y directional strain, are corresponding to the numbers in the loops. The shape of the loops is different from that in proportional loading and different between x and y directions. The slope ratio of load to strain is relatively small when x and y directional strains go in reverse direction, such as 1-2 and 3-4 periods, whereas the value is relatively large when the direction of x and y directional strains is coincident, such as 2-3 and 4-1 in the strain waveform. Thus the loops in nonproportional loading produced incontinuous shapes relating to change in strain going direction between x and y directional strain.

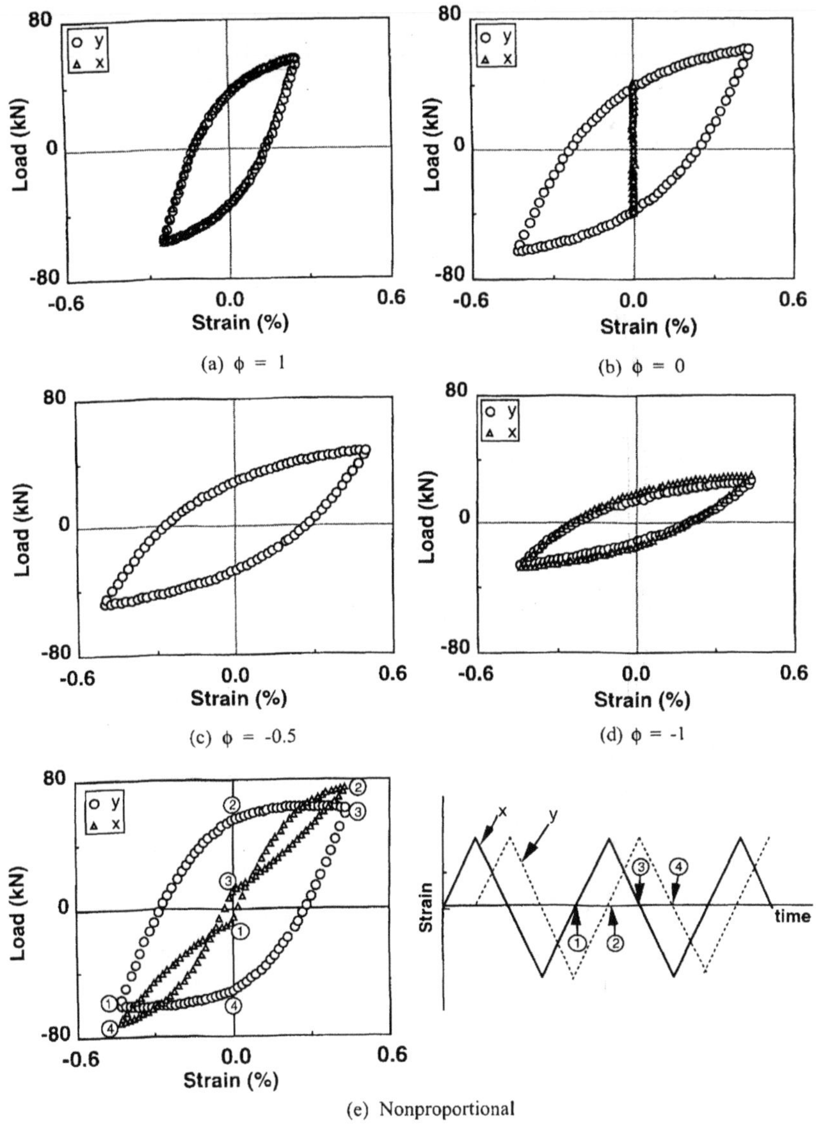

Fig.6 Load-strain hysteresis loop in proportional and nonproportional loading.

Failure appearance

Macroscopic appearances of specimen surface and failure surfaces at the middle of specimen thickness observed by a scanning electron microscope are shown in Fig.7. Since maincracks propagated within the gage area, validity of the specimen design was experimentally verified. Macrocracks propagated only in the x-direction in $\phi = 0$ and -0.5 and both in the x and y directions in $\phi = 1$. The macrocracks initiated both in the x- direction and the maximum shear direction, and connected with each other in $\phi = 0$. These failure appearances might be anticipated, based on the applied strain conditions. The maincrack propagated only in the y direction in the nonproportional loading, where the maximum load amplitude occurred in the x direction. Clear striations indicating that the crack propagated mainly under the Mode I loading, are observed in all failure specimens, both under proportional and nonproportional loading. In $\phi = 1$, a secondary cracks initiated in the normal direction to the failure surface due to contribution by x-directional applied strain. Fractography of $\phi = -1$ is the failure surface normal to the y axis. Significant difference of failure surfaces was not identified.

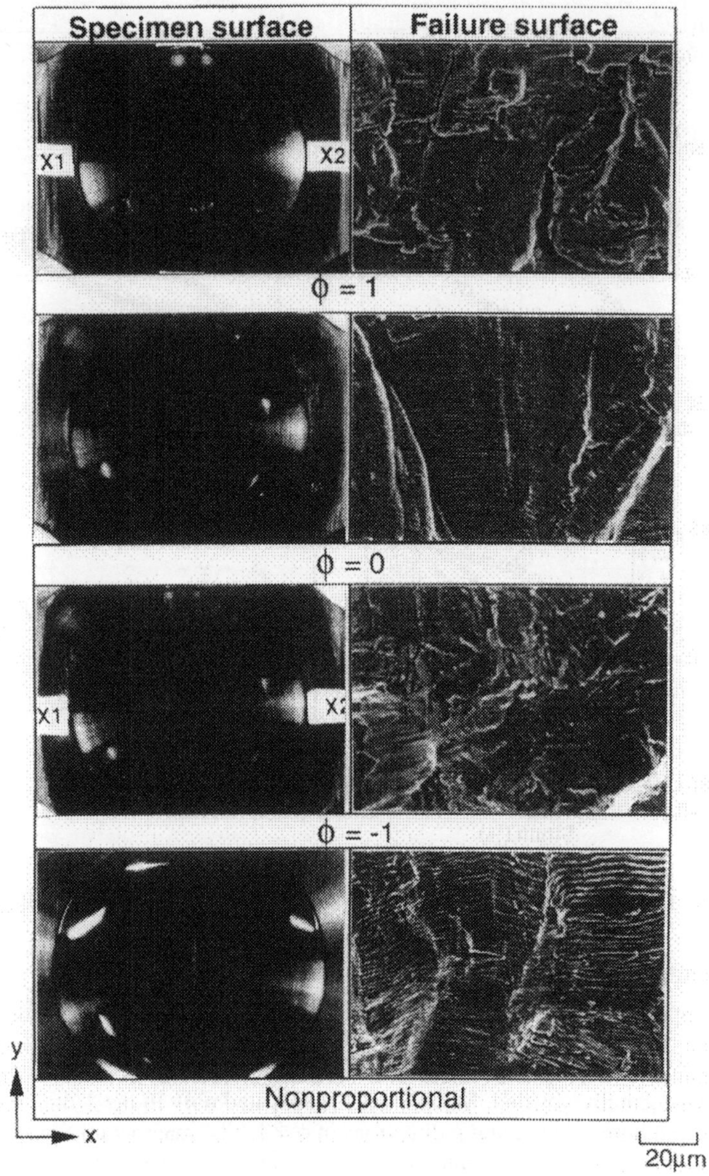

Fig.7 Macrocracks on the specimen surface and fractography.

Fatigue life property

Comparison between fatigue failure life under proportional loading for the same Mises equivalent strain range is shown in Fig.8. Failure life depends on principal strain ratio showing the shortest life in $\phi = 0$ and the longest in $\phi = -1$. Correlation between fatigue failure life and Mises equivalent strain range was shown in Fig.9. Fatigue life data is widely scattered depending on ϕ and loading mode. Fatigue life under nonproportional loading is shorter than that under proportional loading and 90° is the shortest among nonproportional tests. This means that life prediction for nonproportional loading by Mises equivalent strain range based on uniaxial data gives unconservative prediction.

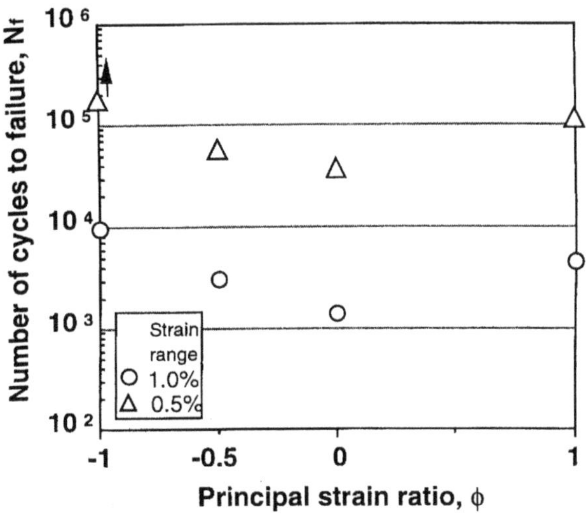

Fig.8 Dependency of biaxial fatigue life on principal strain ratio.

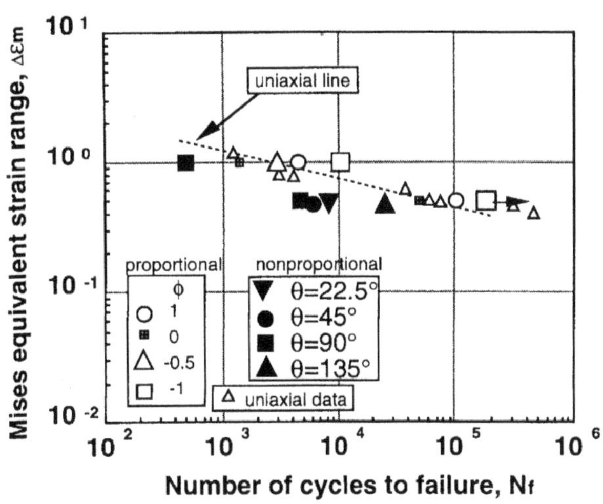

Fig.9 Relation between Mises equivalent strain range and biaxial fatigue life.

NEW BIAXIAL FATIGUE LIFE CRITERION

From the design point of view, development of a biaxial fatigue criterion, which can correlate biaxial fatigue life data with uniaxial data, is important to predict failure life of components subjected to biaxial stress. The authors performed tension/compression and torsion tests under proportional and nonproportional loading conditions and fatigue failure criterion, equivalent shear strain range $\Delta\bar{\gamma}_u$, was previously proposed based on experimental results and strain analysis **(2)**. Biaxial fatigue criterion for proportional loading, where principal strain axis is fixed during a cycle, and nonproportional loading, where principal strain axis rotates during a cycle, are expressed as follows:

$$\Delta\bar{\gamma}_u = \frac{1}{U}\left\{(\Delta\gamma_{max}/2)^2 + 12.0\Delta\varepsilon_n^{\ 2}\right\}^{1/2} \quad \text{(proportional)} \quad (1)$$

$$\Delta\bar{\gamma}_u = \frac{1}{U}\left(\Delta\gamma_{max}/2 + 1.2\Delta\varepsilon_n^{\ 2}\right) \quad \text{(nonproportional)} \quad (2)$$

$$U = 1/\left\{\left(\frac{1+\nu}{2}\right)^2 + 12.0\left(\frac{1-\nu}{2}\right)^2\right\}^{1/2} \quad (3)$$

where $\Delta\gamma_{max}$ is the maximum range of shear strain, $\Delta\varepsilon_n$ is the range of normal strain on the $\Delta\gamma_{max}$ plane and ν is Poisson's ratio. Biaxial fatigue lives both under proportional and nonproportional loading correlated well with $\Delta\bar{\gamma}_u$.

In this study, a unified criterion for the whole range of biaxial stress state under proportional loading is considered. Relationship between failure life and $\Delta\varepsilon_y$ is shown in Fig.10.

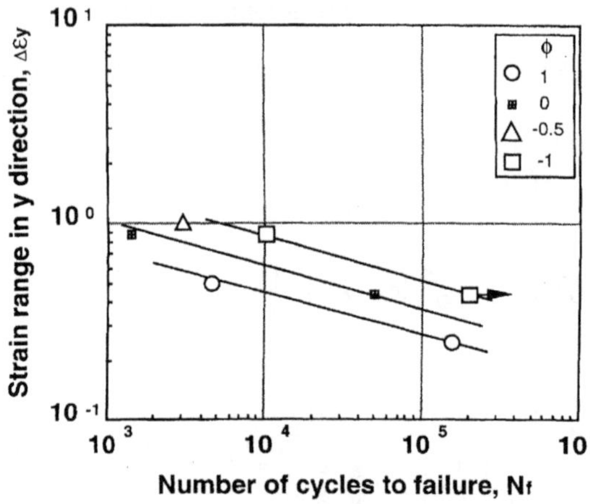

Fig.10 Relationship between principal strain range and biaxial fatigue life.

Although only two data point at each strain ratio was obtained in this study, linear relationship between $\Delta\varepsilon_y$ and N_f can be assumed within this test condition based on the previous study **(2)** in which linear relationship between $\Delta\varepsilon_m$ and N_f existed. It is seen that failure life is not only controlled by the maximum principal strain range but also affected by $\Delta\varepsilon_x$. Based on the experimental results, fatigue failure life criterion can be drawn on ε_x - ε_y plane as a diamond shape shown in Fig.11, where ν is assumed to be 0.5 and an elliptical shape drawn by a broken line represents the Mises equivalent strain criterion.

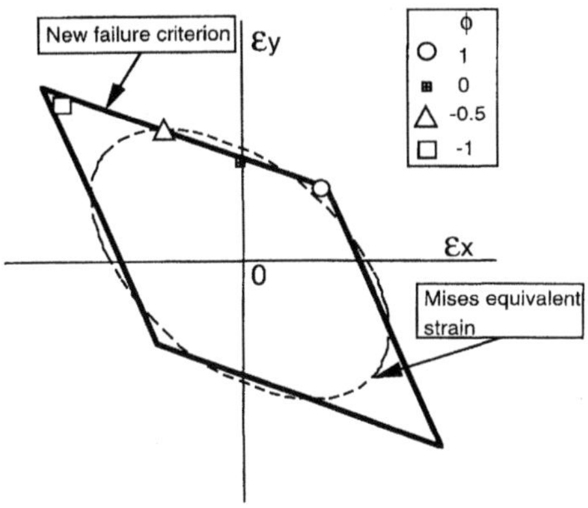

Fig.11 New biaxial fatigue life criterion on ε_x - ε_y diagram.

According to the new criterion, equivalent normal strain range $\Delta\varepsilon_u$ is simply expressed by:

$$\Delta\varepsilon_u = \frac{1}{1+A\nu}\Delta\varepsilon_y - \frac{1}{1+A\nu}\mathrm{sgn}(\phi)\Delta\varepsilon_x \tag{4}$$

where A is a material constant and determined as 0.38 from Fig.11. Although the new criterion gives slightly different results in comparison with the prediction by Eq.(1), it can be applied to the whole range of biaxial stress state under proportional loading. Previously obtained biaxial fatigue data by tension/compression and torsion tests on 304 stainless steel are correlated with $\Delta\varepsilon_u$ in Fig.12.

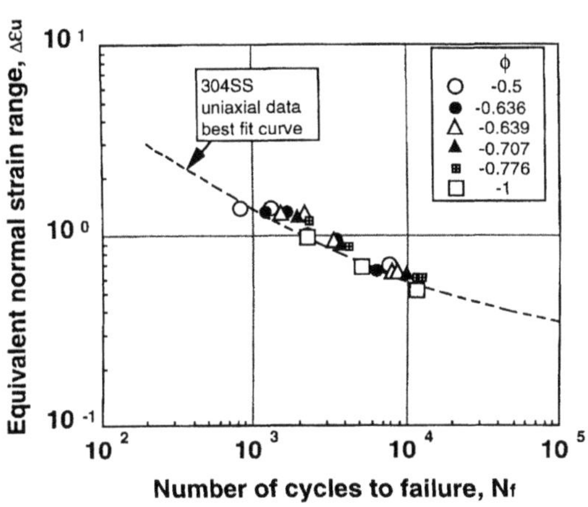

Fig.12 Correlation of fatigue life of 304 stainless steel with equivalent normal strain range.

A broken line represents a best fit curve of fatigue data obtained from uniaxial tests **(6).** Good correlation can be seen between $\Delta\varepsilon_u$ and fatigue life without dependence on principal strain ratio.

The failure life under nonproportional loading conditions was shorter than that in $\phi=0$ in spite of the same $\Delta\varepsilon_y$, and 90° phase difference showed the shortest life within nonproportional loading. This fact also suggests that $\Delta\varepsilon_x$ affected fatigue life under nonproportional loading. Therefore, definition of equivalent normal strain range for nonproportional loading should be taken into account for the difference of life property and strain paths based on new criterion for proportional loading shown in Fig.11. In this study, equivalent normal strain range, $\Delta\varepsilon_u$ for nonproportional loading is proposed as follows:

$$\Delta\varepsilon_u = (\Delta\varepsilon_u)_p + (\Delta\varepsilon_u)_{np} \tag{5}$$

$$(\Delta\varepsilon_u)_p = \frac{1}{1+A\nu}\Delta\varepsilon_{yp} - \frac{1}{1+A\nu}\text{sig}(\phi)\Delta\varepsilon_{xp} \tag{6}$$

$$(\Delta\varepsilon_u)_{np} = \frac{1}{1+A\nu}\left(\Delta\varepsilon_{ynp}{}^2 + \Delta\varepsilon_{xnp}{}^2\right)^{1/2} \tag{7}$$

where $(\Delta\varepsilon_u)_p$ and $(\Delta\varepsilon_u)_{np}$ are proportional and nonproportional part of the equivalent strain, $\Delta\varepsilon_{xp}$ and $\Delta\varepsilon_{yp}$ are x and y directional components of the maximum strain range which is defined as the maximum distance of straight line between two points on the strain path of ε_x and ε_y plane, $\Delta\varepsilon_{yp}$ and $\Delta\varepsilon_{ynp}$ are x and y directional components of strain range which declined 90° from direction of the maximum strain range. Definition of $\Delta\varepsilon_u$ for nonproportional loading condition is shown in Fig.13.

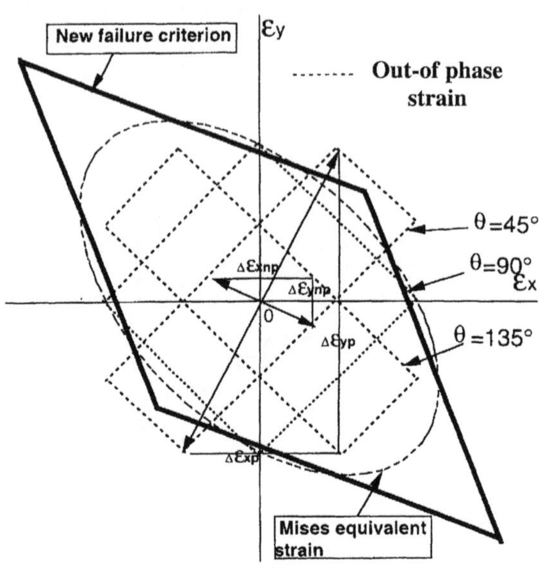

Fig.13 Definition of equivalent normal strain range $\Delta\varepsilon_u$ for nonproportional loading.

Correlation of biaxial fatigue lives of the 316FR stainless steel under proportional and nonproportional loading with $\Delta\varepsilon_u$ is shown in Fig.14. Uniaxial testing data obtained in the CRIEPI is also incorporated. It can be seen that biaxial fatigue lives correlated well with $\Delta\varepsilon_u$ regardless of strain ratio and loading mode. It can thus be concluded that $\Delta\varepsilon_u$ proposed in this study is a useful criterion to predict fatigue failure life of actual components subjected to proportional or nonproportional loading.

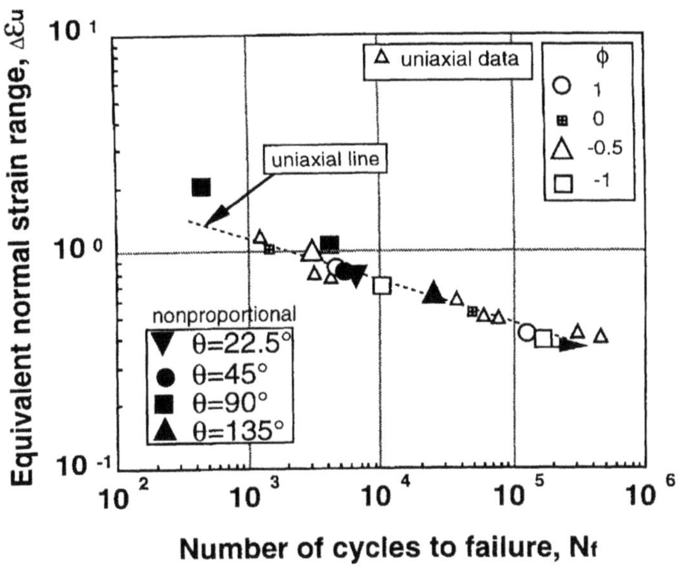

Fig.14 Correlation between $\Delta\varepsilon_u$ and biaxial fatigue life.

CONCLUSIONS

A high-temperature biaxial fatigue testing machine (HTBFM) using a cruciform specimen was developed and biaxial fatigue tests were performed on 316FR stainless steel at 550°C. The main results obtained in this study are summarised as follows.

1. Biaxial fatigue tests under proportional and nonproportional tests were successfully performed by the HTBFM using a cruciform specimen, which was designed by 3D finite element analysis.
2. It was found that biaxial fatigue life could not be correlated with Mises equivalent strain or principal strain range, which provides shorter fatigue life under nonproportional loading than that under proportional loading.
3. Equivalent normal strain range, $\Delta\varepsilon_u$ was proposed as a new biaxial fatigue criterion based on iso-failure line on the applied principal strain diagram. Biaxial fatigue lives both under proportional and nonproportional loading correlated well with $\Delta\varepsilon_u$.

REFERENCES

(1) Nitta A., Ogata T. and Kuwabara K., (1989), Fracture mechanism and life assessment under high-strain biaxial cyclic loading of type 304 stainless steel, Fatigue Fract. Engng. Mater. Struct., Vol.12, No.2, pp.77-92

(2) Ogata T., Nitta A. and Kuwabara K., (1991), Biaxial low-cycle fatigue failure of type 304 stainless steel under in-phase and out-of-phase straining conditions, fatigue under biaxial and multiaxial loading, ESIS 10, Eds. K.Kussmaul, D.L.McDiarmid and D.F.Socie, pp.377-392.

(3) Ogata T., Nitta A. and Blass J.J., (1993), Propagation behaviour of small cracks in 304 stainless steel under biaxial low-cycle fatigue at elevated temperature, *Advances in Multiaxial Fatigue, ASTM STP 1911*, D. L. McDowell and R. Ellis, Eds, pp.313-325.

(4) Sakane M. and Ohnami M., (1991), Creep-fatigue in biaxial stress states using cruciform specimen, same as ref. (2), pp.265-278.

(5) Itoh T., Sakane M., Ohnami M., Takahashi Y. And Ogata T., (1992), Nonproportional multiaxial low cycle fatigue using cruciform specimen at elevated temperature, Proc. 5th Inter. Conf. on Creep Materials, pp.331-339.

(6) Wada Y., Kawakami Y. and Aoto K., (1987), A statistical approach to fatigue life prediction for sus304, 316 and 321 austenitic stainless steels, ASME Pres. Ves. & Piping, Vol.123, pp.37-42.

Acknowledgements

The authors would like to express our gratitude to staffs of MTS Corporation who made tremendous effort to design, manufacture and install the new machine which meets the provided specification. The work has been conducted within a program sponsored by the Ministry of International Trade and Industry in Japan.

HIGH CYCLE MULTIAXIAL FATIGUE ENERGY CRITERION TAKING INTO ACCOUNT THE VOLUME DISTRIBUTION OF STRESSES

Thierry PALIN-LUC and Serge LASSERRE

Ecole Nationale Supérieure d'Arts et Métiers, CER de Bordeaux
Laboratoire Matériaux Endommagement Fiabilité (LA.M.E.F.)
Esplanade des Arts et Métiers, F - 33405 Talence Cédex - France.

ABSTRACT

An energy high cycle multiaxial fatigue criterion based on a new concept using the strain energy density is proposed in this paper. It distinguishes all load types even though no criteria existing in the literature explains the experimental differences between the endurance limits in tension, rotating bending and plane bending. To predict these differences the criterion takes into account the distribution of the strain energy density inside a volume influencing crack initiation at the critical point. By taking into account the effect of the triaxiality of stresses this criterion is available under any fully reversed multiaxial loading. In combined plane bending and torsion, predictions lead to a curve close to the Gough et al. ellipse quadrant for ductile materials and to an ellipse arc for brittle one. Under other combined loadings predictions are also on a curve close to an ellipse quadrant which is load dependent and material brittleness dependent. Predictions are in very good agreement with uniaxial and multiaxial experiments on four materials.

KEY WORDS

High cycle fatigue, multiaxial criterion, strain energy density, multiaxial loading, degree of triaxiality

NOTATION

A	fracture elongation
C_i	critical point, where the fatigue macro-crack initiates
dTa	triaxiality degree of stresses
E	Young modulus
F	analytical function taking into account the stress triaxiality effect on the endurance limit
$R_{p0.02}$	limit of proportionality with a 0.02% plastic strain
$R_{p0.2}$	conventional yield strength with a 0.2% plastic strain
R_m	tensile strength
R_u	true fracture strength

T	loading period
$V^*(C_i)$	volume influencing macro-crack initiation at the critical point
$Wa(M)$	mean value of elastic strain energy density in period T at the point M
$Wa^*(C_i)$	value of Wa at the C_i point, corresponding to σ^* for a sinusoidal fully reversed uniaxial stress state
Wsa	mean (averaged in period T) volumetric strain energy density
Wda	mean (averaged in period T) distortion strain energy density
β	material parameter characteristic of the material triaxiality of stresses sensitivity
$\varepsilon_{ij}^e(M,t)$	tensor of elastic strains at the point M function of time
φ	phase difference between stresses under combined loadings
$\sigma_{ij}(M,t)$	tensor of stresses at the point M function of time
σ^D	endurance limit
σ_{Trac}^D	endurance limit under fully reversed tension on smooth specimen
$\sigma_{RotBend}^D$	endurance limit under fully reversed rotating bending on smooth specimen
σ_{PlBend}^D	endurance limit under fully reversed plane bending on smooth specimen
σ^*	stress limit, below σ^D, such that there is no observable micro-crack
υ	Poisson ratio
$\varpi_a(C_i)$	damaging part of the elastic strain energy density inside $V^*(C_i)$
$\varpi_a{}^D(C_i)$	value of $\varpi_a(C_i)$ at the endurance limit

INTRODUCTION

Since 1951 and the work of Gough and Pollard **(1)** many high cycle multiaxial fatigue criteria have been proposed but none of them predicts the well-known experimental difference observed on all metallic materials between the endurance limits in tension, four point plane bending and rotating bending **(2, 3).** This paper presents a new concept and a criterion able to predict these observations under any fully reversed loadings. Furthermore, this criterion is phase independent for combined tension torsion and combined bending torsion but phase dependent under biaxial tension; this is in agreement with the Simbürger **(4)** and Froustey **(5)** experiments. After having presented the new concept used to establish the proposed calculation method, the criterion itself is presented. Then its predictions are compared with experimental data on four materials.

A NEW CONCEPT

It has already been proven that micro-cracks exist on components loaded at their endurance limit even if there is no macro-crack. Indeed, Vivensang **(6)** carried out SEM observations on smooth specimens in 35CD4 annealed steel loaded in four points rotating bending. She observed that persistent slip bands already exist after 50000 cycles and micro-cracks after only 300000 cycles at the surface of smooth specimens loaded at their endurance limit. SEM observations **(7, 8)** with the interrupted test technique on smooth specimens in Spheroidal Graphite (SG) cast iron loaded under fully reversed plane bending prove that even at the endurance limit micro-cracks initiate very early (50000 cycles). This study shows that a new limit, called σ^*, can be defined below the usual endurance limit of the material, σ^D. At a considered point a stress amplitude below this new limit does not initiate observable damage at the microscopic scale (no micro-cracks) in the matrix of the SG cast iron. Between σ^* and σ^D a stress amplitude only contributes to the initiation of micro damage, which could develop if, either near this point or in the course of time, there is a stress amplitude higher than the endurance limit. The usual endurance limit is not a limit of no damage initiation but is a limit of no damage propagation (micro-cracks) **(7, 8)**.

In Fig.1a we can see that the distribution of stresses at a moment of maximum loading in cycle is the same in plane bending and in rotating bending. In rotating bending, however, all the points lying on a circle centred on the middle of the specimen cross-section support the same stress during a cycle. In plane bending there is no axisymmetry, there are only two points supporting the greatest stresses. That is why it is important to reason on a complete loading cycle such as proposed by Tsybanyov **(9)**.

In their fatigue criterion, Froustey et al. **(10)** are working on a complete cycle of stresses; our proposal is based on their work. They use the mean of elastic strain energy density in time at one cycle, *Wa*, defined by (1) whatever the point *M* in the mechanical part.

$$Wa(M)=\frac{1}{T}\int_0^T \frac{1}{2}\sigma_{ij}(M,t)\varepsilon_{ij}^e(M,t)dt \tag{1}$$

$\sigma_{ij}(M,t)$ and $\varepsilon_{ij}^e(M,t)$ are respectively the tensor of stresses and the tensor of elastic strains at the considered point M function of time. Usually the endurance limit is low enough to consider that the material remains elastic at the macroscopic scale **(11)**. Thus, *Wa* can be considered as the mean value on one cycle of the total strain energy density at the considered point. Fig.1b illustrates the *Wa* distributions on the cross-section of a smooth specimen loaded in tension, rotating bending and plane bending. These distributions are very different, they are loading dependent. In order to take into account these differences we reason upon a volume around the critical point; this volume is defined below. The critical points C_i with regard to fatigue are those where *Wa* is maximum as proposed by Froustey et al. **(10)**.

From σ^* and by analogy with a sinusoidal tension the corresponding mean value of the strain energy density, *Wa**, can be calculated by (2), where *E* is the Young's modulus of the material.

$$Wa^* = \frac{\sigma^{*2}}{4E} \tag{2}$$

We postulate that the part of $Wa(M)$ exceeding $Wa^*(C_i)$ in some volume is the damaging part of the strain energy density. Around each critical point C_i the volume influencing crack initiation at this point is noted $V^*(C_i)$, it is the set of points M where $Wa(M)$ is higher than $Wa^*(C_i)$ - see Eq. 3.

$$V^*(C_i) = \{\text{points } M(x,y,z) \text{ around } C_i \text{ such that } Wa(M) \geq Wa^*(C_i)\} \quad (3)$$

From $V^*(C_i)$, $\varpi_a(C_i)$ is defined by (4), it is the mean value of the strain energy density around the critical point C_i in volume $V^*(c_i)$

$$\varpi_a(c_i) = \frac{1}{V^*(C_i)} \iiint_{V^*(C_i)} [Wa(x,y,z) - Wa^*(c_i)]\,dv \quad (4)$$

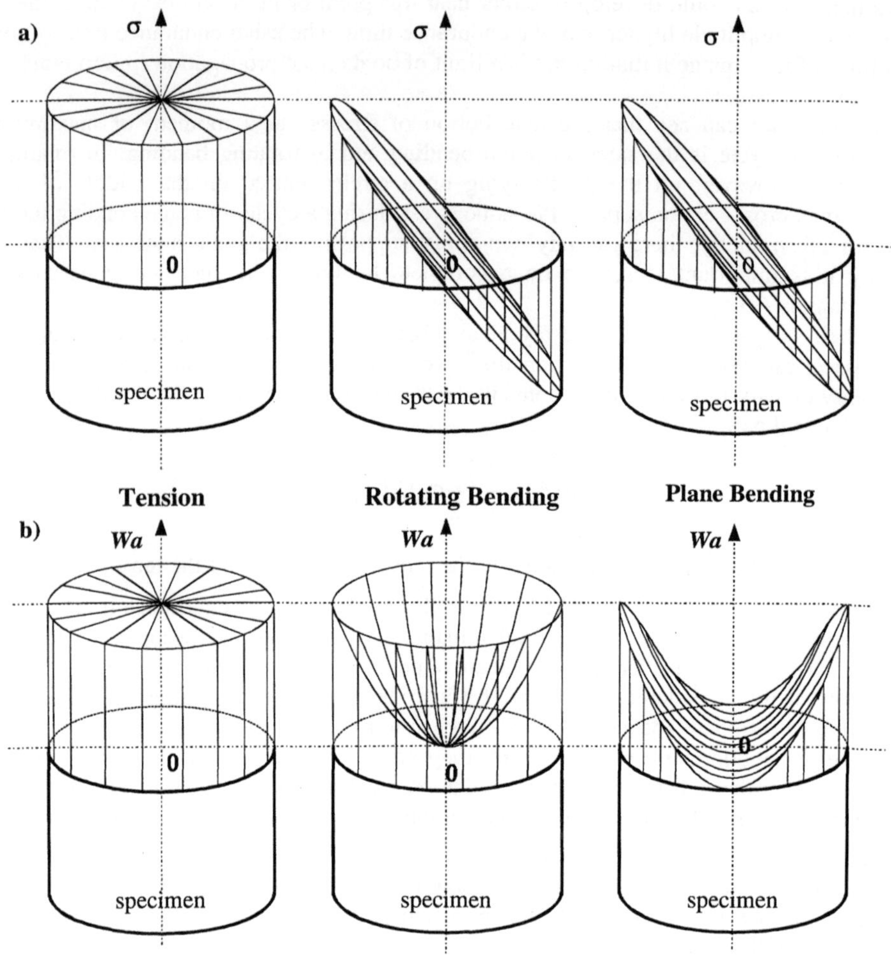

Fig.1 The stress distribution at maximum loading in cycle (1a) and Wa distribution (1b) on the cross-section of a smooth specimen loaded in tension, rotating bending and plane bending.

UNIAXIAL STRESS STATE

At the endurance limit at the critical point C_i, the quantity $\varpi_a(C_i)$ is supposed to be constant. If we note $\varpi_a^D(\text{Uniax})$ its value at the endurance limit for uniaxial stress state our criterion can be written by inequality (5). Failure occurs if this inequality is not satisfied.

$$\varpi_a(C_i) < \varpi_a^D(\text{Uniax}) \tag{5}$$

To apply this criterion, the following parameters have to be identified: σ^*, $Wa^*(\text{Uniax})$ and $\varpi_a^D(\text{Uniax})$. As there is no stress gradient along the longitudinal axis of a smooth cylindrical specimen under tension, the volume V^* can be reduced to the surface S^* inside the specimen cross-section, Eqs (3) and (4) become:

$$S^* = \{\text{points } M(x,y,z) \text{ around } C_i \text{ such that } Wa(M) \geq Wa^*(C_i)\} \tag{6}$$

$$\varpi_a(C_i) = \frac{1}{S^*(C_i)} \iint_{S^*(C_i)} [Wa(x,y,z) - Wa^*(C_i)]ds \tag{7}$$

The *Wa* distributions are axisymmetric in tension and in four point rotating bending (Fig.1b), for this reason these two sinusoidal loadings are taken in reference to identify σ^*. In tension, all the points of the cross-section of the specimen have the same *Wa* value (Fig.2), expression (7) becomes:

$$Wa(Trac) = \frac{\sigma_{Trac}^{2}}{4E} \Rightarrow \varpi_{aTrac} = Wa(Trac) - Wa^* = \frac{\sigma_{Trac}^{2} - \sigma^{*2}}{4E} \tag{8}$$

In four point rotating bending S^* is a crown shape, the iso-Wa lines are circular as shown in Fig.2. For such a loading on smooth cylindrical specimens ϖ_a is given by (9) where $\sigma_{RotBend}$ is the maximum stress due to rotating bending on the cross-section $(0 \leq r \leq R)$ and ρ^* is the radius of the circle representing the iso-Wa* line (Fig.2).

$$Wa(RotBend) = \frac{\sigma_{RotBend}^{2} \cdot r^2}{4E \cdot R^2} \Rightarrow \varpi_{aRotBend} = \frac{\sigma_{RotBend}^{2}}{8E} \cdot \left(1 - \left(\frac{\rho^*}{R}\right)^2\right) \quad \text{if } \rho^* < R \tag{9}$$

At the endurance limit, $\varpi_a(C_i)$ is supposed to be constant whatever the uniaxial stress state at the critical point C_i: $\varpi_a^D(\text{Tract}) = \varpi_a^D(\text{RotBend})$. Thus Eq.(10) is obtained from (8) and (9); it is a convenient expression for design. From (2) and (8) it is easy to prove that at the endurance limit $Wa^*(\text{Uniax})$ is given by (11). $\varpi_a^D(\text{Uniax})$ can be calculated by (12).

$$\sigma^* = \sqrt{2\sigma_{\text{Trac}}^{D\ 2} - \sigma_{\text{RotBend}}^{D\ 2}} \tag{10}$$

$$Wa^*(\text{Uniax}) = \frac{2\sigma_{Trac}^{D\ 2} - \sigma_{RotBend}^{D\ 2}}{4E} \tag{11}$$

$$\varpi_a^D(\text{Uniax}) = \frac{\sigma_{RotBend}^{D\ 2} - \sigma_{Trac}^{D\ 2}}{4E} \tag{12}$$

$\sigma_{RotBend}^{D}$ can be considered as a material parameter if the radius of the specimen is larger than about 5 mm as shown by Pogorotskii and Karpenko (from Papadopoulos and Panoskaltsis (**12**)). σ_{Trac}^{D} is not dependent of the size of the specimen, thus σ^* can be identified as a material parameter. Equations (10) and (11) are available if $\sigma_{RotBend}^{D} / \sigma_{Trac}^{D} < \sqrt{2}$; according to the authors this condition is true on all metallic materials, usually this ratio is less than 1.3 (**12**).

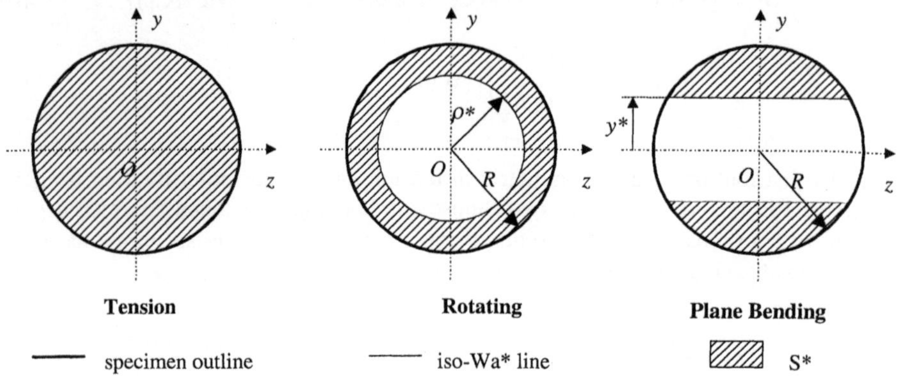

Fig.2 Iso-Wa lines and S* surfaces on the cross-section of a specimen loaded in tension, four points rotating bending or plane bending.

MULTIAXIAL STRESS STATES

The influence of the triaxiality of stresses on the endurance limit has already been proven by several works. We propose to take into account this influence by using the *F* function defined by Froustey et al. (**10**). By referring to the work of De Leiris (**13**) these authors define the degree of triaxiality, dTa, for a fully reversed loading by expression (15) where Wsa and Wda are respectively mean in-cycle volumetric and distortion strain energy density (16). For any periodic loading it is easy to prove that $Wa = Wsa + Wda$.

$$dTa = \frac{Wsa}{Wsa + Wda} \tag{15}$$

$$Wsa = \left(\frac{1-2\upsilon}{6E}\right)\frac{1}{T}\int_0^T I_{1a}^2(t)dt \quad \text{and} \quad Wda = \left(\frac{1+\upsilon}{E}\right)\frac{1}{T}\int_0^T J_{2a}(t)dt \tag{16}$$

where $I_{1a}(t)=\sigma_{kk}(t)$, $J_{2a}(t)=\frac{1}{2}s_{ij}(t)s_{ij}(t)$ by noting $\sigma_{ij}(t)=\frac{\sigma_{kk}(t)}{3}\cdot\delta_{ij}+s_{ij}(t)$

Based on much experimental data in high cycle fatigue, Froustey et al. **(10)** have proposed to relate, at the endurance limit, the value of $Wa(c_i,_{load})$, whatever the loading, to the value of Wa in torsion, $Wa(c_i,_{tors})$, by the function *F* (see Eq. 17) depending on the degree of triaxiality of the stresses at the critical point C_i, $dTa(c_i,_{load})$ and a new material dependent parameter β.

$$F(dTa(c_i),\beta)=\frac{Wa(c_i,_{load})}{Wa(c_i,_{tors})}=\frac{1}{1-dTa(c_i)}\cdot\left\{1-\frac{1}{\beta}\cdot\ln\left[1+dTa(c_i)\cdot\left(e^{\beta}-1\right)\right]\right\} \quad (17)$$

The β parameter is representative of the triaxiality sensitivity of the material. β is equal to zero for a XC18 annealed steel and is around 3 for a spheroidal graphite cast iron. The evolution of the function *F* is illustrated by Fig.3. The identification of the β parameter has to be done by applying equation (17) with the endurance limits in rotating bending and in torsion. It becomes (18) where the only unknown is β; υ is the Poisson ratio. The endurance limit in torsion is the only other experimental data needed to apply this proposal.

$$\frac{Wa^{D}(c_i,_{RotBend})}{Wa^{D}(c_i,_{Tors})}=\frac{1}{1-\frac{1-2\upsilon}{3}}\cdot\left\{1-\frac{1}{\beta}\cdot\ln\left[1+\frac{1-2\upsilon}{3}\cdot\left(e^{\beta}-1\right)\right]\right\} \quad (18)$$

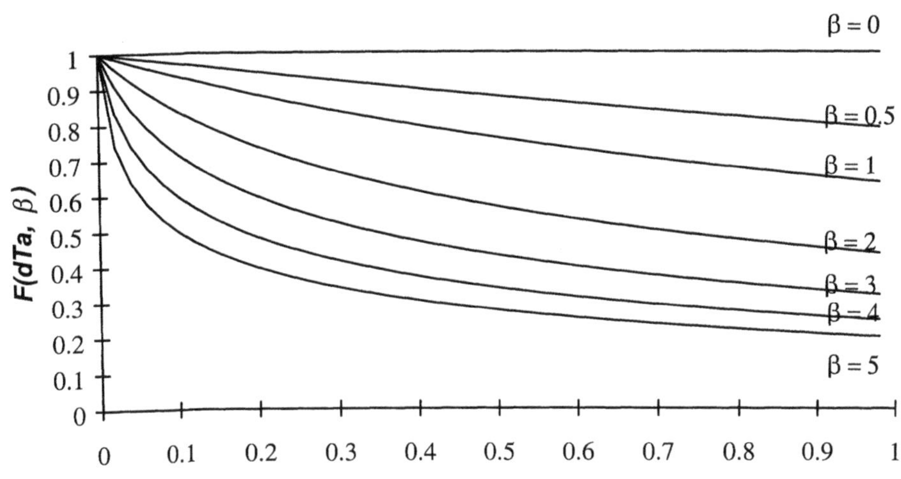

Fig.3 Evolution of the function *F*.

In order to take into account the triaxiality influence, we suppose that Wa* is loading dependent and verifies equation (19). It can be noted that in tension, rotating bending and plane bending, $dTa(c_i, \text{load})$ is the same, and so Wa* has the same value. With (19) Wa* is defined whatever the loading, so *V** is also defined.

$$\frac{Wa^*(\text{load})}{Wa^*(\text{Tors})} = F(dTa(c_i, \text{load}), \beta) \tag{19}$$

The influence of triaxiality has also to be taken into account in the definition of the limit value of $\varpi_a^D(c_i)$. By analogy with the previous assumption, we postulate that for any loading at the endurance limit the value of $\varpi_a^D(c_i)$, signified by $\varpi_a^D(\text{load})$, satisfies (20).

$$\frac{\varpi_a^D(\text{load})}{\varpi_a^D(\text{Tors})} = F(dTa(c_i, \text{load}), \beta) \Rightarrow \frac{\varpi_a^D(\text{load})}{\varpi_a^D(\text{Uniax})} = \frac{F(dTa(c_i, \text{load}), \beta)}{F(dTa(\text{Uniax}), \beta)} \tag{20}$$

With this last point the criterion can be applied on any fully reversed loading. Its use can be summarised as described in the following paragraph.

USE OF THIS CALCULATION METHOD

Two static properties of the material are necessary: E and υ. Only three experimental endurance limits under fully reversed loadings are needed: in tension, σ_{Trac}^D, in rotating bending, $\sigma_{\text{RotBend}}^D$, and in torsion, τ_{To}^D. From $\sigma_{\text{RotBend}}^D$ and τ_{To}^D, the β material parameter is identified by solving equation (18). At the endurance limit on any mechanical part the terms of the tensor of stresses are solutions of Eq.(20). In this equation $\varpi_a^D(c_i, \text{load})$ is defined by:

$$\varpi_a^D(c_i, \text{load}) = \frac{1}{V^*(C_i)} \iiint_{V^*(C_i)} [Wa(x, y, z, \text{load}) - Wa^*(\text{load})] dv \tag{21}$$

where $V^*(c_i) = \{\text{points } M(x, y, z) \text{ such that } Wa(M) \geq Wa^*(\text{load})\}$

and $Wa^*(\text{load}) = Wa^*(\text{Uniax}) \dfrac{F(dTa(\text{load}), \beta)}{F(dTa(\text{Uniax}), \beta)}$

The use of this criterion in a design department can be synthesised as shown in Fig.4.

APPLICATION OF THE CRITERION UNDER COMBINED LOADING

The predictions of this proposal are presented below for several combined loading conditions usually used in laboratory tests.

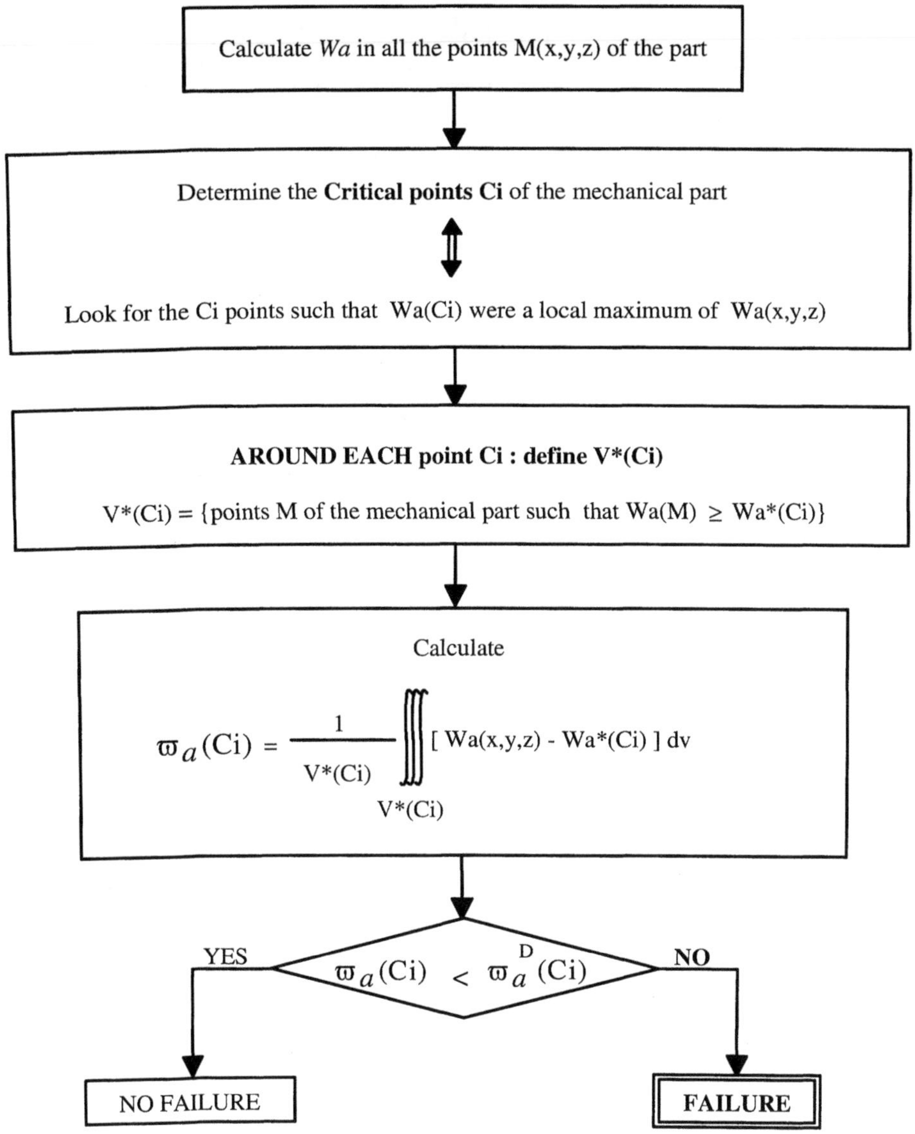

Fig.4 Failure prediction algorithm for use in a design department.

Combined rotating bending and torsion on a smooth cylindrical specimen

On a smooth cylindrical specimen loaded at its endurance limit $\left(\sigma^D_{RotBend+To}, \tau^D_{RotBend+To}\right)$ with $\sigma^D_{RotBend+To} / \tau^D_{RotBend+To} = k$, the Wa distribution on the specimen cross-section is given by (22).

$$W_a(\text{RotBend+To}) = \left(\frac{r}{R}\right)^2 \cdot \left[\frac{\sigma^{D\ 2}_{\text{RotBend+To}}}{4E} + \left(\frac{1+\upsilon}{2E}\right)\tau^{D\ 2}_{\text{RotBend+To}}\right] \quad (22)$$

The $Wa*(c_i)$ value under combined rotating bending and torsion is given by:

$$W_a*(\text{RotBend+To}) = \frac{2\sigma^{D\ 2}_{\text{Trac}} - \sigma^{D\ 2}_{\text{RotBend}}}{4E} \cdot \frac{F(dTa(\text{RotBend+To}),\beta)}{F(dTa(\text{Uniax}),\beta)} \quad (23)$$

$\varpi^D_a(\text{RotBend+To})$ is given by the following expression calculated by using (22), (23) and its definition (21),

$$\varpi^D_a(\text{RotBend+To}) = \frac{\sigma^{D\ 2}_{\text{RotBend+To}}}{8E}\left(1+\frac{2(1+\upsilon)}{k^2}\right) + \frac{2\sigma^{D\ 2}_{\text{Trac}} - \sigma^{D\ 2}_{\text{RotBend}}}{8E} \cdot \frac{F(dTa(\text{RotBend+To}),\beta)}{F(dTa(\text{Uniax}),\beta)} - Wa*(\text{RotBend+To}) \quad (24)$$

and its limit value is defined by (25).

$$\varpi^D_a(\text{RotBend+To}) = \frac{\sigma^{D\ 2}_{\text{RotBend}} - \sigma^{D\ 2}_{\text{Trac}}}{4E} \cdot \frac{F(dTa(\text{RotBend+To}),\beta)}{F(dTa(\text{Uniax}),\beta)} \quad (25)$$

From (24) and (25) and by using (23) the endurance limits $\left(\sigma^D_{RotBend+To}, \tau^D_{RotBend+To}\right)$ can be calculated. They are given by (26) whatever the ratio *k*.

$$\left.\begin{aligned} \sigma^D_{\text{RotBend+To}} &= \sqrt{\frac{\sigma^{D\ 2}_{\text{RotBend}}}{1+2(1+\upsilon)/k^2} \cdot \frac{F(dTa(\text{RotBend+To}),\beta)}{F(dTa(\text{Uniax}),\beta)}} \\ \tau^D_{\text{RotBend+To}} &= \frac{\sigma^D_{\text{RotBend+To}}}{k} \end{aligned}\right\} \quad (26)$$

These predictions are phase independent because the degree of triaxiality (27) is also phase independent.

$$dTa(\text{RotBend+To}) = \frac{(1-2\upsilon)\cdot k^2}{3\cdot k^2 + 6\cdot(1+\upsilon)} \quad (27)$$

Furthermore, these predictions lie on a curve all the closer to the Gough et al. (**1**) ellipse quadrant, $\sigma^{D\ 2}_{\text{RotBend+To}} / \sigma^{D\ 2}_{\text{RotBend}} + \tau^{D\ 2}_{\text{RotBend+To}} / \tau^{D\ 2}_{\text{To}} = 1$, than the β parameter is close to zero (ductile material). β differs from zero when the material is brittle ; in this case predictions are close to the Gough et al. ellipse arc.

Plane bending and torsion, tension and torsion

This proposal is loading type dependent; its predictions are different for each of these combined loadings, as illustrated by Fig.5. This method leads to three curves close to ellipse quadrants, which are not phase dependent. This is in agreement with the experiments.

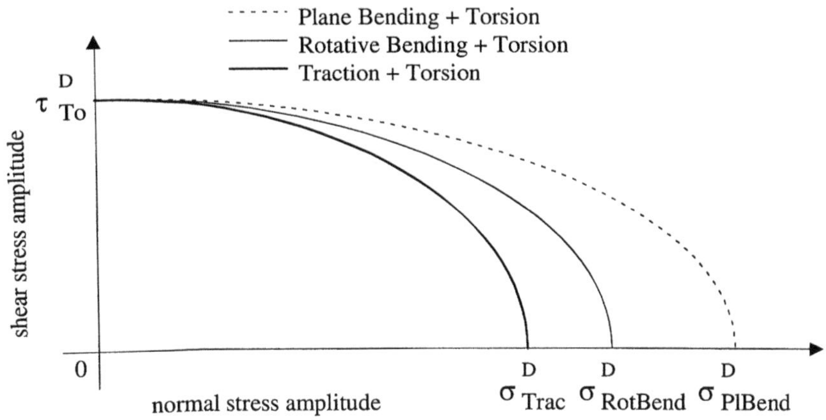

Fig.5 Prediction of the criterion under combined torsion with bending and tension.

Biaxial tension on thin walled tube (tension and internal/external pressure)

The longitudinal stress is $\sigma_l \sin\omega t$ and the tangent one is $\sigma_t \sin(\omega t + \varphi)$ with $\sigma_l/\sigma_t = \lambda$. It must be pointed out that in biaxial tension of a thin walled tube the degree of triaxiality defined by (15) is phase dependent (see Eq. (29)). Under this loading the endurance limits predicted by the criterion are phase dependent (30). They lie on a curve close to an ellipse arc as shown in Fig.6.

$$dTa(\text{BiaxTrac}) = \left(\frac{1-2\upsilon}{3}\right)\cdot\left(\frac{\lambda^2 + 2\cdot\lambda\cdot\cos\varphi + 1}{\lambda^2 - 2\cdot\upsilon\cdot\lambda\cdot\cos\varphi + 1}\right) \tag{29}$$

$$\sigma_l^D = \sigma_{Trac}^D \sqrt{\frac{1}{1+1/\lambda^2 - 2(\upsilon\cos\varphi)/\lambda}\cdot\frac{F(dTa(\text{BiaxTrac}),\beta)}{F(dTa(\text{Uniax}),\beta)}} \quad \text{and} \quad \sigma_t^D = \frac{\sigma_l^D}{\lambda} \tag{30}$$

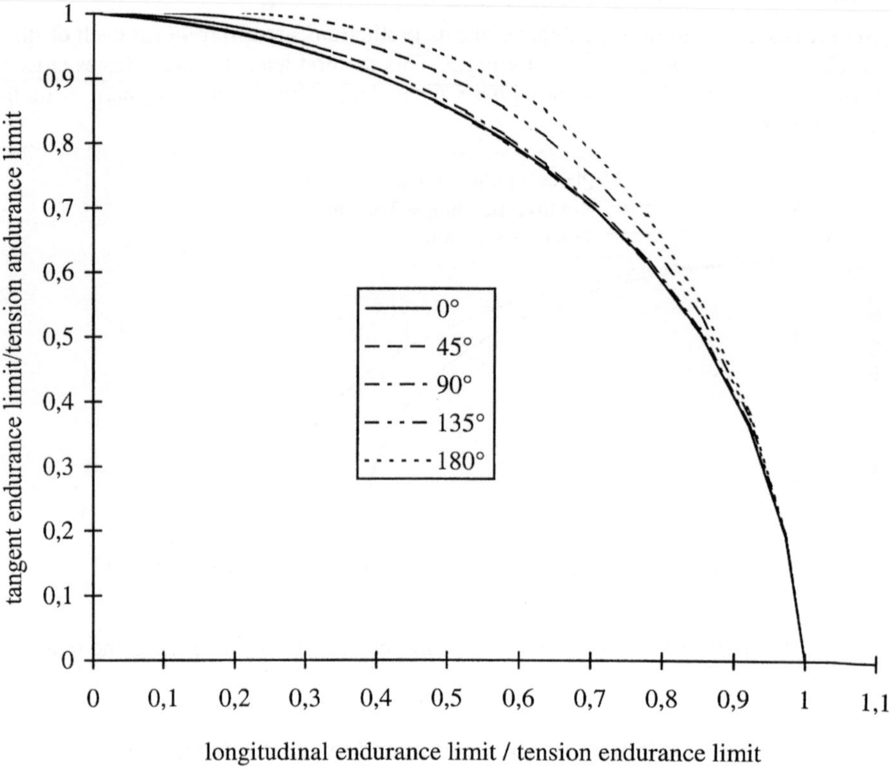

Fig.6 Predictions of the criterion under biaxial tension with phase influence.

COMPARISON BETWEEN EXPERIMENT AND PREDICTION

In order to test the accuracy of the predictions of our proposal, a comparison between experiment and prediction has been made for four materials and 14 experimental endurance limits of smooth cylindrical specimens. The materials are: 30NDC16 quenched and tempered steel **(5)**, XC18 annealed steel **(14)**, 35CD4 quenched and tempered steel and a spheroidal graphite cast iron (AFNOR standard close to FGS800-2) **(7)** and **(15)**. Their mechanical properties are summarised in Table 1.
For an objective and easy comparison the Relative Error of Prediction of the criterion, *REP*, is defined by (31).

$$\mathrm{REP}(\%) = \frac{\sigma^{D}_{\mathrm{Experiment}} - \sigma^{D}_{\mathrm{Prediction}}}{\sigma^{D}_{\mathrm{Experiment}}} \times 100 \tag{31}$$

All the *REP* are shown in Table 2 with the experimental data. This table proves that our criterion is in very good agreement with the experiments. The absolute value of the *REP* is always less than 10%.

Table 1 Static mechanical properties of the tested materials.

Material	*E* (GPa)	υ	$R_{p0.02}$ (MPa)	$R_{p0.2}$ (MPa)	R_m (MPa)	R_u (MPa)	A (%)
30NCD16 quenched & tempered	200	0.29	895	1080	1200	-	-
XC18 annealed	210	0.3	350	-	520	1530	24
35CD4 quenched & tempered	200	0.3	1015	1019	1123	-	13
FGS 800-2	164.9	0.275	320	462	795	815	9

Table 2 Experimental results and Relative Error of Prediction, *REP*, of the criterion. The italic values are used to identify the different parameters of the criterion for each material.

Material	β	Loading	σ_D (MPa)	τ_D (MPa)	σ_D/τ_D	φ (degree)	***REP* *(%)***
30NCD16	0.96	*Tension*	*560*	-	-	-	-
		Rotating Bending	*658*	-	-	-	-
		Torsion	-	*428*	-	-	-
		Plane Bending	690	-	-	-	**-4.5**
		Plane Bending + Torsion	519	291	1.78	0	**-8.1**
		Plane Bending + Torsion	514	288	1.78	90	**-9.1**
		Rotating Bending + Torsion	337	328	1.03	-	**-8.3**
		Rotating Bending + Torsion	482	234	2.06	-	**-9.3**
XC18	≈ 0	*Tension*	*273*	-	-	-	-
		Rotating Bending	*310*	-	-	-	-
		Torsion	-	*186*	-	-	-
		Plane Bending	332	-	-	-	**-6.3**
		Plane Bending + Torsion	246	138	1.783	0	**1.6**
		Plane Bending + Torsion	246	138	1.783	45	**1.6**
		Plane Bending + Torsion	264	148	1.783	90	**8.3**
35CD4	1.33	*Tension*	*558*	-	-	-	-
		Rotating Bending	*581*	-	-	-	-
		Torsion	-	*384*	-	-	-
		Plane Bending	620	-	-	-	**4.3**
FGS 800-2	3.09	*Tension*	*245*	-	-	-	-
		Rotating Bending	*280*	-	-	-	-
		Torsion	-	*220*	-	-	-
		Plane Bending	294	-	-	-	**-9.2**
		Plane Bending + Torsion	228	132	1.732	0	**-7.2**
		Plane Bending + Torsion	245	142	1.732	90	**0.2**
		Plane Bending + Torsion	199	147	1.35	0	**-10.2**

CONCLUSION AND PROSPECTS

Based on the volume distribution of the mean value on a cycle of the strain energy density Wa, this criterion is the first predicting the experimental differences between tension, rotating bending and plane bending in high cycle multiaxial fatigue. These differences are explained by taking into account the mean value of Wa over a volume V^* corresponding to the set of points influencing crack initiation at the critical point. Furthermore, this approach is based on a new limit σ^* below the conventional endurance limit. The predictions of this criterion are in very good agreement with uniaxial and multiaxial experimental data of smooth specimens for several materials. Under combined loadings, predictions are on a curve close to the ellipse quadrant or the ellipse arc of Gough et al. depending on the material brittleness.

Other SEM observations and experiments have to be done to confirm that σ^* is a material parameter under uniaxial stress state; nevertheless this concept with the energy approach is promising. This method has also to be developed to take into account the mean stress effect and has to be tested with endurance limits of notched specimens. A post-processor for a finite element software is studied now to be able to know the strain energy distribution around a notch. After these two steps of development this very promising criterion should be able to be used in design departments.

REFERENCES

(1) Gough H.J., Pollard H.V. and Glenshaw W.J., (1951), Some experiments on the resistance of metals to fatigue under combined stresses, Aeronaut. Research Council Reports and Memoranda, London, p.141

(2) Massonnet Ch., (1955), Le dimensionnement des pièces de machines soumises à la fatigue. Contribution expérimentale à l'étude de l'effet de l'échelle et des entailles, Rev. Univ. Mines, Paris, 9, T. XI, No. 6, pp.204-222.

(3) Barrault J. and Lasserre S., (1980), Limites de fatigue de l'acier 35CD4 en flexion rotative et en flexion plane, Rev. Mécanique Matériaux et Electricité, Sept., pp.275-278.

(4) Simbürger A., (1975), Festigkeitsverhalten zäher Werkstoffe bei einer mehrachsigen, phasenverschobenen Schwingbeanspruchung mit körperfesten und veränderlichen Hauptspannungsrichtungen, Laboratorium für Betriebsfestigkeit, Darmstadt, Germany, Bericht, Nr. FB-121, p.89

(5) Froustey C., (1987), Fatigue multiaxiale en endurance de l'acier 30NCD16, PhD. thesis, ENSAM CER de Bordeaux, France, p.131

(6) Vivensang M., (1994), Comportement en fatigue de deux nuances d'acier 35CD4. Cumul d'endommagement. Aspect microstructural de l'endommagement, PhD thesis, ENSAM CER de Bordeaux, France, p.197

(7) Palin-Luc T., (1996), Fatigue multiaxiale d'une fonte GS sous sollicitations combinées d'amplitude variable, PhD thesis, ENSAM CER de Bordeaux, France, p.261

(8) Palin-Luc T., Lasserre S. and Bérard J-Y., (1997), Damage evolution of a spheroidal graphite cast iron loaded around its endurance limit. Proc. EUROMAT 97, (Eds Sarton L.A.J.L. and Zeedijk H.B.), 21-23 April, Maastricht -NL, Vol. 1, pp. 511-514.

(9) Tsybanyov G.V., (1994), An energy approach to fatigue tests and crack initiation stage determination, from Problemy Prochnosti, Ukranian Acad. Sci., Kiev. Plenum Publishing Corp., 2, pp. 12-27.

(10) Froustey C., Lasserre S. and Dubar L., (1992), Validité des critères de fatigue multiaxiale à l'endurance en flexion-torsion, Mat-Tech 92, IITT-International, France, pp.79-85

(11) Lemaitre J. and Chaboche J-L., (1988), Mécanique des matériaux solides, (Ed. Bordas), Paris, p.544

(12) Papadopoulos I.V. and Panoskaltsis V.P., (1994), Gradient dependent multi-axial high-cycle fatigue criterion. Proc. 4th Int. Conf. Biaxial/Multiaxial Fatigue, (Ed. SF2M), St Germain en Laye, France, Vol.1, pp. 461-476.

(13) De Leiris H., (1969), Triaxialité des contraintes et critère de non fragilité. Bulletin ATMA, pp. 481-491.

(14) Lasserre S. and Froustey C., (1992), Multiaxial fatigue of steel - testing out of phase and in blocks : validity and applicability of some criteria. Int. J. Fatigue, 14, No. 2, pp. 113-120.

(15) Bennebach M., (1993), Fatigue d'une fonte GS. Influence de l'entaille et d'un traitement de surface, PhD. thesis, ENSAM CER de Bordeaux, France, p.157

Acknowledgements

This work was carried out as part of a research contract with the Materials Engineering Department of RENAULT. RENAULT is gratefully acknowledged for enabling the authors to do this work.

DISLOCATION STRUCTURE AND NON-PROPORTIONAL HARDENING OF TYPE 304 STAINLESS STEEL

Masao SAKANE*, Takamoto ITOH**, Seiji KIDA*, Masateru OHNAMI*
and Darrell SOCIE***

* Department of Mechanical Engineering, Faculty of Science and Engineering, Ritsumeikan University,
1-1-1 Nojihigashi Kusatsu-shi Shiga, 525-0077, Japan.

** Department of Mechanical Engineering, Faculty of Engineering, Fukui University, Japan

*** Department of Mechanical and Industrial Engineering, University of Illinois, USA

ABSTRACT

This paper describes the microstructure of Type 304 stainless steel after cyclic loading at room temperature under tension-torsion nonproportional strain paths. The degree of cyclic nonproportional hardening is correlated with changes in the dislocation substructure. Dislocation cells, dislocation bundles, twins and stacking faults are all observed. The type of microstructure formed and resultant stress response is dependent on the degree of nonproportional loading and strain range. Cyclic stress range was uniquely correlated with mean cell size.

KEY WORDS

Dislocation structure, nonproportional loading, stacking fault, additional hardening

NOTATION

$\varepsilon_1(t)$	Maximum principal strain at time t
$\varepsilon_3(t)$	Minimum principal strain at time t
$\varepsilon_I(t)$	Maximum absolute value of the principal strain at time t : Max $[\lvert\varepsilon_1(t)\rvert, \lvert\varepsilon_3(t)\rvert]$
$\varepsilon_{I\,max}$	Maximum value of $\varepsilon_I(t)$ in a cycle
$\Delta\varepsilon_I$	Maximum principal strain range under nonproportional straining
$\Delta\varepsilon_{NP}$	Nonproportional strain range
$\sigma_1(t)$	Maximum principal stress at time t
$\sigma_3(t)$	Minimum principal stress at time t
$\sigma_I(t)$	Maximum absolute value of the principal stress at time t : Max $[\lvert\sigma_1(t)\rvert,\lvert\sigma_3(t)\rvert]$
$\Delta\sigma_I$	Maximum principal stress range under nonproportional straining
$\xi(t)$	Angle between the principal strain directions of $\varepsilon_I(t)$ and $\varepsilon_{I\,max}$
f_{NP}	Nonproportional factor
α	Material constant which expresses the amount of additional hardening
N_f	Number of cycles to failure

INTRODUCTION

Many practical applications such as the nuclear vessel of a fast breeder reactor have nonproportional stresses and strains under the combination of thermal and mechanical loading. Type 304 stainless steel is known as a material, which shows a significant additional cyclic hardening under nonproportional loading in comparison with proportional loading. Recent studies have shown that the degree of the additional cyclic hardening is material dependent. Doong et. al. **(1)** reported the relationship between the microstructure and additional cyclic hardening behaviour of 1100-aluminum alloy, oxygen free pure copper and Type 304 and 310 stainless steels. They reported that no additional hardening occurred in aluminum alloy but significant additional hardening in stainless steel. Nonproportional cyclic hardening was reported for pure copper. They discussed the microstructure change with proceeding cycles in detail for a limited number of strain paths. Cailletaud et. al. **(2)** compiled much of the published data and concluded that the main parameter governing the degree of nonproportional hardening in solid solution materials is the ease of cross slip. Itoh et. al. **(3, 4)** studied nonproportional cyclic hardening of Type 304 stainless steel, pure copper, pure nickel, pure aluminum and 6061-T6 aluminum and reported that the degree of additional cyclic hardening is related to the stacking fault energy of the material. For a material with a low stacking fault energy such as Type 304 stainless steel, planar slip occurs and results in a large amount of additional cyclic hardening. This is caused by the interaction of many slip systems. Materials with a high stacking fault energy such as pure aluminum and 6061-T6 aluminum alloy deform by wavy slip. These materials do not show additional cyclic hardening during nonproportional loading The difference in the additional hardening behaviour between high and low stacking fault energy materials is be related to the microstructure of the material but extensive and systematic studies have not yet been reported.

Several investigators have examined the dislocation structure for room temperature tests. Doquet **(5)** reported twin deformation as a primary deformation mechanism under nonproportional loading for binary Co33Ni. She reported that the increase in the amount of twin deformation is a cause of additional cyclic hardening during nonproportional loading. Jiao et. al. **(6)** examined alloy 800 H and observed deformation twins and suggested that the formation of twins depends not only on the shear stress but also on the normal stress acting on the slip plane. McDowell et. al. **(7)** found that the heterogeneity of α-martensite and other planar slip deformation products (e.g. α' martensite) are a function of the nonproportionality in 304 stainless steel. They found that the homogeneity and morphology of the deformation products is of key importance. Cailletaud et. al. **(1)** observed ladders, veins or dislocation cell structures with loose outlines in uniaxial specimens but walls, mazes, cells, and, above all, abundant micro-twinning for nonproportionally loaded specimens of Type 316 stainless steel. Twinning is not an easy deformation mode in 316 steel at room temperature. The critical shear stress needed to induce twinning was reached because of the additional hardening during the nonproportional tests. Doong et. al. **(1)** found single slip structures under proportional loading of both 304 and 310 stainless steels. Multi-slip structures such as cells and labyrinths were found for nonproportional loading. At high temperatures, Nishino et. al. **(8)** observed the dislocation structure of Type 304 stainless steel cyclically loaded at 823 K and have discussed the relationship between the dislocation structure and hardening behaviour. They concluded that anisotropic hardening is caused by the directionally developed cell formation and isotropic hardening by the formation of round-shaped cells.

Microstructural studies of additional nonproportional cyclic hardening have been limited to a small number of strain paths so that the results of these studies are rather qualitative. Little quantitative discussion has been reported. This paper studies the microstructure and cyclic stress-strain relationships obtained at room temperature under 14 nonproportional strain paths for Type 304 stainless steel, and will discuss the relationship between stress response and cell structure quantitatively.

EXPERIMENTAL PROCEDURE

The material tested is Type 304 stainless steel, which received a solution treatment at 1373 K for one hour. Hollow cylindrical specimens with 9 mm I.D., 12 mm O.D., and 4.6 mm gage length were employed in this study. Strain controlled cyclic loading tests at a Mises' effective strain rate of 0.1 %/sec were carried out at ambient temperature. Testing details are reported by Itoh et. al. **(9)**. Figure 1 shows the 14 proportional and nonproportional loading histories employed, where ε and γ are the axial and shear strains, respectively. Case 0 is a push-pull proportional test that is the basic data for examining the microstructure. Case 0 testing was carried out at strain ranges between 0.5 and 1.5 %. Case 5 is also a proportional test as is Case 0, but is a combined push-pull and reversed torsion test. The other Cases are nonproportional tests in which the severity of nonproportional loading is determined by the strain history. In all the tests except for Case 0, axial and shear strain ranges were 0.5 and 0.8 % Mises' equivalent strain.

One cycle is defined here as a full straining for both axial and shear strains. All the tests except Cases 3, 4 and 13 were counted as one cycle and these tests were counted as two cycles for a full straining along the strain path chosen in Fig.1. The number of cycles to failure (N_f) was defined as the cycle at which the axial stress amplitude decreased to 5 percent of the saturation stress in tension.

After the cyclic loading tests, thin foils of 3 mm diameter were cut from specimens away from cracks by a wire cutter to observe the microstructure. They were polished down to about 0.2 mm in thickness with emery papers and were jet-electropolished in acetate perchlorate for observation by the transmission electron microscope (TEM). A JEOL JEM-100C (100kV) was used to observe the microstructure and diffraction pattern.

DEFINITION OF PRINCIPAL STRAIN AND STRESS RANGES AND NONPROPORTIONAL PARAMETER

During nonproportional loading, stress and strain amplitudes vary with time, so that the principal strain and stress ranges must be defined. In a previous paper **(9)**, the authors have proposed a definition of the maximum principal strain and stress ranges for nonproportional loadings and this paper follows that definition.

The maximum principal strain range, $\Delta\varepsilon_I$, is defined as,

$$\Delta\varepsilon_I = \mathrm{Max}\left[\varepsilon_{I\max} - \cos(\xi(t))\cdot\varepsilon_I(t)\right] \qquad (1)$$

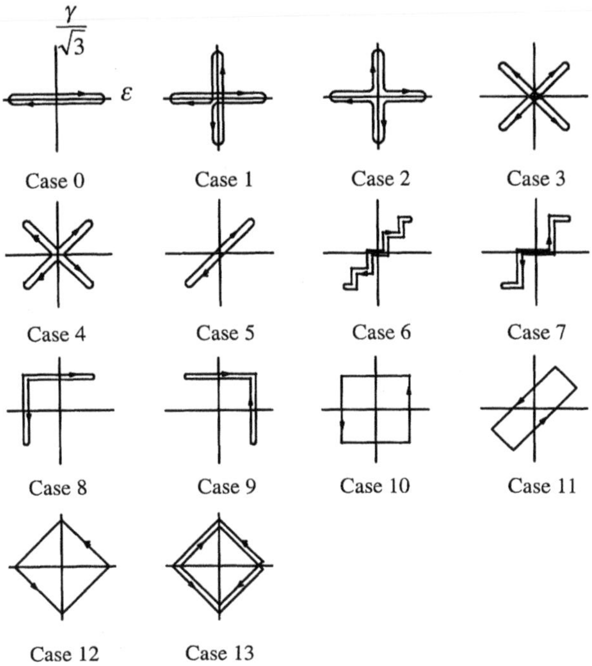

Fig.1 Proportional and nonproportional loading paths.

In this equation, $\varepsilon_I(t)$ is the maximum absolute value of principal strain at time t and

$$\begin{aligned} \varepsilon_I(t) &= |\varepsilon_1(t)| \quad \text{for} \quad |\varepsilon_1(t)| \geq |\varepsilon_3(t)| \\ \varepsilon_I(t) &= |\varepsilon_3(t)| \quad \text{for} \quad |\varepsilon_1(t)| \leq |\varepsilon_3(t)| \end{aligned} \tag{2}$$

where $\varepsilon_1(t)$ and $\varepsilon_3(t)$ are the maximum and minimum principal strains at time t, respectively.

Figure 2 is a polar figure of $\varepsilon_I(t)$ schematically showing $\varepsilon_I(t)$, $\xi(t)$ and $\Delta\varepsilon_I$. In Eq. (1), ε_{Imax} is the maximum value of $\varepsilon_I(t)$ in a cycle and $\xi(t)$ is the angle between ε_{Imax} and $\varepsilon_I(t)$ directions. Thus, $\Delta\varepsilon_I(t)$ is determined by the two strains, $\varepsilon_I(A)$ and $\varepsilon_I(B)$, and by the angle between the two strain directions in Fig.2, where A is the time giving ε_{Imax} and B the time maximising the strain range in Eq. (1).

The maximum principal stress range, $\Delta\sigma_I(t)$, has a similar definition to $\Delta\varepsilon_I$,

$$\begin{aligned} \Delta\sigma_I &= \sigma_I(A) - \cos(\xi(B)) \cdot \sigma_I(B) \\ \sigma_I(t) &= |\sigma_1(t)| \quad \text{for} \quad |\sigma_1(t)| \geq |\sigma_3(t)| \\ \sigma_I(t) &= |\sigma_3(t)| \quad \text{for} \quad |\sigma_1(t)| \leq |\sigma_3(t)| \end{aligned} \tag{3}$$

where $\sigma_1(t)$ and $\sigma_3(t)$ are the maximum and minimum principal stresses, respectively. The two times A and B correspond with those defined for the maximum principal strain range.

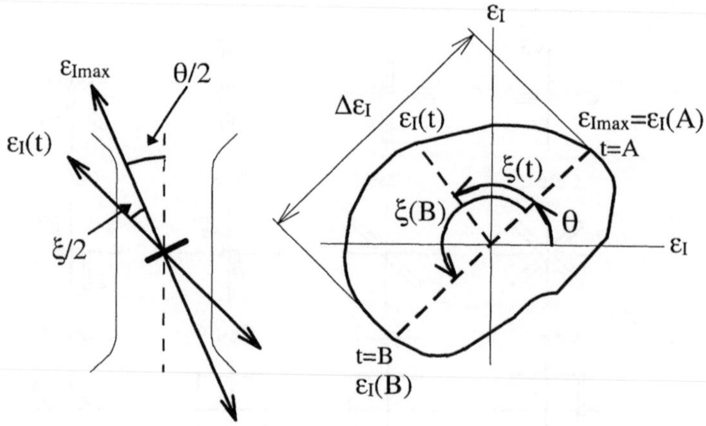

Fig.2 Definition of maximum principal strain range under nonproportional loading.

A nonproportional factor, f_{NP}, was proposed by the authors to express the severity of nonproportional loading **(6)**.

$$f_{NP} = \frac{\pi}{2 \cdot T \cdot \varepsilon_{Imax}} \int_0^T \left(|\sin \xi(t)| \cdot \varepsilon_I(t) \right) dt \tag{4}$$

where T is the time for a cycle shown in Fig.1. The value of f_{NP} is zero under proportional loading and is the range of $0 \leq f_{NP} \leq 0$ under nonproportional loading. As shown in Eq. 4, f_{NP} is a function of only the applied strain history to avoid the need to compute stresses and plastic strains.

The authors **(9)** have proposed the following strain parameter to correlate nonproportional LCF lives,

$$\Delta \varepsilon_{NP} = (1 + \alpha \cdot f_{NP}) \cdot \Delta \varepsilon_I \tag{5}$$

where α is a material parameter related to the additional hardening of the material under nonproportional loading, and α is defined as the ratio of stress range under nonproportional circular loading in $\gamma/\sqrt{3}$ plot to that under proportional loading at the same Mises equivalent strain. The value of α becomes larger for lower stacking fault energy materials **(1-4)**. Murakami et. al. **(10)** showed that this parameter will decrease with increasing temperatures. For Type 304 stainless steel at room temperature, α takes the value of 0.9. Benallal and Marquis **(11)** show a small strain range dependence of α but here we take it as a constant. The term $(1+\alpha \cdot f_{NP})$ accounts for the additional cyclic hardening observed during nonproportional cyclic loading and is similar to damage parameters that are based on the product of stress and strain range. The advantage of this parameter is that it does not require a sophisticated transient cyclic plasticity model to obtain the stress ranges.

EXPERIMENTAL RESULTS AND DISCUSSION

A complete tabular listing of all test data is available in **(9)** but the test data is not presented here because of the space of the paper. Fatigue lives for Cases 0-13 significantly depend on the strain history. Rotating principal strain directions in tests such as Cases 8-10, 12 and 13 yields the largest reduction in fatigue lives by as much as a factor of 10. In Cases 6-9, steps in the path can have a large influence on fatigue lives when the number of steps is small and the path length is large. Thus, Case 6 shows a small reduction in fatigue life, as the strain history is nearly proportional loading because of the small step length. Figure 3 correlates LCF lives with principal stress range. The figure shows that a significant additional hardening occurs under nonproportional loadings. Greater additional cyclic hardening results in smaller fatigue lives. Thus, an estimate of additional hardening is necessary for predicting fatigue lives under nonproportional loading in the LCF regime. Figure 4 correlates the nonproportional LCF lives with equivalent strain given in Eq.5. Most of the data are correlated within a factor of two scatter band.

Figure 5 shows the relationship between the stabilised axial and shear stresses for twelve of the fourteen loading histories. The stress response for Case 8 is the mirror image of Case 9 and was omitted from the figure. In the figures, dashed lines are the results at $\Delta\varepsilon = 0.5$ %, and solid lines are the results at $\Delta\varepsilon = 0.8$ %. The shear stress scale has been plotted as $\sqrt{3}\tau$ so that stresses can easily be compared on Mises basis. Comparing the equivalent stresses for Cases 1 and 2 with Case 0 at $\Delta\varepsilon = 0.5$ and 0.8 % shows significant nonproportional hardening due to the change of principal strain direction at zero strain. The normal stress–shear stress relationship of Case 1 is different from that of Case 2 which shows that a fully reversed straining has a different influence on stress response from a zero-to-maximum straining, and the former strain history causes greater additional hardening than the latter one. This additional hardening also results in a lower fatigue life for Case 2. The shape of the τ–σ plot of Cases 3 and 4 is similar to that of Cases 1 and 2, respectively, after giving a rotation of 45 degrees to the former two cases. However, the stress amplitude of Cases 3 and 4 is larger than that of Cases 1 and 2 because the shear and axial strains are applied simultaneously resulting in a cyclic strain range in the former two cases that is larger by about 1.4 times.

A simple method for visualising degree of nonproportionality is useful when interpreting the stresses in Fig.5. If an ellipse is drawn so as to circumscribe the entire stress path, nonproportionality can be thought of as the ratio of the minor axis to the major axis. In Case 5, the minor axis of an ellipse circumscribing the loading history is small corresponding to a low degree of nonproportionality. The degree of nonproportionality increases in going from Case 5 to Case 10. This is easily visualised as an increasingly circular ellipse circumscribing the stress history. The size of the ellipse also increases as the degree of nonproportionality increases. Case 5 is proportional loading where the normal stress amplitude σ is same as the shear stress amplitude $\sqrt{3}\,\tau$ since the normal strain equals the shear strain on Mises basis. Comparing the stress response in Fig.5 of Case 5 with that of Cases 6, 7 and 8 shows that the normal and shear stress amplitudes are larger as the degree of nonproportionality increases. Fatigue lives tend to decrease as these stresses increase.

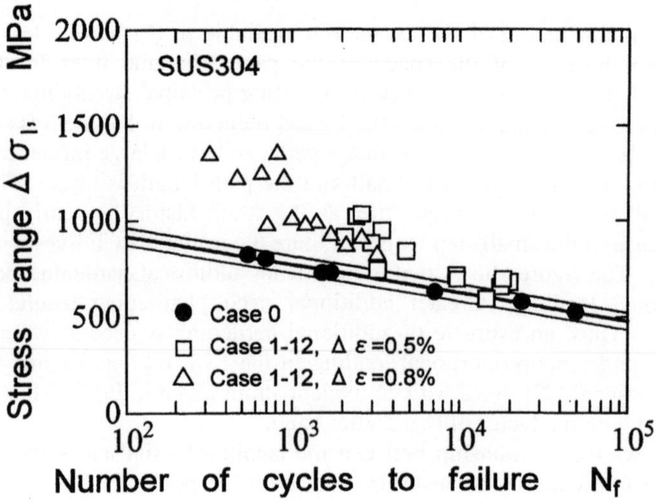

Fig.3 Correlation of nonproportional fatigue lives with principal stress.

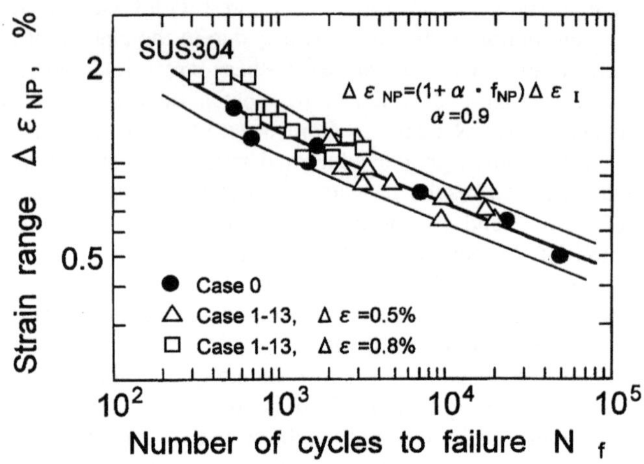

Fig.4 Correlation of nonproportional fatigue lives with nonproportional strain in Eq.5.

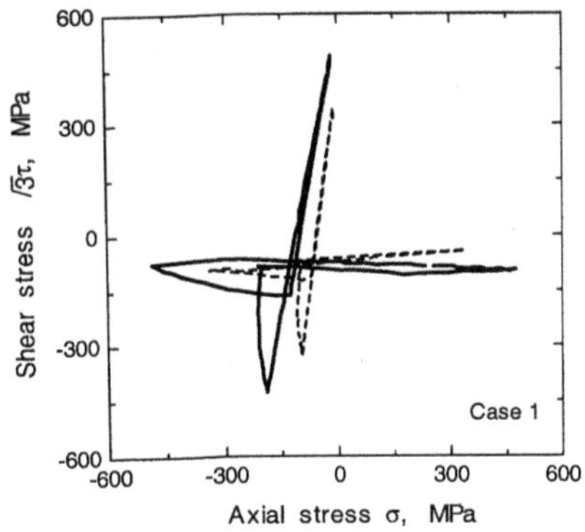

(a) Case 1

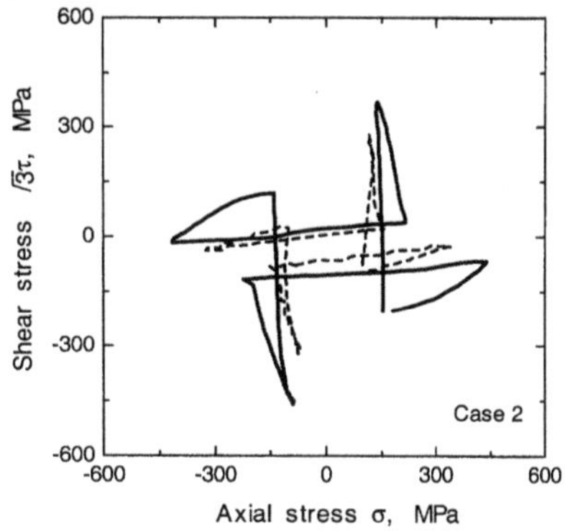

(b) Case 2

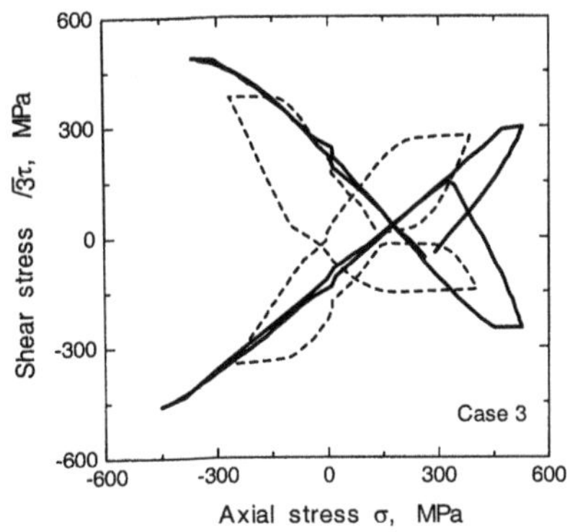

(c) Case 3

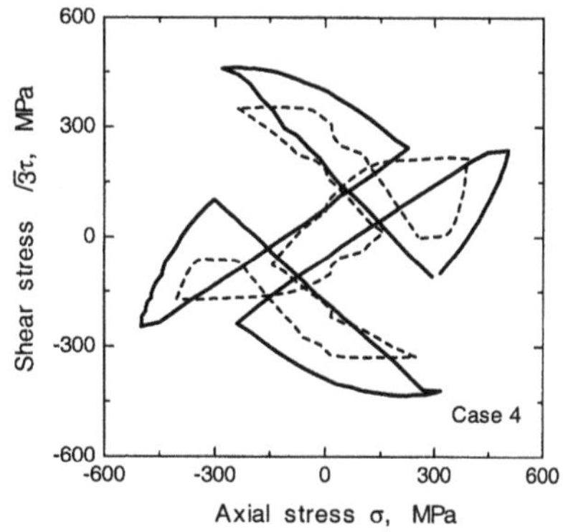

(d) Case 4

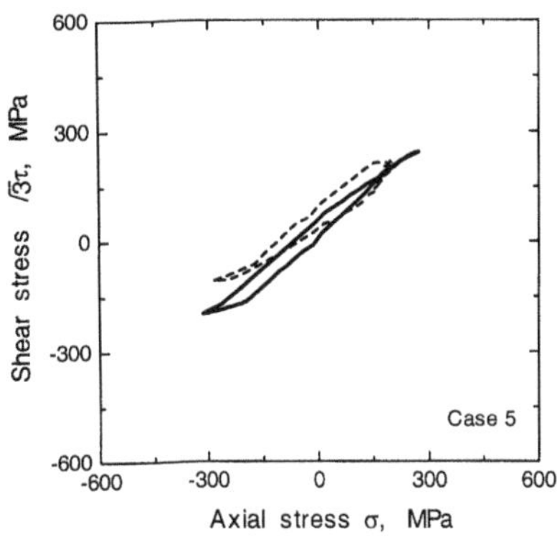

(e) Case 5

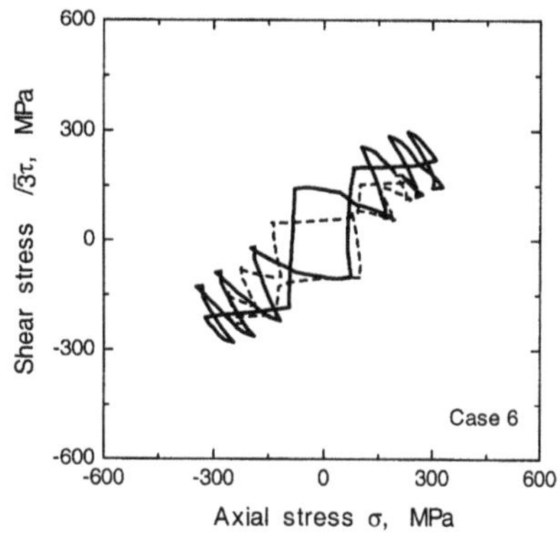

(f) Case 6

Fig.5 Correlation between normal and shear stresses under nonproportional loading (dotted line : $\Delta\varepsilon$=0.5%, solid line : $\Delta\varepsilon$=0.8%).

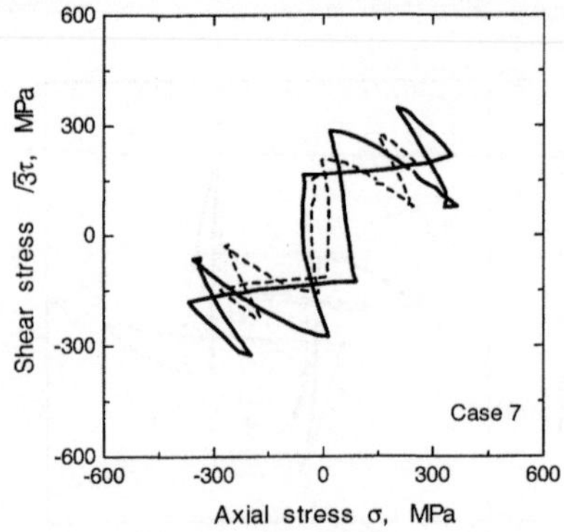

(g) Case 7

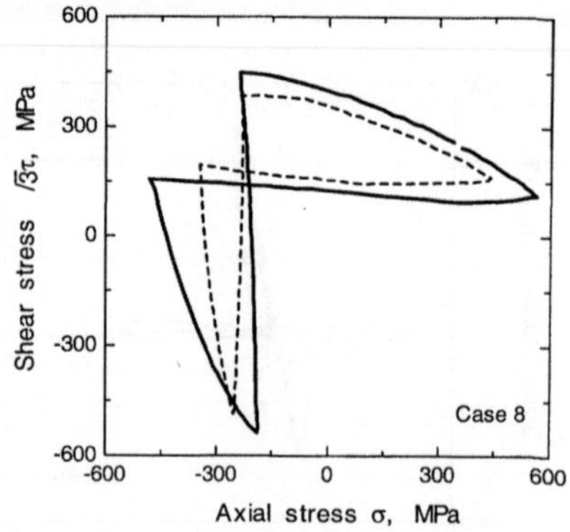

(h) Case 8

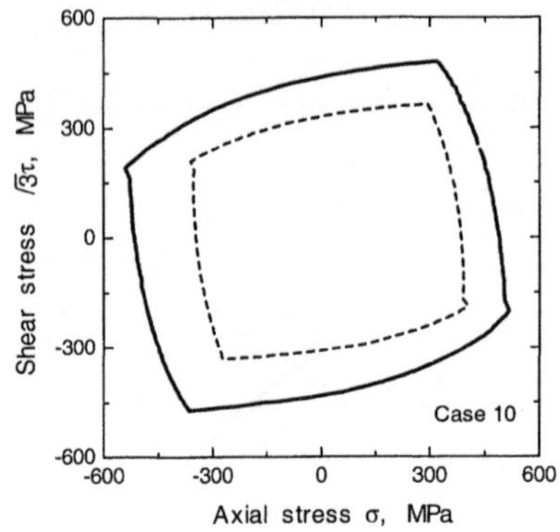

(i) Case 10

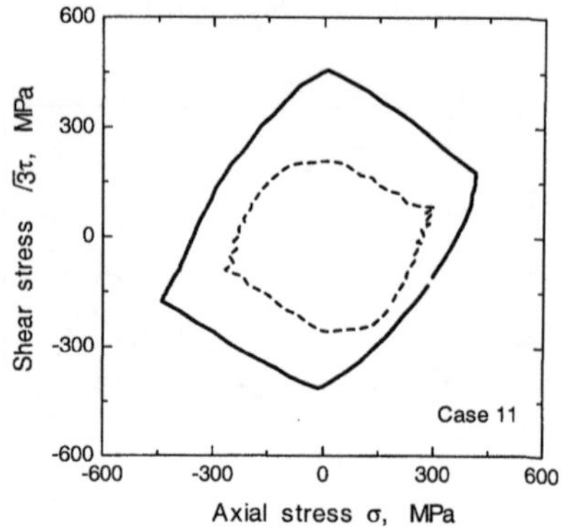

(j) Case 11

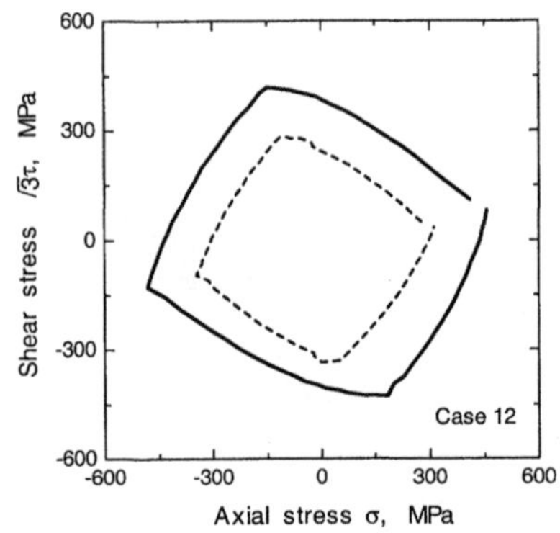

(k) Case 12

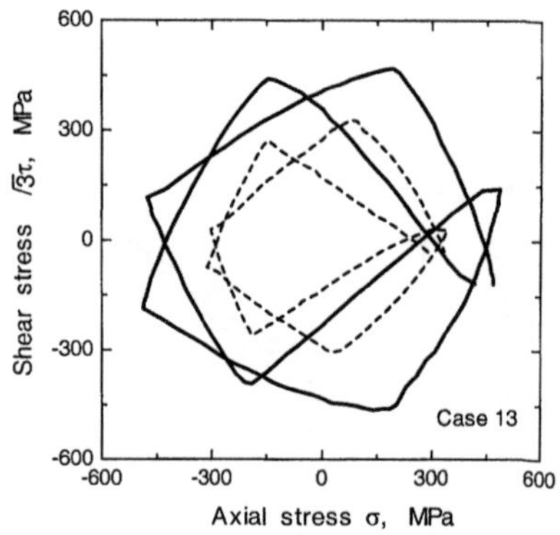

(l) Case 13

Fig.5 Correlation between normal and shear stresses under nonproportional loading. (dotted line : $\Delta\varepsilon$=0.5%, solid line : $\Delta\varepsilon$=0.8%)

Comparison of the stress response between Cases 5, 11 and 12 illustrates the effect of loading phase between the normal and shear strains. A linear stress-strain relationship is obtained on τ–σ plot for Case 5, but a box τ–σ relationship is found for Cases 11 and 12. The normal and shear stress ranges in Cases 11 and 12 are significantly larger than those of Case 5 and the stresses in Case 12 are somewhat larger than in comparison with Case 11. The principal strain amplitude of Case 12 is smaller than that of Case 11. This indicates that the 90-degrees phase difference has a greater hardening effect than a 45-degrees phase difference. Case 13 exhibits a much more significant additional hardening than Case 12, which indicates that the turn around of straining increases the additional hardening effect.

Cases 8, 9 and 10 give the largest additional hardening among the 13 strain paths. A 40 - 60 % normal stress increase is found in these three strain paths in comparison with the Case 0 test. Socie **(12)** has reported the circular strain path has a more pronounced additional hardening than the box strain path for this material where a 90 % stress increase was found in the circular strain path.

OBSERVATIONS OF DISLOCATION STRUCTURE

Additional hardening has been reported to have a close connection with dislocation structure **(9)**, but there have been few systematic and quantitative studies of the relationship between the microstructure and additional hardening. Figures 6(a)-(j) show the microstructure observed by TEM. Figure 6(a) shows the dislocation structure before testing where the dislocation density is very low and no specific substructure is identified.

A cell structure is observed in Case 0, Fig.6(b), where the mean cell diameter is around 1μm. Cell formation was also observed in the specimens cyclically loaded at large strain ranges (>1%) in Case 0. Dislocation bundles which indicates the cluster of dislocations were observed at low strain ranges (<0.8 %) in Case 0.

Cell structures, twins and stacking faults were observed in Case 1, Fig.6(c), but only twins and stacking faults were observed in Case 3, Fig.6(d). No clear cell formation was found in Case 3, and dislocation bundles were observed. Many stacking faults occurred before cell formation and they appear to hinder the cell formation in Case 3. The number of stacking faults in Case 3 is larger than that in Case 1. The phasing of the applied strains produces larger stress and strain ranges for Case 3.

Nishino et. al. **(8)** reported that a ladder or maze structure was a common structure for Type 304 stainless steel in proportional straining and a cell structure was primarily found in the nonproportional straining like Case 1 at high temperature. At room temperature, however, cell structures formed and no ladder or maze structures were observed in Case 0 loading. This difference in dislocation structure between room and elevated temperatures results from the difference in the thermal activation. At elevated temperatures, dislocations glide more easily to form a structure of low elastic energy by the assistance of thermal activation so that a ladder or maze structure, which is a lower elastic energy microstructure than the cell, was found. In Case 5, which is a proportional straining, Fig.6(e), cell structures were observed. A twin boundary was also observed at the centre of the photograph. Case 7, Fig.6(f), exhibits a clear twin boundary at the centre of the photograph. Dislocation walls are observed in the right of the photograph, and a cell structure is found in the left of the photograph. In Case 7, columnar cells were formed and Fig.6(f), shows the two different sections of the columnar grains; the left is the normal section to columnar axes and the right is the parallel section of them. In Case 7, the maximum shear strain direction changed its direction, so that dislocations would easily rearrange to columnar structure by the cross steps.

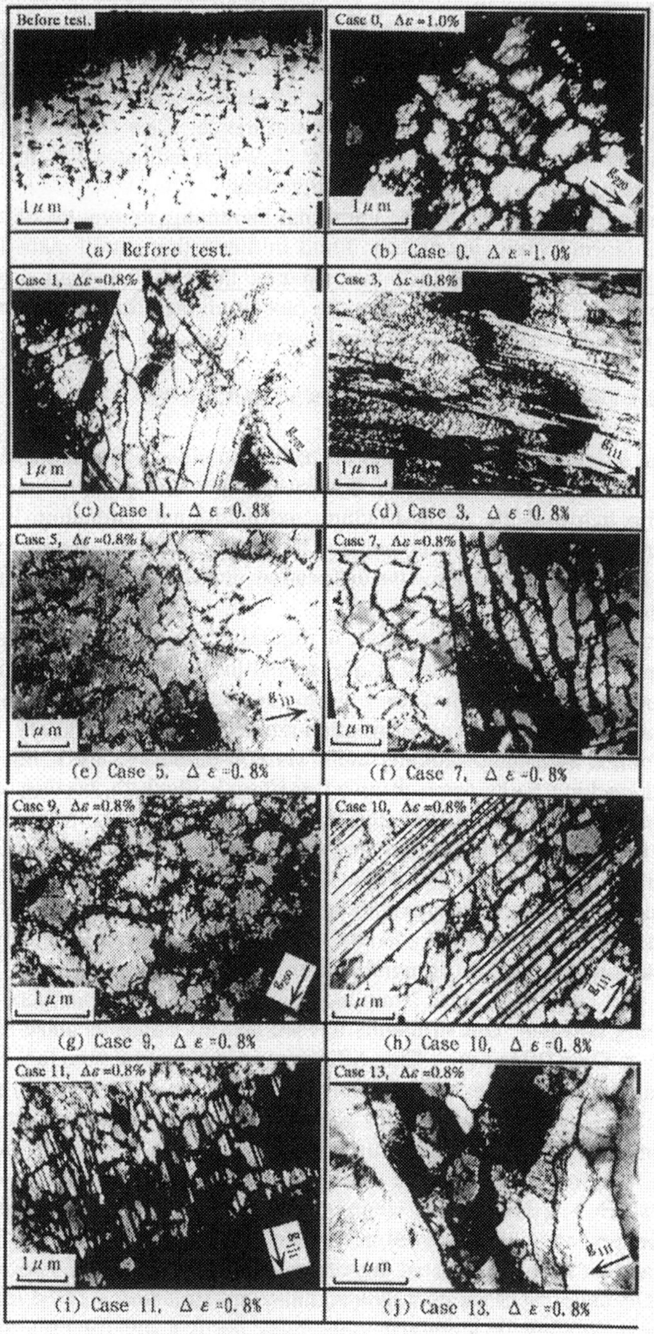

Fig.6 Microstructure observed by TEM.

Cases 5 - 10 all have the same maximum shear strain ranges. Cases 8 - 10 have a rectangular or box strain history. In these strain paths, the maximum shear strain direction rotates continuously, so that many slip systems operate. The additional hardening was most significant in these strain paths. Cases 11 -13 are also rectangular strain paths but the phasing of the strains is such that the maximum shear strain range is smaller than Cases 5 - 10. In Case 9, Fig.6(g), cell boundaries are not clear, but many dislocations exist even in the cells. The maximum shear stress direction rotates continuously in Case 9, so many slip planes operate and interact, and which results in the significant additional hardening.

In Cases 10 and 11, Figs.6(h) and (i), many stacking faults were observed. Since Type 304 stainless steel is a material of low stacking fault energy, slip is planar and there are many partial dislocations which make a stacking fault between them. Long stacking faults exceeding several subgrains in length were formed in Case 10 with short stacking faults formed within cells in Case 11. The long stacking faults were formed by the severe box nonproportional straining and which hindered the cell formation, while, in Case 11, the cells were formed earlier than the stacking faults and stacking faults were stopped by the cell boundaries.

For Case 13, Fig.6(j), the principal stress range in Case 13 is larger than that of Case 9, so that cell structure of Case 13 is different from Case 9. For Case 13, fine cells are found and they are rather close to subgrain since the cell boundaries are rigid and misorientation angle between cells is rather large. This strain path made resulted in clear cells and rigid cell boundaries.

Figure 7 is a microstructure map showing the cell, dislocation bundle and stacking fault boundaries as functions of maximum principal strain range and nonproportional factor for all the strain paths. In the figure, solid symbols indicate tests, in which only cells were observed, while open symbols represent tests in which cells and other dislocation structures were found. Asterisks indicate tests where stacking faults were observed and the number at the data indicates the strain path number shown in Fig.1. This figure shows that stacking faults were observed in almost all the tests and did not depend on the principal strain range and nonproportional factor. Type 304 is a low stacking fault energy material and a dislocation easily splits into partial dislocations, making a stacking fault between them. A partial dislocation glides on the slip plane, and a stacking fault arises between the partial dislocations. Many stacking faults seem to be generated by this mechanism.

There is a critical combination of strain range and nonproportional factor for forming cells indicated by the solid line. In the region above the line, the microstructure is only cells but other microstructures together with cells were observed for the test conditions below the solid line.

Figure 8 shows the relationship between the mean cell size and the maximum principal stress range for all the tests where the cell structure was observed. The mean cell size was determined by the Heyn method (JIS G0552), observing 3 or 4 locations of each specimen. Maximum principal stress range and mean cell size can be approximated by a straight line for all of the strain histories. The relationship is,

$$\Delta\sigma = m \times d^{n} \tag{8}$$

The values of m and n are 975 MPa and -0.57, respectively when d is measured in μm. The value of exponent is close to -1/2, so that the Hall-Petch relationship holds in proportional and nonproportional loadings. As shown in Fig.7, various microstructures are formed under nonproportional loading. However, the result in

Fig.8 indicates that the additional hardening in nonproportional loading is mainly caused by reduction of cell size. The severe interaction of slip systems under nonproportional loading reduces the cell size and results in the additional hardening. The results also imply that microstructures other than cell structure have almost no influence on the additional hardening.

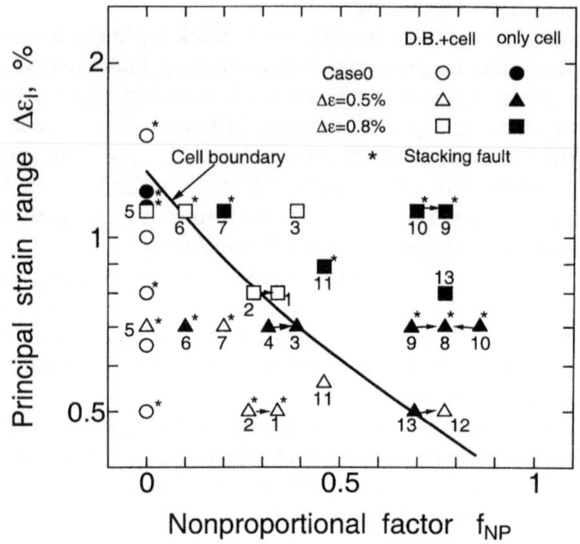

Fig.7 Relationship between microstructure, maximum principal strain range and nonproportional factor.

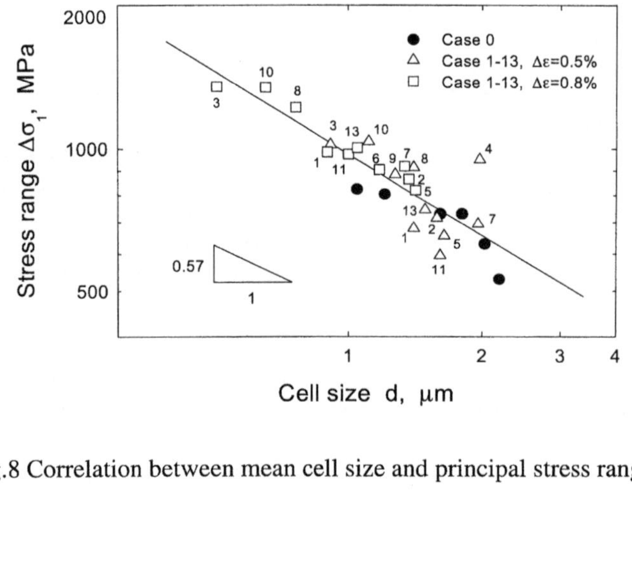

Fig.8 Correlation between mean cell size and principal stress range

CONCLUSIONS

(1) Dislocation substructures observed under nonproportional loading were associated with cells, stacking faults, twins and bundles.

(2) A microstructure map was proposed that show conditions for forming cells and stacking faults as functions of the maximum principal strain range and a nonproportional factor. There exists a critical boundary for forming cells. Stacking faults were observed in almost all the proportional and nonproportional tests.

(3) The principal stress range was uniquely correlated with the mean cell size and is independent of the strain loading path which indicates that the additional hardening was mainly associated with a reduction of cell size.

REFERENCES

(1) Doong S. H., Socie D. F. and M.Robertson I., (1990), Dislocation substructure and nonproportional hardening, Trans. ASME, Eng. Mater. Technol., Vol. 112 - 4, pp.456-465.

(2) Cailletaud G., Doquet V. and Pineau A., (1991), Cyclic multiaxial behaviour of an austenitic stainless steel: microstructural observations and macromechanical modelling, Fatigue Under Biaxial and Multiaxial Loading, ESIS 10, ed. Kussmaul et. al., pp.131-149.

(3) Itoh T., Sakane M., Ohnami M. and Ameyama K., (1992), Effect of stacking fault energy on cyclic constitutive relation under nonproportional loading, JSMS, Japan, Vol.41-468, pp.1361-1367 (in Japanese).

(4) Itoh T., Sakane M., Ohnami M. and Ameyama K., (1992), Additional hardening due to nonproportional cyclic loading - a contribution of stacking fault energy, MECAMAT'92, Proc. Int. Seminar on Multiaxial Plasticity, Cachan, France (Benallal et. al., eds.), pp.43-50.

(5) Doquet V., (1992), Deformation twinning and cyclic behaviour of a coni alloy under multiaxial loading, MECAMAT'92, Proc. Int. Seminar on Multiaxial Plasticity, Cachan, France, pp.51-66.

(6) Jaio F., Osterle W., Portella P.D. and Ziebs J., (1995), Biaxial path dependence of low cycle fatigue behaviour and microstructure of alloy 800 h at room temperature, Materials Science and Engineering, Vol.A196, pp.19-24.

(7) McDonnell D.L., Stahl D.R., Stock S.R. and Antolovich S.D., (1988), biaxial path dependence of deformation substructure of type 304 stainless steel, Metallurgical Transactions A, Vol.19A, pp.1277-1293.

(8) Nishino S., Hamada N., Sakane M., Ohnami M., Matsumura N. and Tokizane M., (1986), microstructural study of cyclic strain hardening behaviour in biaxial stress state at high temperature, Fatigue Fract. Eng. Mater. Struct., Vol.9-1, pp.65-77.

(9) Itoh T., Sakane M., Ohnami M. and Socie D. F., (1995), Nonproportional low cycle fatigue criterion for type 304 stainless steel, Trans. ASME, Eng. Mater. Technol., Vol.117-3, pp.285 - 292.

(10) Murakami S., Kawai M., Aoki K. and Ohmi Y., (1989), Temperature dependence of multiaxial non-proportional cyclic behaviour of type 316 stainless steel, Trans. ASME, Eng. Mater. Technol., Vol.111-1, pp.32-39.

(11) Benallal A. and Marquis D., (1987), Constitutive equations for nonproportional cyclic elasto-viscoplasticity Trans. ASME, Eng. Mater. Technol., Vol.109-4, 326-336.

(12) Socie D.F., (1987), Multiaxial fatigue damage models Trans. ASME, Eng. Mater. Technol., Vol.109-3, pp.293-298.

Acknowledgement

The authors express their gratitude to Dr. Kei Ameyama, the associate professor of Ritsumeikan University, for assisting with the TEM observations.

III. VARIABLE AMPLITUDE AND RANDOM LOADING

COMPARISON OF VARIANCE AND DAMAGE INDICATOR METHODS FOR PREDICTION OF THE FRACTURE PLANE ORIENTATION IN MULTIAXIAL FATIGUE

Włodzimierz BĘDKOWSKI*, Bastien WEBER**, Ewald MACHA*
and Jean-Louis ROBERT***

* Department of Mechanics and Machine Design - Technical University of Opole,
ul. St. Mikołajczyka 5, 45-271 OPOLE, Poland

** SOLLAC- LEDEPP,17, avenue des Tilleuls, 57191 FLORANGE CEDEX, France

***INSA Lyon, Laboratoire de Mécanique des Solides - Bât. 304
20, avenue Albert Einstein, 69621 VILLEURBANNE CEDEX, France

ABSTRACT

Two methods that enable prediction of the fracture plane orientation are presented and compared in this paper. The first one is a statistical approach, which is based on the variance of an equivalent stress. It is assumed that the fracture plane is the one where the variance of a linear combination of the shear and normal stresses acting on this plane is maximum. The second one uses the so-called damage indicator of a multiaxial fatigue criterion, which is based on the research of the critical plane. The formulation of the criterion involves shear and normal stress amplitudes and mean normal stress. The fracture plane is the critical plane; That is to say the one where the damage indicator is the highest. A comparison of the two methods against experimental results is made for biaxial cyclic and random stress states.

KEY WORDS

Biaxial fatigue, random loadings, critical fracture plane, fracture plane orientation, variance method, damage indicator method, fatigue criteria

NOTATION

σ_{-1} fatigue limit under reversed tensile test (R= -1),

τ_{-1} fatigue limit under reversed torsion test (R= -1),

σ_0 fatigue limit under zero to maximum tensile test (R= 0),

$\sigma_{-1}(N)$ fatigue strength corresponding to N cycles under reversed tensile test (R= -1),

$\tau_{-1}(N)$ fatigue strength corresponding to N cycles under reversed torsion test (R= -1),

$\sigma_0(N)$ fatigue strength corresponding to N cycles under zero to maximum tensile test (R= 0),
$\vec{h}$ unit vector normal to a the critical plane (expected fracture),
$\vec{h}_r$ unit vector normal to the experimental fracture plane,
$\vec{h}_t$ unit vector normal to the theoretical (calculated) critical plane,
$\tau_{ha}(t)$ alternate shear stress acting on the plane defined by $\vec{h}$,
$\tau_{hs}(t)$ shear stress acting on the fracture plane,
$\sigma_{hh}(t),\sigma_h(t)$ normal stress acting on the fracture plane,
$\sigma_{eq}(t)$ equivalent stress,
$\mu_{\sigma eq}$ variance of the equivalent stress,
$[\sigma(t)]$ stress states tensor,
$\sigma_{ij}(t)$ stress state components (i,j=x,y,z)
$\sigma_{hha}(t)$ alternate normal stress acting on the plane defined by $\vec{h}$,
σ_{hhm} mean normal stress during the cycle acting on the plane defined by $\vec{h}$,
E_h damage indicator of the plane whose unit normal vector is $\vec{h}$,
E fatigue function of the multiaxial criterion,
R stress ratio,
ψ,θ,φ Euler's angles,
$\hat{l}_n, \hat{m}_n, \hat{n}_n$ expected direction cosines of principal stress $\sigma_1(t) \geq \sigma_2(t) \geq \sigma_3(t)$, n=1,2,3
$\vec{s}$ unit vector on the plane with normal $\overset{\perp}{h}$ defined by the mean direction of the maximum shear stress $\tau_{hsmax}(t)$,
μ_{xst}, covariance of the stress state components, s,t=1, ... , 6,
R_m ultimate tensile strength,
B,K,F material constants,
a_k,a_s,a_t coefficients dependent on direction cosines $\hat{l}_n, \hat{m}_n, \hat{n}_n$, n=1,3 and on constant K
$\theta(N),\alpha(N),\beta(N)$ coefficients in the damage indicator $E_h(t)$ dependent on N,
N number of cycles to failure,
Mxyz system of axes connected with the considered point M of the material,
Mhuv system of axes connected with the critical plane,
$\vec{h}_r \cdot \vec{h}_t$ dot product of vectors $\vec{h}_r$ and $\vec{h}_t$.

INTRODUCTION

Most mechanical components or structures are designed with special attention to the fatigue calculations. The industrial purpose is to improve their service safety to avoid mechanical failures and to lower the cost of maintenance. This is why many industrial engineers and researchers have worked during the last decades both on experiments and theoretical fatigue models to improve the understanding and modelling of the fatigue behaviour of materials. The major objective in fatigue research is to assess the fatigue life of the component submitted to variable loading. This induces various states of stress in any point of the structure. The fatigue assessment is realised everywhere at order to find out the critical area of the component. An important point of fatigue behaviour models is the determination of the crack plane orientation

because it is required to calculate the fatigue life. The so-called critical plane is the one where the crack occurs or will be expected. The material life can then be established through the normal and shear stresses that are applied on it.
The aim of this paper is to present and discuss two different methods for determining the crack plane orientation. One is a statistical approach that is developed by the Department of Mechanics and Machine Design of the Technical University of Opole (Poland) **(1)**. The other is a stress-based approach that is proposed by the Laboratory of Solid Mechanics of INSA Lyon (France) **(2)**.
The Polish point of view is the so-called variance method. An equivalent stress is calculated with respect to a multiaxial fatigue criterion. The assumption is made that the fracture plane is the one where the variance of the equivalent stress is maximum.
The French method uses a multiaxial fatigue criterion that defines a damage indicator E_h for any physical plane. The plane for which E_h reaches the highest value both in time and space domain is assumed to be the critical plane, i.e. the fracture plane. The steps of the two procedures are detailed in the next sections.
The validation of the two methods against some experimental cyclic and random biaxial fatigue tests results is realised. The predicted crack planes orientations are compared with experimental ones. An extension of the two methods to multiaxial random stress states is then proposed. Tests carried out in the Polish laboratory allow comparison of the assessed orientation of the fracture plane with the one observed on cruciform specimen submitted to biaxial random tensile-compressive loads.

PRESENTATION OF THE VARIANCE METHOD

Multiaxial stress states due to the action of various external loads exist at the considered point of the machine or structure. In the proposed algorithm the multiaxial stress states history is reduced to an uniaxial equivalent one by a maximum shear and normal stresses criterion **(1)**. It is assumed that the fracture is influenced only by those stress components, which act on the expected fracture plane(s). The generalised criterion is formulated as:

$$\max_{t} \{B\tau_{hs}(t) + K\sigma_h(t)\} = F \tag{1}$$

where $\tau_{hs}(t)$ and $\sigma_h(t)$ are respectively shear stress in $\vec{s}$ direction and normal stress acting on the fracture plane, whose unit normal vector is denoted as $\vec{h}$. The direction $\vec{s}$ agrees with the mean direction of the maximum shear stress $\tau_{hsmax}(t)$. $\tau_{hs}(t)$ and $\sigma_h(t)$ are functions of stress components $\sigma_{ij}(t)$ (i, j = x, y, z). B, K and F are constants of the criterion. K is expressed as:

$$K = \sqrt{\left(\frac{\sigma_{-1}}{2\tau_{-1} - \sigma_{-1}}\right)^2 - 1} \tag{2}$$

where σ_{-1} and τ_{-1} are the material fatigue limits determined during completely reversed tensile and torsion tests respectively (R= -1).
In particular case of the criterion (1) when we assume that the expected fracture plane is determined by the mean position of one of the two planes of maximum shear stress and B=1, the equivalent stress $\sigma_{eq}(t)$ is defined as:

$$\sigma_{eq}(t) = \frac{1}{K+1}\Big\{\Big[\hat{l}_1^2 - \hat{l}_3^2 + K(\hat{l}_1 + \hat{l}_3)^2\Big]\sigma_{xx}(t) +$$
$$+\Big[\hat{m}_1^2 - \hat{m}_3^2 + K(\hat{m}_1 + \hat{m}_3)^2\Big]\sigma_{yy}(t) + \Big[\hat{n}_1^2 - \hat{n}_3^2 + K(\hat{n}_1 + \hat{n}_3)^2\Big]\sigma_{zz}(t) +$$
$$+ 2\Big[\hat{l}_1\hat{m}_1 - \hat{l}_3\hat{m}_3 + K(\hat{l}_1 + \hat{l}_3)(\hat{m}_1 + \hat{m}_3)\Big]\sigma_{xy}(t) + \tag{3}$$
$$+ 2\Big[\hat{l}_1\hat{n}_1 - \hat{l}_3\hat{n}_3 + K(\hat{l}_1 + \hat{l}_3)(\hat{n}_1 + \hat{n}_3)\Big]\sigma_{xz}(t) +$$
$$+ 2\big[\hat{m}_1\hat{n}_1 - \hat{m}_3\hat{n}_3 + K(\hat{m}_1 + \hat{m}_3)(\hat{n}_1 + \hat{n}_3)\big]\sigma_{yz}(t)\Big\}$$

where $\hat{l}_n$, $\hat{m}_n$, $\hat{n}_n$ (n=1,2,3) are mean direction cosines of principal stresses written so that $\sigma_1(t) \geq \sigma_2(t) \geq \sigma_3(t)$. These direction cosines are used in the description of the expected fatigue fracture plane, which is determined using direction cosines of its normal $\vec{h}$ and tangential $\vec{s}$ vectors:

$$\hat{l}_h = \frac{\hat{l}_1 + \hat{l}_3}{\sqrt{2}} \; ; \; \hat{m}_h = \frac{\hat{m}_1 + \hat{m}_3}{\sqrt{2}} \; ; \; \hat{n}_h = \frac{\hat{n}_1 + \hat{n}_3}{\sqrt{2}}$$
$$\hat{l}_s = \frac{\hat{l}_1 - \hat{l}_3}{\sqrt{2}} \; ; \; \hat{m}_s = \frac{\hat{m}_1 - \hat{m}_3}{\sqrt{2}} \; ; \; \hat{n}_s = \frac{\hat{n}_1 - \hat{n}_3}{\sqrt{2}} \tag{4}$$

From equation (3) it appears that $\sigma_{eq}(t)$ is linearly depending on the stress state components $\sigma_{ij}(t)$. This can be simply written as:

$$\sigma_{eq}(t) = \sum_{k=1}^{6} a_k X_k(t) \tag{5}$$

where:

$$X_1(t) = \sigma_{xx}(t), \quad X_2(t) = \sigma_{yy}(t), \quad X_3(t) = \sigma_{zz}(t),$$
$$X_4(t) = \sigma_{xy}(t), \quad X_5(t) = \sigma_{xz}(t), \quad X_6(t) = \sigma_{yz}(t),$$

For stationary and ergodic random stress state the variance of the equivalent stress can be calculated as:

$$\mu_{\sigma_{eq}} = \sum_{s=1}^{6}\sum_{t=1}^{6} a_s a_t \mu_{xst} \tag{6}$$

where μ_{xst} are the components of the (6x6) covariance matrix of the variables X_k and a_s, a_t are the same coefficients as a_k in Eq.(5) - suitably chosen for μ_{xst}.

In general case the variance $\mu_{\sigma eq}$ depends on the direction cosines $\hat{l}_n$, $\hat{m}_n$, $\hat{n}_n$ (n=1,2,3) which have to fulfil 6 conditions of orthogonality. Practically, the direction cosines are expressed as functions of the three Type 1 Euler's angles ψ,θ,φ (see Fig.1). By this way, the variance of the equivalent stress can be written as:

$$\mu_{\sigma_{eq}} = f(\psi, \theta, \varphi, K, \mu_{xst}) \tag{8}$$

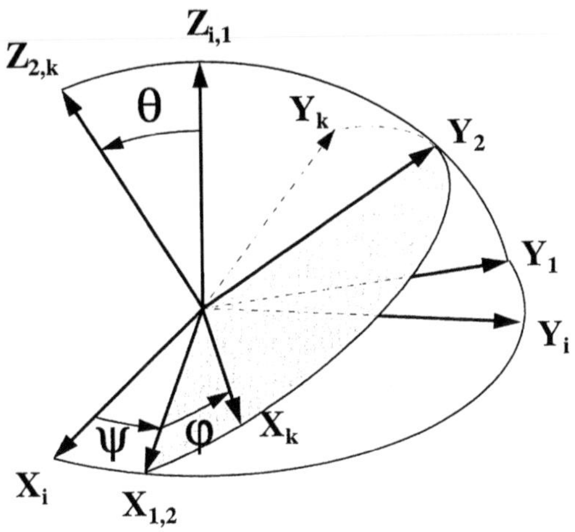

Fig.1 Type 1 Euler's angles.

$$\begin{aligned}
l_1 &= \cos\psi\cos\varphi - \cos\theta\sin\psi\sin\varphi \\
m_1 &= \sin\psi\cos\varphi + \cos\theta\cos\psi\sin\varphi \\
n_1 &= \sin\theta\sin\varphi \\
l_3 &= \sin\theta\sin\psi \\
m_3 &= -\sin\theta\cos\varphi \\
n_3 &= \cos\theta
\end{aligned} \tag{7}$$

As K and μ_{xst} are constants, $\mu_{\sigma eq}$ depends on the three parameters ψ, θ, φ.

The determination of the maximum value of function (8) is generally not possible in an analytical way. Then it is numerically calculated. In the case of multiaxial stationary and ergodic random stress history, the variance matrix components μ_{xst} are calculated from the representative parts of the stress histories and then the set of critical Euler's angles (ψ_c, θ_c, φ_c) that gives the highest value to the variance of the equivalent stress is searched.

PRESENTATION OF THE DAMAGE INDICATOR METHOD

Fracture plane assessment under cyclic stress states

The method, which is developed in this section, is a deterministic one with respect to the former. First, it is presented for a cyclic stress state $[\sigma(t)]$ that is known at the considered point M of the material where the fatigue damage assessment is realised. The criterion defines a time dependent damage indicator $E_h(t)$ for any plane which unit normal vector is denoted as $\vec{h}$. This damage indicator is a linear combination of the alternate shear stress $\tau_{ha}(t)$, the alternate normal stress $\sigma_{hha}(t)$ and the mean normal stress σ_{hhm} **(2)** as:

$$E_h(t) = \frac{1}{\theta(N)}\left[\tau_{ha}(t) + \alpha(N)\sigma_{hha}(t) + \beta(N)\sigma_{hhm}\right] \tag{9}$$

The damage indicator E_h corresponding to the considered plane is defined as the maximum value of $E_h(t)$ during the cycle.

$$E_h = \max_t \left[E_h(t)\right] \tag{10}$$

In this criterion, the normal and shear stress are distinguished as it is well recognised they do not have the same influence in fatigue. Concerning the normal stress, mean and alternate stresses are also separated because they do not have the same incidence on the fatigue behaviour of materials, as the tensile-compressive constant life diagram (Haigh diagram) demonstrates these differences of influence. The mean shear stress does not appear in the formulation of the damage indicator as it is generally assumed to have no practical influence on the fatigue behaviour. The coefficients α and β describe the respective contribution of the stresses components to fatigue damage.

The critical plane (i.e. fracture plane) whose unit normal vector is denoted $\vec{h}_c$ is the one for which the damage indicator is the highest. This maximum value is the fatigue function E of the criterion. Finding the critical plane requires assessment of all the possible planes passing through the point where the fatigue analysis is realised.

$$E = E_{h_c} = \max_t \left[E_h\right] \tag{11}$$

A plane passing through one point M is determined by its unit normal vector $\vec{h}$ defined by the two angles γ and φ with respect to body frame (x, y, z) as shown on Fig.2.

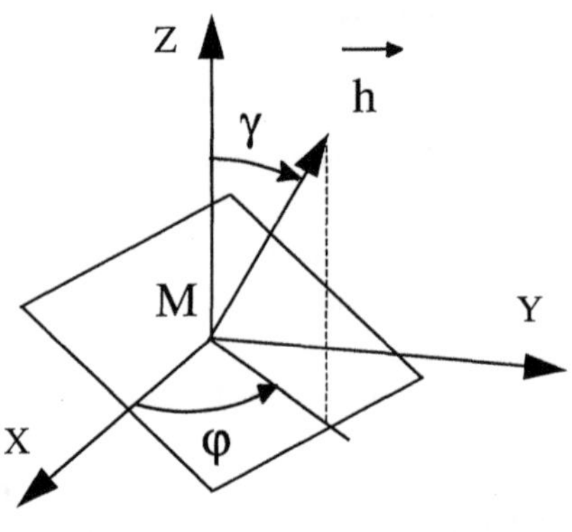

Fig.2 Orientation of one plane with respect to the body frame by the definition of its unit normal vector $\vec{h}$.

E is generally used in order to check whether a given multiaxial stress states cycle reaches the fatigue limit or the fatigue strengths (corresponding to N cycles) of the material. This is expressed as:

$$E = 1 \tag{12}$$

α, β and θ are the three parameters of the criterion. They are determined by stating that the criterion is checked (E=1) for the three fatigue limits of the material, or its three fatigue strengths corresponding to N cycles when the criterion is used as a N cycles fatigue criterion. In other words, it means that the criterion may be utilised as well for endurance limit as for finite fatigue lives **(2, 3, 4)**.

The alternate normal stress $\sigma_{hha}(t)$ is calculated with:

$$\sigma_{hha}(t) = \sigma_{hh}(t) - \sigma_{hhm} \tag{13}$$

where σ_{hhm} is the mean value during the stress cycle of the normal stress $\sigma_{hh}(t)$ that is expressed as:

$$\sigma_{hh}(t) = {}^t\{\vec{h}\}[\sigma(t)]\{\vec{h}\} \tag{14}$$

The alternate shear stress $\tau_{ha}(t)$ is defined by a geometrical method. During a cycle, the tip of the shear stress vector acting on the plane with normal vector $\vec{h}$ makes a closed loop. The smallest circle surrounding to this loop is built (Fig.3).

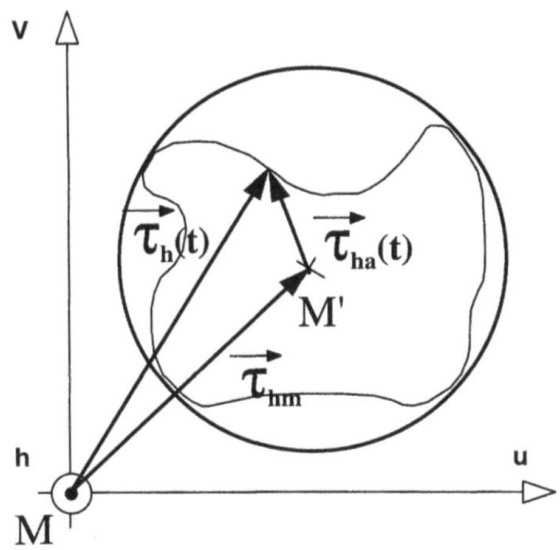

Fig.3 Definition of the alternate shear stress vector $\vec{\tau}_{ha}(t)$ on the plane with normal vector $\vec{h}$.

The centre of the circle gives the mean component $\vec{\tau}_{hm}$ of the shear stress vector $\vec{\tau}_h(t)$ during the cycle. The alternate shear stress vector $\vec{\tau}_{ha}(t)$ is obtained by:

$$\vec{\tau}_{ha}(t) = \vec{\tau}_h(t) - \vec{\tau}_{hm} \tag{15}$$

Fracture plane assessment under multiaxial random loading

The concept of a plane by plane damage accumulation is used for this purpose. The multiaxial random stress states history is decomposed into cycles by the way of the definition of a counting variable and the application of the rainflow counting procedure to this variable. The normal stress acting on one particular plane is used as the counting variable **(3)**. A multiaxial cycle is identified and extracted from the multiaxial stress history when a cycle of the counting variable is obtained from the corresponding part of the multiaxial history.

The criterion allows calculation of the life of the material through equation (11), plane after plane, as shear and normal components of stresses are known for any plane. A damage law such as Miner's rule allows the determination of the corresponding damage induced by the multiaxial cycle on that plane. The calculations are made for all the possible planes passing through point M and for all the stress cycles. A damage accumulation is performed plane by plane for the whole multiaxial stress state history (**4**). Finally, the most damaged plane is assumed to be the critical one, i.e. the fracture plane.

COMPARISON OF BOTH METHODS AGAINST EXPERIMENTAL RESULTS

An experimental verification is realised in order to validate the methods. It is based on the results obtained from fatigue tests of specimens under biaxial cyclic and random states of stress. They are issued from experiments found in literature connected with fatigue fracture planes. Collected tests results are those for which fatigue material data are complete - that is necessary for the application of the two methods - and fracture planes orientations are precisely described.

Cyclic biaxial stress states tests

Six sets of experimental results (corresponding to 86 tests) were collected and are given in Table 1. Rotvel (**5**) made his experiments with cylindrical 0.35% carbon steel specimens. Biaxial sinusoidal tension-compression stress states were generated with different values of mean stress and for some cases out of phase. Nishihara and Kawamoto (**6**) obtained the orientation of the fracture planes under complex bending and torsion cyclic tests. Various ratios of stress amplitudes and many different dephasings are provided. Round specimens were used and several materials are investigated: 0.51% carbon hardened steel, 0.1% carbon mild steel, 3.87% carbon cast iron and 3.81% Cu duraluminium. Achtelik et al. (**7**) tested grey cast iron Zl 250 (3.32% C) round specimens under bending-torsion stress states.

Table 1 Cyclic stress states and fatigue data (**5,6,7**).

Material : carbon steel 0.35% C, σ_{-1}=215.8 MPa, σ_0=349.9 MPa, τ_{-1}=138.5 MPa, Rm=570 MPa		
Test number	Stress states	
	$\sigma_{xx}(t)$	$\sigma_{xy}(t)$
1	$227.6 \sin(\omega t)$	$1.96 \sin(\omega t)$
2	$-2.94 + 224.6 \sin(\omega t)$	$6.87 \sin(\omega t + \pi)$
3	$52 + 233.5 \sin(\omega t)$	$41.2 + 191.3 \sin(\omega t)$
4	$-11.8 + 228.6 \sin(\omega t)$	$-24.5 + 117.7 \sin(\omega t)$
5	$-7.8 + 156 \sin(\omega t)$	$11.77 + 121.6 \sin(\omega t + \pi)$
6	$79.5 + 155 \sin(\omega t + \pi)$	$118.7 \sin(\omega t)$

Material : hardened steel 0.51% C, σ_{-1}=313.9 MPa, σ_0=485.8 MPa, τ_{-1}=196.2 MPa, Rm=694 MPa		
Test number	Stress states	
	$\sigma_{xx}(t)$	$\sigma_{xy}(t)$
HNK50	0.0	$225.63 \sin(\omega t)$
HNK53	$353.16 \sin(\omega t)$	0.0
HNK54	0.0	$201.11 \sin(\omega t)$
HNK55	$323.73 \sin(\omega t)$	0.0
HNK59	$294.30 \sin(\omega t + \pi/2)$	$147.15 \sin(\omega t)$

HNK60	274.68 sin(ωt)	137.34 sin(ωt)
HNK63	264.87 sin(ωt + π/2)	132.44 sin(ωt)
HNK67	162.85 sin(ωt + π/2)	196.69 sin(ωt)
HNK69	154.45 sin(ωt + π/2)	184.23 sin(ωt)
HNK74	162.85 sin(ωt)	195.69 sin(ωt)
HNK75	308.03 sin(ωt)	63.86 sin(ωt)
HNK76	141.85 sin(ωt)	171.28 sin(ωt)
HNK79	344.33 sin(ωt)	71.32 sin(ωt)
HNK83	344.33 sin(ωt + π/2)	71.32 sin(ωt)
HNK84	157.65 sin(ωt + π/3)	190.31 sin(ωt)
HNK86	308.03 sin(ωt + π/3)	63.86 sin(ωt)
HNK89	255.06 sin(ωt)	127.53 sin(ωt)
HNK90	264.87 sin(ωt + π/3)	132.44 sin(ωt)
HNK91	255.06 sin(ωt + π/3)	127.53 sin(ωt)
HNK94	147.15 sin(ωt + π/3)	177.56 sin(ωt)
HNK96	141.95 sin(ωt + π/6)	171.18 sin(ωt)
HNK97	152.35 sin(ωt + π/6)	183.94 sin(ωt)
HNK98	264.87 sin(ωt + π/6)	132.44 sin(ωt)
HNK99	255.06 sin(ωt + π/6)	127.53 sin(ωt)

Material : soft steel 0.1% C, σ_{-1}=235.4 MPa, σ_0=325.7 MPa, τ_{-1}=137.3 MPa, Rm=382 MPa

Test number	Stress states	
	$\sigma_{xx}(t)$	$\sigma_{xy}(t)$
LNK5	194.30 sin(ωt)	0.0
LNK11	0.0	142.25 sin(ωt)
LNK12	187.12 sin(ωt)	93.59 sin(ωt)
LNK16	101.34 sin(ωt)	122.33 sin(ωt)
LNK18	235.64 sin(ωt)	48.85 sin(ωt)
LNK22	235.83 sin(ωt + π/2)	117.92 sin(ωt)
LNK24	208.07 sin(ωt + π/2)	104.08 sin(ωt)
LNK27	112.62 sin(ωt + π/2)	135.97 sin(ωt)
LNK28	244.76 sin(ωt + π/2)	50.72 sin(ωt)
LNK29	235.64 sin(ωt + π/2)	48.85 sin(ωt)
LNK31	201.11 sin(ωt + π/3)	100.55 sin(ωt)
LNK32	194.24 sin(ωt + π/3)	97.12 sin(ωt)
LNK35	245.25 sin(ωt)	0.0
LNK36	105.16 sin(ωt + π/3)	126.84 sin(ωt)
LNK40	108.89 sin(ωt + π/3)	131.45 sin(ωt)

Material : cast iron 3.87% C, σ_{-1}=96.1 MPa, σ_0=142.3 MPa, τ_{-1}=91.2 MPa, Rm=185 MPa

Test number	Stress states	
	$\sigma_{xx}(t)$	$\sigma_{xy}(t)$
CNK4	103.0 sin(ωt)	0.0
CNK6	96.19 sin(ωt)	0.0
CNK7	83.38 sin(ωt)	41.59 sin(ωt)
CNK12	95.16 sin(ωt)	19.72 sin(ωt)
CNK16	104.18 sin(ωt + π/2)	21.58 sin(ωt)
CNK19	99.57 sin(ωt + π/2)	20.6 sin(ωt)
CNK23	56.31 sin(ωt)	67.98 sin(ωt)
CNK30	93.68 sin(ωt + π/3)	46.89 sin(ωt)
CNK33	67.59 sin(ωt + π/3)	81.62 sin(ωt)
CNK36	0.0	98.1 sin(ωt)
CNK38	75.05 sin(ωt + π/2)	90.64 sin(ωt)
CNK39	71.32 sin(ωt + π/2)	86.13 sin(ωt)

Material : duraluminium 3.81% Cu, σ_{-1}=156 MPa, σ_0=257.1 MPa, τ_{-1}=100 MPa, Rm=443 MPa

Test number	Stress states	
	$\sigma_{xx}(t)$	$\sigma_{xy}(t)$
D-30 2	0.0	98.1 sin(ωt)
D-30 5	0.0	127.53 sin(ωt)
D-30 6	156.96 sin(ωt)	0.0
D-30 7	196.2 sin(ωt + π/2)	0.0
D-30 8	181.29 sin(ωt + π/2)	37.57 sin(ωt)
D-30 12	152.55 sin(ωt)	76.32 sin(ωt)
D-30 15	138.7 sin(ωt + π/2)	69.36 sin(ωt)
D-30 16	124.88 sin(ωt)	62.49 sin(ωt)
D-30 17	163.14 sin(ωt)	33.75 sin(ωt)
D-30 19	117.92 sin(ωt + π/2)	58.96 sin(ωt)
D-30 20	82.6 sin(ωt)	99.67 sin(ωt)
D-30 22	199.44 sin(ωt)	41.3 sin(ωt)
D-30 23	199.44 sin(ωt + π/2)	41.3 sin(ωt)
D-30 24	82.6 sin(ωt + π/2)	99.67 sin(ωt)

Material : grey cast iron 3.32% Cu, σ_{-1}=143 MPa, σ_0=212.7 MPa, τ_{-1}=110 MPa, Rm=278.8 MPa

Test number	Stress states	
	$\sigma_{xx}(t)$	$\sigma_{xy}(t)$
Zla 1	168.0 sin(ωt)	0.0
Zla 2	164.0 sin(ωt)	0.0
Zla 3	160.0 sin(ωt)	0.0
Zlb 1	0.0	142.0 sin(ωt)
Zlb 2	0.0	130.0 sin(ωt)

Zlb 3	0.0	132.0 sin(ωt)
Zlc 1	149.9 sin(ωt)	74.95 sin(ωt)
Zlc 2	121.62 sin(ωt)	60.81 sin(ωt)
Zlc 3	118.79 sin(ωt)	59.4 sin(ωt)
Zld 1	176.67 sin(ωt)	51.0 sin(ωt)
Zld 2	155.88 sin(ωt)	45.0 sin(ωt)
Zld 3	152.42 sin(ωt)	44.0 sin(ωt)
Zle 1	118.0 sin(ωt)	102.2 sin(ωt)
Zle 2	108.0 sin(ωt)	93.53 sin(ωt)
Zle 3	106.0 sin(ωt)	91.78 sin(ωt)

Random biaxial stress states tests

Some biaxial random tension-compression fatigue tests have been carried out by W.Będkowski **(8)**. Low carbon steel (10HNAP) thin walled cruciform specimens were used. Table 2 gives the chemical composition of this steel. Ten different random histories were generated by a random signals generator. The track of the fracture plane with the (O, x, y) free surface plane was observed through angle α_r as shown in Fig.4. The unit vector $\vec{h}_r$ normal to this fracture plane is such that: $(\vec{h}_r, \vec{x}) = \alpha_r \pm 90^o$.

Table 2 Chemical composition of the 10HNAP steel.

Elements	C	Mn	Si	P	S	Cr	Cu	Ni
Content [%]	0.115	0.71	0.41	0.082	0.028	0.81	0.30	0.50

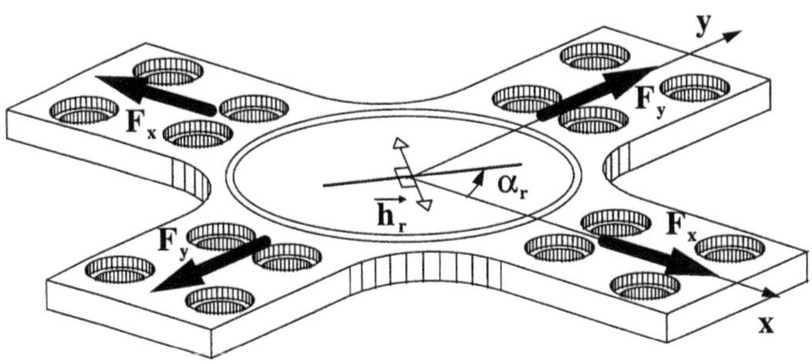

Fig.4 Cruciform specimen.

FRACTURE PLANE ORIENTATION RESULTS

The experimental fracture plane is defined by its unit normal vector $\vec{h}_r$. The predicted fracture planes are defined by the theoretical unit normal vector $\vec{h}_{t1}$ and $\vec{h}_{t2}$ for variance and damage indicator methods respectively. In the case where several assessed fracture planes are obtained, the most similar to that obtained experimentally is assumed. The suitability of the predicting methods is measured by the closeness of the theoretical and experimentally observed fracture planes, that is to say by the closeness of the corresponding unit normal vectors $\vec{h}_r$ and $\vec{h}_{t1,t2}$. The dot product of these vectors is calculated to express the agreement (or disagreement) between assessments and tests results.

The direction cosines of vectors $\vec{h}_r$, $\vec{h}_{t1}$ and $\vec{h}_{t2}$ are reported in Table 3 for biaxial cyclic tests and in Table 4 for biaxial random tests.

Table 3 Experimental and predicted fracture plane orientations (cyclic stress states).

Test number	Real cosines directions			Theoretical variance cosines directions				Theoretical damage indicator cosines directions			
	l_r	m_r	n_r	l_{t1}	m_{t1}	n_{t1}	$\vec{h}_r \cdot \vec{h}_{t1}$	l_{t2}	m_{t2}	n_{t2}	$\vec{h}_r \cdot \vec{h}_{t2}$
1	1.0	0.0	0.0	0.990	0.143	0.0	0.990	0.845	0.074	-0.530	0.845
2	1.0	0.0	0.0	0.990	0.143	0.0	0.990	0.842	0.526	0.122	0.842
3	1.0	0.0	0.0	0.990	0.143	0.0	0.990	0.848	0.0	-0.530	0.848
4	1.0	0.0	0.0	0.990	0.143	0.0	0.990	0.839	0.0	-0.545	0.839
5	1.0	0.0	0.0	0.990	0.143	0.0	0.990	0.829	0.559	0.0	0.829
6	1.0	0.0	0.0	0.990	0.143	0.0	0.990	0.848	0.530	0.0	0.848
HNK50	0.71	0.71	0.0	0.791	0.612	0.0	0.996	0.982	0.191	0.0	0.832
HNK53	1.0	0.0	0.0	0.992	0.126	0.0	0.992	0.833	-0.337	0.438	0.833
HNK54	0.71	0.71	0.0	0.791	0.791	0.0	0.996	0.982	0.191	0.0	0.832
HNK55	1.0	0.0	0.0	0.992	0.126	0.0	0.992	0.835	0.543	0.087	0.835
HNK59	1.0	0.0	0.0	0.992	0.124	0.0	0.992	0.875	-0.485	0.0	0.875
HNK60	0.92	0.38	0.0	0.965	0.264	0.0	0.988	0.978	-0.208	0.0	0.821
HNK63	1.0	0.0	0.0	0.992	0.124	0.0	0.992	0.876	-0.466	0.122	0.876
HNK67	0.91	0.41	0.0	0.848	0.530	0.0	0.990	0.988	0.156	0.0	0.963
HNK69	0.88	0.47	0.0	0.848	0.53	0.0	0.995	0.985	0.174	0.0	0.948
HNK74	0.82	0.57	0.0	0.755	0.655	0.0	0.992	0.375	0.927	0.0	0.836
HNK75	0.98	0.19	0.0	0.998	0.069	0.0	0.991	0.927	-0.375	0.0	0.837
HNK76	0.83	0.56	0.0	0.755	0.655	0.0	0.993	0.375	0.927	0.0	0.830
HNK79	0.98	0.19	0.0	0.998	0.069	0.0	0.991	0.927	-0.374	0.035	0.837
HNK83	1.0	0.0	0.0	0.992	0.126	0.0	0.992	0.837	0.483	0.259	0.837
HNK84	0.92	0.39	0.0	0.881	0.473	0.0	0.995	0.999	0.052	0.0	0.939
HNK86	1.0	0.0	0.0	0.992	0.126	0.0	0.992	0.837	0.483	0.259	0.837
HNK89	0.93	0.37	0.0	0.965	0.264	0.0	0.995	0.982	0.191	0.0	0.843
HNK90	0.99	0.14	0.0	0.983	0.181	0.0	0.999	0.978	-0.208	0.0	0.939
HNK91	0.99	0.14	0.0	0.983	0.181	0.0	0.999	0.978	-0.208	0.0	0.939
HNK94	0.93	0.37	0.0	0.881	0.473	0.0	0.994	0.999	0.053	0.0	0.948
HNK96	0.85	0.53	0.0	0.892	0.452	0.0	0.998	1.0	0.0	0.0	0.850
HNK97	0.92	0.39	0.0	0.892	0.452	0.0	0.997	1.0	0.0	0.0	0.920
HNK98	0.96	0.28	0.0	0.968	0.251	0.0	1.0	0.978	-0.208	0.0	0.881
HNK99	0.96	0.28	0.0	0.968	0.251	0.0	1.0	0.978	-0.208	0.0	0.881
LNK5	1.0	0.0	0.0	0.997	0.077	0.0	0.997	0.810	0.487	0.326	0.810
LNK11	0.71	0.71	0.0	0.764	0.645	0.0	0.996	0.990	0.139	0.0	0.802
LNK12	0.93	0.37	0.0	0.953	0.303	0.0	0.998	0.970	-0.242	0.0	0.813
LNK16	0.87	0.49	0.0	0.874	0.486	0.0	0.999	0.999	-0.052	0.0	0.843
LNK18	0.98	0.20	0.0	0.961	0.276	0.0	0.997	0.680	0.730	0.07	0.813
LNK22	0.99	0.14	0.0	0.997	0.077	0.0	0.998	0.875	0.485	0.0	0.934
LNK24	0.99	0.14	0.0	0.997	0.077	0.0	0.998	0.875	0.485	0.0	0.934
LNK27	0.78	0.63	0.0	0.826	0.564	0.0	0.999	0.999	0.139	0.0	0.860

LNK28	1.0	0.0	0.0	0.997	0.077	0.0	0.997	0.810	0.487	0.326	0.810
LNK29	1.0	0.0	0.0	0.997	0.077	0.0	0.997	0.810	0.487	0.326	0.810
LNK31	0.99	0.14	0.0	0.975	0.222	0.0	0.996	0.970	-0.242	0.0	0.927
LNK32	0.98	0.20	0.0	0.975	0.222	0.0	1.0	0.970	-0.242	0.0	0.903
LNK35	1.0	0.0	0.0	0.997	0.077	0.0	0.997	0.810	0.487	0.326	0.810
LNK36	0.93	0.37	0.0	0.861	0.509	0.0	0.989	0.999	0.017	0.0	0.936
LNK40	0.99	0.14	0.0	0.860	0.512	0.0	0.923	0.999	0.017	0.0	0.992
CNK4	1.0	0.0	0.0	0.848	0.530	0.0	0.848	0.991	-0.052	-0.122	0.991
CNK6	1.0	0.0	0.0	0.848	0.530	0.0	0.848	0.991	-0.052	-0.122	0.991
CNK7	0.91	0.41	0.0	0.986	-0.166	0.0	0.848	0.865	0.499	-0.052	0.992
CNK12	0.98	0.20	0.0	0.729	0.684	0.0	0.851	0.946	0.326	0.0	0.992
CNK16	1.0	0.0	0.0	0.848	0.530	0.0	0.848	0.991	-0.052	-0.122	0.911
CNK19	1.0	0.0	0.0	0.848	0.530	0.0	0.848	0.991	-0.052	-0.122	0.911
CNK23	0.83	0.56	0.0	0.999	0.032	0.0	0.867	0.731	0.682	0.0	0.989
CNK30	0.96	0.28	0.0	0.976	-0.216	0.0	0.876	0.982	0.191	0.0	0.996
CNK33	0.84	0.54	0.0	0.996	0.087	0.0	0.887	0.891	0.454	0.0	0.994
CNK36	0.71	0.71	0.0	0.974	0.226	0.0	0.857	0.809	0.588	0.0	0.992
CNK38	0.79	0.61	0.0	0.983	0.183	0.0	0.888	0.857	0.515	0.0	0.991
CNK39	0.80	0.60	0.0	0.983	0.183	0.0	0.896	0.857	0.515	0.0	0.995
D-30 2	1.0	0.0	0.0	0.801	-0.598	0.0	0.801	0.996	0.087	0.0	0.996
D-30 5	1.0	0.0	0.0	0.801	-0.598	0.0	0.801	0.996	0.087	0.0	0.996
D-30 6	0.97	0.24	0.0	0.990	0.141	0.0	0.994	0.846	0.529	0.070	0.947
D-30 7	0.98	0.20	0.0	0.990	0.141	0.0	0.998	0.846	0.529	0.070	0.935
D-30 8	0.88	0.47	0.0	0.999	0.045	0.0	0.900	0.896	0.545	0.0	0.994
D-30 12	0.87	0.49	0.0	0.86	0.510	0.0	0.998	0.574	0.819	0.0	0.900
D-30 15	1.0	0.0	0.0	0.990	0.141	0.0	0.990	0.882	-0.469	0.052	0.882
D-30 16	0.82	0.57	0.0	0.860	0.510	0.0	0.996	0.574	0.819	0.0	0.937
D-30 17	0.80	0.60	0.0	0.943	0.334	0.0	0.995	0.719	0.695	0.0	0.992
D-30 19	1.0	0.0	0.0	0.990	0.141	0.0	0.990	0.882	-0.469	0.052	0.882
D-30 20	1.0	0.0	0.0	0.903	0.429	0.0	0.903	1.0	0.0	0.0	1.0
D-30 22	0.82	0.57	0.0	0.943	0.334	0.0	0.964	0.719	0.695	0.0	0.986
D-30 23	0.85	0.53	0.0	0.990	0.045	0.0	0.873	0.839	0.545	0.0	1.0
D-30 24	1.0	0.0	0.0	0.857	-0.515	0.0	0.857	0.999	-0.052	0.0	0.999
Zla 1	1.0	0.0	0.0	0.960	0.281	0.0	0.960	0.916	-0.370	0.156	0.916
Zla 2	1.0	0.0	0.0	0.960	0.281	0.0	0.960	0.916	-0.370	0.156	0.916
Zla 3	1.0	0.0	0.0	0.960	0.281	0.0	0.960	0.916	-0.370	0.156	0.916
Zlb 1	0.71	0.71	0.0	0.878	0.479	0.0	0.963	0.342	0.940	0.0	0.910
Zlb 2	0.71	0.71	0.0	0.878	0.479	0.0	0.963	0.342	0.940	0.0	0.910
Zlb 3	0.71	0.71	0.0	0.878	0.479	0.0	0.963	0.342	0.940	0.0	0.910
Zlc 1	0.88	0.47	0.0	0.779	0.627	0.0	0.980	0.682	0.731	0.0	0.944
Zlc 2	0.88	0.47	0.0	0.779	0.627	0.0	0.980	0.682	0.731	0.0	0.944
Zlc 3	0.91	0.42	0.0	0.994	0.108	0.0	0.950	0.682	0.731	0.0	0.928
Zld 1	0.95	0.30	0.0	0.854	0.520	0.0	0.967	0.777	0.629	0.0	0.927
Zld 2	0.96	0.29	0.0	0.854	0.520	0.0	0.971	0.777	0.629	0.0	0.929
Zld 3	0.97	0.23	0.0	1.0	0.0	0.0	0.970	0.988	-0.156	0.0	0.922
Zle 1	0.81	0.59	0.0	0.690	0.724	0.0	0.986	0.574	0.819	0.0	0.948
Zle 2	0.80	0.60	0.0	0.690	0.724	0.0	0.986	0.574	0.819	0.0	0.950
Zle 3	0.82	0.57	0.0	0.690	0.724	0.0	0.978	0.574	0.819	0.0	0.937

Table 4 Experimental and predicted fracture plane orientations (random stress states).

Sequences	Experimental fracture plane angle α_r (degrees)	Theoretical variance method angle α_{t1} (degrees)	Theoretical damage indicator method angle α_{t2} (degrees)
GP9302	72.0 -73.5 -11.0	71.2 -73.0	30.0 (60.0) -30.0(-60.0)
GP9305	-62.0	-71.9 72.3	30.0 (60.0) -30.0 (-60.0)
GP9307	-73.5 -67.0	-73.6	30.0 (60.0) -30.0 (-60.0)
GP9308	71.0	72.8 -71.4	30.0 (60.0) -30.0 (-60.0)
GP9310	-25.0	-17.9 17.9	-30.0 30.0
GP9312	-72.5	-72.8 71.4	-60.0 60.0
GP9313	-71.0	-75.4 68.7	-60.0 60.0
GP9314	-70.0	-73.5 70.6	-60.0 60.0
GP9315	68.0 -70.5	70.0 -74.2	65.0 -65.0
GP9619	-65.0 -42.0	-69.7 74.5	-65.0 65.0

DISCUSSION

Figs. 5 and 6 give a summary of the previous results for constant amplitude and random stress states respectively. The angle deviation between experimentations and predictions are plotted for both theoretical methods and for each group of tests.

In the case of cyclic stress states tests, the variance method gives very good predictions of the fracture plane orientation. It provides the best predictions for the tests carried on 0.35% C steel, 0.51% C hardened steel, 0.1% C soft steel and grey cast iron (3.32% Cu). The damage indicator (deterministic) method gives the most suitable predicted results for the cast iron (3.87% C) and the two assessing methods are equivalent for the duraluminium.

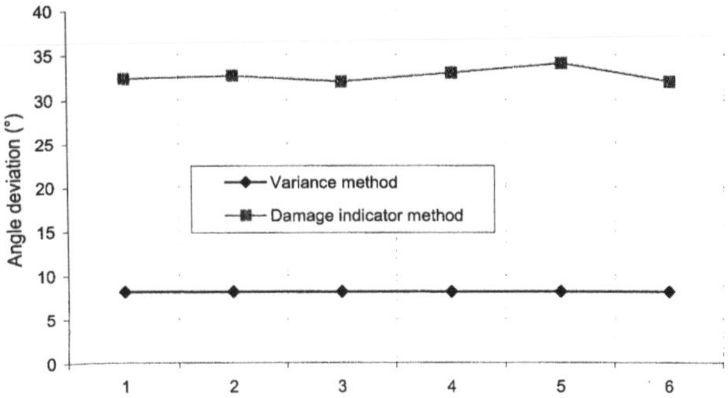
40
35
30
25
20
15
10
5
0
Angle deviation (°)
Variance method
Damage indicator method
1
2
3
4
5
6

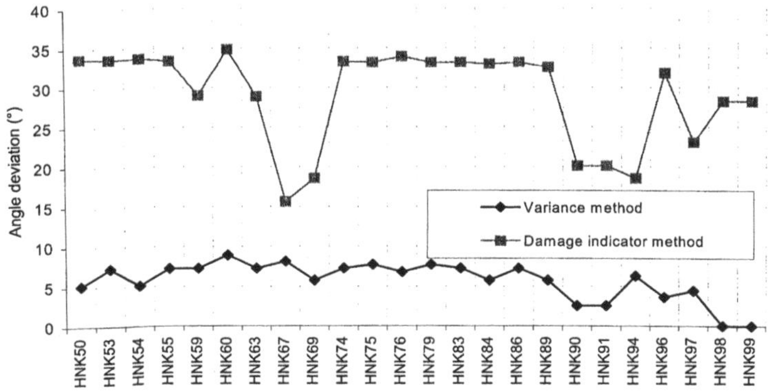
40
35
30
25
20
15
10
5
0
Angle deviation (°)
Variance method
Damage indicator method
HNK50
HNK53
HNK54
HNK55
HNK59
HNK60
HNK63
HNK67
HNK69
HNK74
HNK75
HNK76
HNK79
HNK83
HNK84
HNK86
HNK89
HNK90
HNK91
HNK94
HNK96
HNK97
HNK98
HNK99

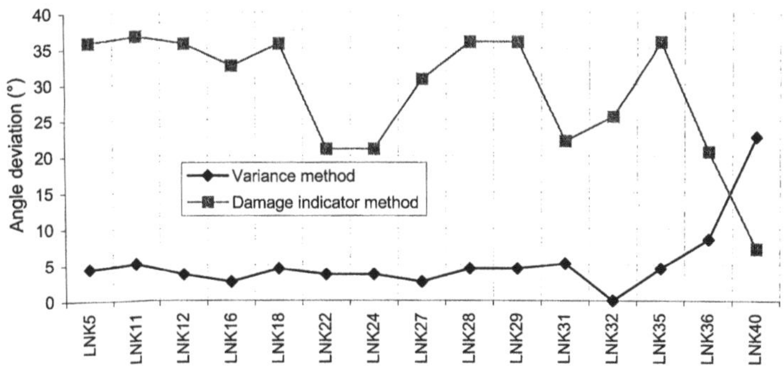
40
35
30
25
20
15
10
5
0
Angle deviation (°)
Variance method
Damage indicator method
LNK5
LNK11
LNK12
LNK16
LNK18
LNK22
LNK24
LNK27
LNK28
LNK29
LNK31
LNK32
LNK35
LNK36
LNK40

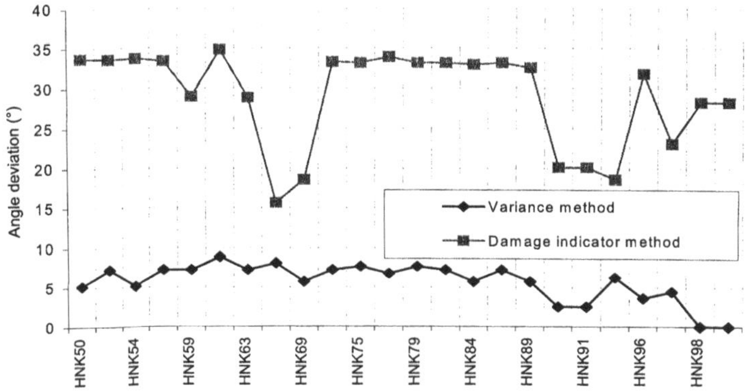
40
35
30
25
20
15
10
5
0
Angle deviation (°)
Variance method
Damage indicator method
HNK50
HNK54
HNK59
HNK63
HNK69
HNK75
HNK79
HNK84
HNK89
HNK91
HNK96
HNK98

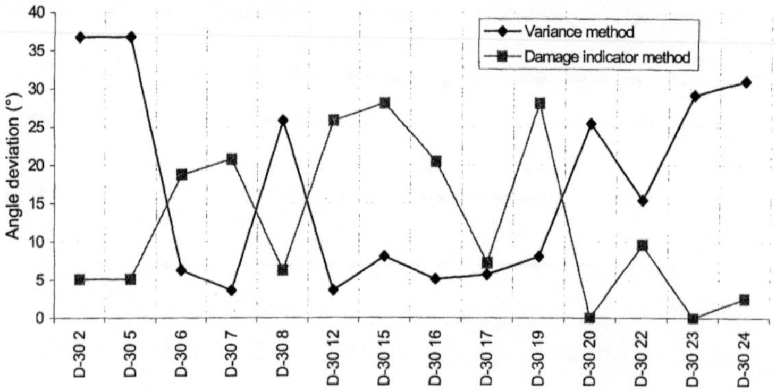

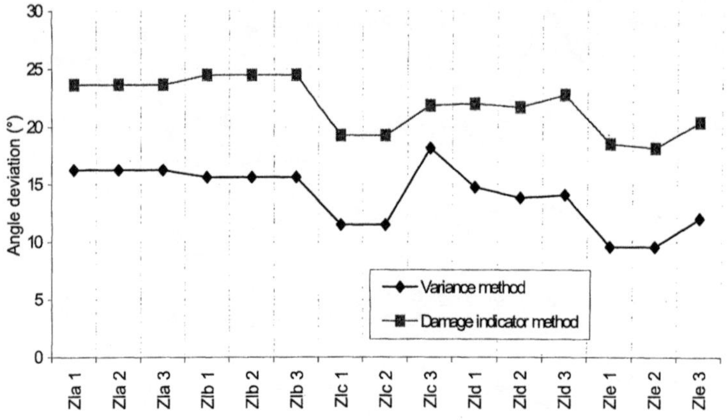

Fig.5 Angles deviations between predicted and experimental fracture plane orientations for constant amplitude multiaxial stress states.

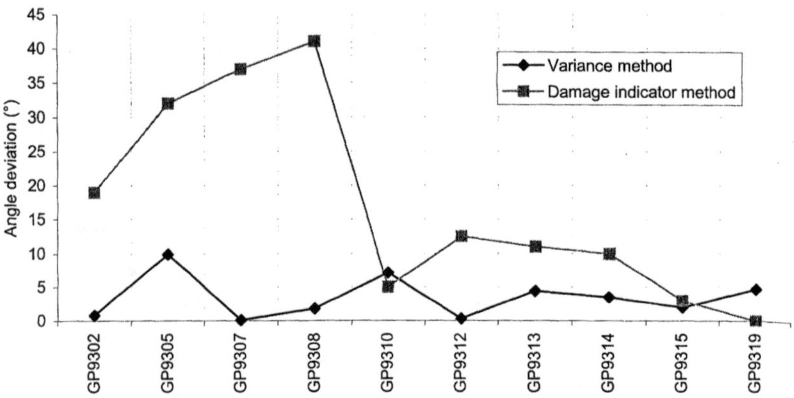

Fig.6 Angles deviations between predicted and experimental fracture plane orientations for random stress states histories.

The overall mean value of the dot product is about 0.961 for the variance method and 0.908 for the deterministic one. It indicates average deviation angles of 16.1° and 24.7° respectively of the predicted fracture planes against the real ones.
Possible explanations of the deviation between predicted and experimental fracture planes may be caused by heterogeneous defaults of metals, which become crack initiation sites because of local stress concentrations and / or mechanical weakness of the material. Such defects are responsible of modification of the crack plane direction because of local changes of stress states. The second explanation is an average effect. In the case of many equally critical planes each one quite close to each other, the macroscopic fracture plane can be observed as the average fracture plane. Figure 7 shows for instance the distribution of the damage indicator all over the possible material planes (whose unit normal vector is defined with angles φ and γ) for a non-proportional bending-torsion cyclic fatigue test. All the critical planes make a ring and because of many activated slipping planes, the real fracture plane has a unit normal vector close to the average of the set of unit vectors of all the critical planes.
This phenomenon presented in Fig.7 for the damage indicator method is also available for the statistical variance method.

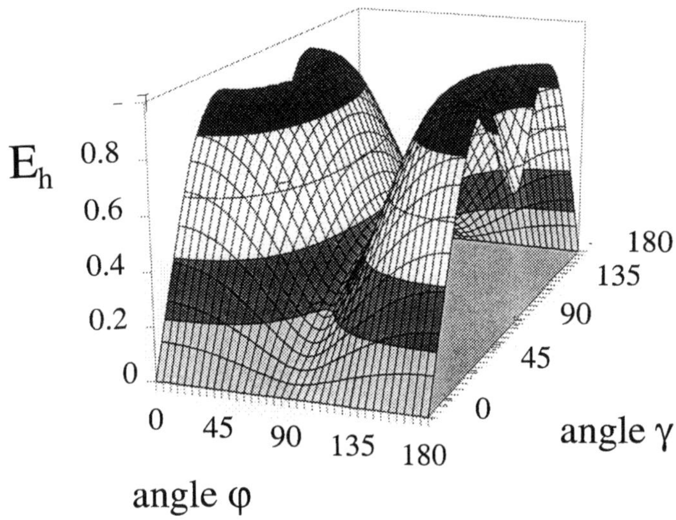

Fig.7 Distribution of the damage indicator.

In the case of biaxial random stress histories, the mean angle deviations between predicted and experimental fracture planes are closed to 3.3° for the variance method and 20.0° for the damage indicator method when only the most critical plane is considered. For this method some other critical fracture planes may appear and are reported mentioned between brackets in Table 4. They correspond to other critical plane (Fig.8), even if the accumulated damage is less than the one obtained on the principal damaged planes. When these planes are also considered the mean deviation between predictions and experiments is represented by a 7.4° angle. More than one critical plane is also observed in the case of variance method - see several extremal points of $\mu_{\sigma eq}$ in Fig.9.

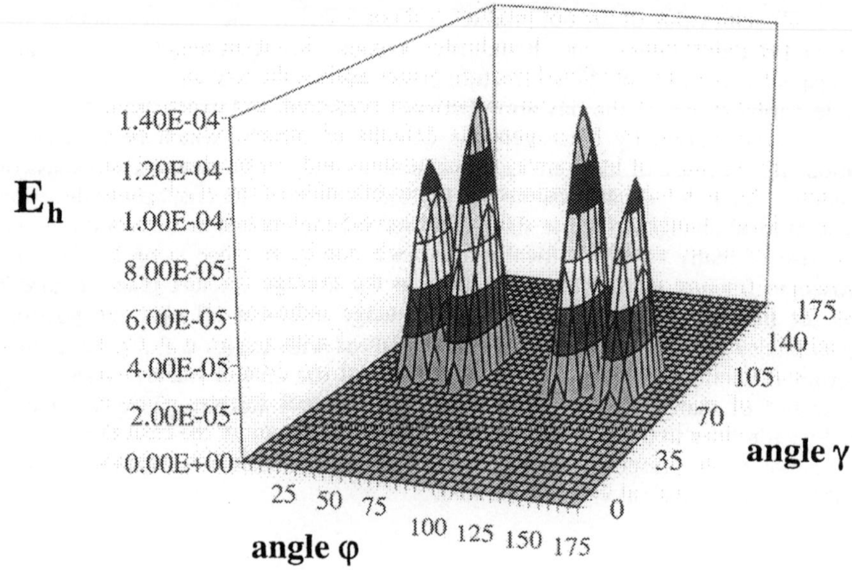

Fig.8 Plane by plane accumulated damage.

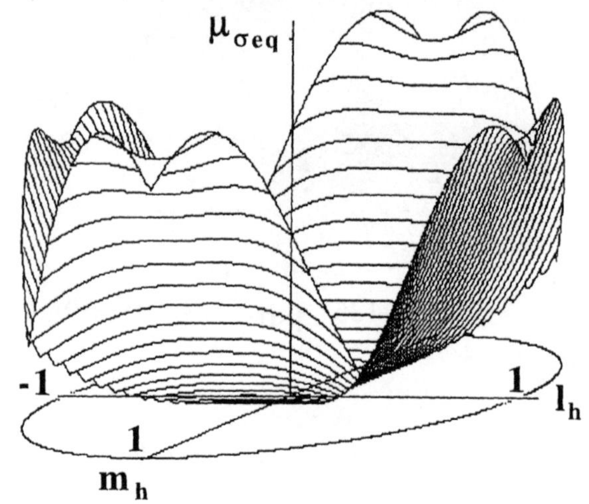

Fig.9 Distribution of the variance function.

CONCLUSION

Two statistical and deterministic methods for predicting fracture plane orientation have been described and predictions are compared with experimental results concerning cyclic or random biaxial stress states. The statistical approach uses the variance of an equivalent stress, which is a linear combination of the normal and shear stresses acting on a plane. It is assumed that the fracture plane is the one where

the variance of the equivalent stress is the highest. The deterministic method uses a critical plane criterion that defines a damage indicator for any plane. It is a function of the alternate and mean components of the normal and shear stresses acting on this plane. The fracture plane corresponds to the most damaged one.
In the case of low mean stress the variance approach is a very promising method for the fracture plane assessment.
As a matter of fact, this stage is a preliminary step for fatigue life assessment. The accuracy of the fatigue life prediction methods strongly depends on the ability to determine the critical plane of the material.

REFERENCES

(1) Będkowski W. and Macha E., (1992), Fatigue fracture plane under multiaxial random loadings - Prediction by variance of equivalent stress based on the maximum shear and normal stresses, Mat.-wiss. u. Werkstofftech. 23, pp.82-94.

(2) Robert J.L., (1992), Contribution à l'étude de la fatigue multiaxiale sous sollicitations périodiques ou aléatoires, Thesis of the National Institute of Applied Sciences (INSA) of Lyon. Order number 92ISAL0004.

(3) Kenmeugne B., (1996), Contribution à la modélisation du comportement en fatigue sous sollicitations multiaxiales d'amplitude variable, Thesis of the National Institute of Applied Sciences (INSA) of Lyon, Order number 96ISAL0064.

(4) Weber B., Clement J.C., Kenmeugne B. and Robert J.L., (1997), On a global stress-based approach for fatigue assessment under multiaxial random loading, Int. Conf. on Engineering Against Fatigue, Sheffield (England), A.A. Balkema Publishers, pp.407-414.

(5) Rotvel F., (1970), Biaxial fatigue tests with zero mean stresses using tubular specimens, Int. J. of Mech. Sc., Pergamon Press, vol. 12, pp.597-615.

(6) Nishihara T. and Kawamoto M., (1945), The strength of metals under combined alternating bending and torsion with phase difference, Memoirs of the College of Engineering, Kyoto Imperial University, vol. XI, n°4, pp.95-112.

(7) Achtelik H., Jakubowska I. and Macha E., (1983), Actual and estimated directions of fatigue fracture plane in ZI250 grey cast iron under combined alternating bending and torsion, Studia Geotechnica et Mechanica, vol. V, n°2, pp.9-30.

(8) Będkowski W., (1994), Determination of the critical plane and effort criterion in fatigue life evaluation for materials under multiaxial random loading - Experimental verification based on fatigue tests of cruciform specimens, Proc. 4th Int. Conf. on Biaxial/Multiaxial Fatigue, Paris (France), pp.435-447.

CRITICAL FRACTURE PLANE UNDER MULTIAXIAL RANDOM LOADING BY MEANS OF EULER ANGLES AVERAGING

Andrea CARPINTERI*, Ewald MACHA**, Roberto BRIGHENTI*
and Andrea SPAGNOLI*

* Dept. Civil Engineering, University of Parma, Viale delle Scienze, 43100 Parma, Italy
** Dept. Mechanics and Machine Design, Technical University of Opole, ul. Mikołajczyka 5, 45-233 Opole, Poland

ABSTRACT

Several authors have experimentally observed that the position of the fatigue fracture plane strongly depends on the directions of the principal stresses or strains. The expected principal stress directions under multiaxial random loading are obtained herein by averaging the instantaneous values of the three Euler angles through some suitable weight functions, in order to take into account the main factors influencing the fatigue fracture behaviour. Then the correlation between such theoretical principal directions and the experimental fracture plane is examined for some biaxial random fatigue tests.

KEY WORDS

Critical plane approach, fatigue fracture plane, multiaxial random loading, principal stress directions, weight function method.

NOTATION

A	matrix of the principal direction cosines
$A(t_k)$	matrix of the principal direction cosines at time instant t_k
c	constant coefficient, with $0 < c \leq 1$
f	generic frequency
G(f)	power spectral density function matrix
$G_{ij}(f)$, $i, j = 1, \ldots, 6$	power spectral density function (PSDF)
l_n, m_n, n_n, $n = 1, 2, 3$	principal direction cosines (eigenvectors of the stress tensor)
$\hat{l}_n, \hat{m}_n, \hat{n}_n$, $n = 1, 2, 3$	expected principal direction cosines
$m_\sigma = -1/m$	coefficient depending on the slope m of the S-N curve for uniaxial tension-compression with loading ratio R = -1
$m_\tau = -1/m^*$	coefficient depending on the slope m^* of the S-N curve for reversed torsion (R = -1)

N	number of time instants being considered, i.e. number of the stress tensor realisations
N_a	S-N fatigue life for stress amplitude σ_a
N_{t_k}	S-N fatigue life for $\sigma_a = \sigma_1(t_k)$
PDF	probability density function
PSDF	power spectral density function
$t_1, t_2, \ldots, t_k, \ldots, t_N$	time instants
W	summation of the weights $W(t_k)$, from t_1 to t_N
$W(t_k)$	weight function at time instant t_k
$x = [x_1, \ldots, x_6]$	vector of the variables
$\hat{x} = [\hat{x}_1, \ldots, \hat{x}_6]$	vector of the mean values
$X(t) = [X_1(t), \ldots, X_6(t)]$	six-dimensional vectorial stochastic process representing the components of the stress tensor
σ_a	stress amplitude
σ_{af}	fatigue limit stress for uniaxial tension-compression with loading ratio $R = -1$
σ_n, $n = 1, 2, 3$	principal stresses (eigenvalues of the stress tensor), with $\sigma_1 \geq \sigma_2 \geq \sigma_3$
$\sigma_{xx}(t), \sigma_{yy}(t), \sigma_{zz}(t)$	normal stresses
$\sigma_{xy}(t), \sigma_{xz}(t), \sigma_{yz}(t)$	shear stresses
τ_{af}	fatigue limit shear stress for reversed torsion ($R = -1$)
ϕ, θ, ψ	Euler angles
$\phi(t_k), \theta(t_k), \psi(t_k)$	Euler angles at time instant t_k
$\hat{\phi}, \hat{\theta}, \hat{\psi}$	expected values of the Euler angles
η_{cal}	calculated angle between the expected direction $\hat{1}$ of the maximum principal stress and the longitudinal axis of the specimen
η_{exp}	mean value of the experimental angle between the normal vector to the fracture plane and the longitudinal axis of the specimen

INTRODUCTION

Several models of crack initiation and propagation under multiaxial cyclic loading have been proposed (**1-3**). One group of such models is based on the critical plane approach (**4**), according to which the expected fracture plane needs to be determined in order to calculate the fatigue life.

From a review of many test results obtained under multiaxial stress state, caused by in- or out-of-phase cyclic loadings, the fatigue fracture plane position appears to greatly depend on the directions of the maximum principal stress or strain and the maximum shear stress or strain (**5-7**). However, position changes of the principal axes are often neglected, although such directions change in most cases of fatigue loadings.

The main damage mechanisms and factors which affect the position of the principal stress or strain axes under multiaxial fatigue loading can be expressed by averaging, through some suitable weight functions **(8)**, the instantaneous values of the parameters defining the principal directions. As an example, each element of the instantaneous 3 x 3 matrix of the principal direction cosines could be averaged to determine the expected position of the principal axes. In a general case, no averaging procedure gives us an orthogonal matrix, because only 3 out of 9 direction cosines are independent.
In order to solve such a problem, the authors have recently proposed **(9,10)** to carry out a weighted average of the 3 Euler angles, which define the principal stress axes. This theoretical procedure is here applied to some biaxial random fatigue tests **(11)**, to examine how the expected principal directions are correlated to the position of the experimental fracture plane.

MULTIAXIAL RANDOM LOADING

Let us consider a six-dimensional vectorial process:

$$X(t) = [X_1(t),...,X_6(t)] \tag{1}$$

where $X_i(t), i = 1,...,6$, are unidimensional stochastic processes representing the components of the stress tensor for a given multiaxial random loading. In particular, we assume that the first three elements of the above vectorial process represent the normal stresses ($\sigma_{xx}(t), \sigma_{yy}(t), \sigma_{zz}(t)$), whereas the other elements represent the shear stresses ($\sigma_{xy}(t), \sigma_{xz}(t), \sigma_{yz}(t)$).
According to the correlation theory, the stationary and ergodic vectorial process $X(t)$ is usually described through its mean value, $\hat{x} = [\hat{x}_1,...,\hat{x}_6]$, and its covariance matrix, $\mu_x(\tau)$, with $\tau = t_k - t_h$, for *h*, *k* = 1,2,...,*N* and *h* < *k*, where *N* is the number of time instants being considered. By assuming the above process to be Gaussian, the joint probability density function (PDF) is given by **(12)**

$$f_{x_1,...,x_6}(x_1,...,x_6,\tau) = \frac{1}{\sqrt{(2\pi)^6 \cdot |\mu_x(\tau)|}} \exp[-0.5 \cdot \sigma_x \cdot \mu_x^{-1}(\tau) \cdot \sigma_x^T] \tag{2}$$

where

$$\mu_x(\tau) = \begin{bmatrix} \mu_{x11}(\tau) & \dots & \mu_{x16}(\tau) \\ \dots & \dots & \dots \\ \mu_{x61}(\tau) & \dots & \mu_{x66}(\tau) \end{bmatrix} \tag{3}$$

$$\sigma_x = [x_1 - \hat{x}_1,...,x_6 - \hat{x}_6] \tag{4}$$

with

- $\mu_x(\tau)$ positive definite covariance matrix of random variables $X_1,...,X_6$
- $|\mu_x(\tau)|$ determinant of the matrix $\mu_x(\tau)$
- $\mu_x^{-1}(\tau)$ inverse of covariance matrix $\mu_x(\tau)$
- σ_x row vector of variables $x_1,...,x_6$ and mean values $\hat{x}_1,...,\hat{x}_6$
- σ_x^T column vector (σ_x transposed).

In practice, the joint PDF expressed by Eq.(2) can also be assumed if the acting loadings present probability distribution different to the normal one, but the law of large numbers can be applied because of the large number of loadings **(13)**.

Generally speaking, if the joint probability density function (Eq.(2)) of the random tensor must be determined, six mean values $\hat{x}$ and twenty-one elements of the symmetric covariance matrix $\mu_x(\tau)$ should be known. The remaining fifteen elements are determined from $\mu_{ij}(\tau) = \mu_{ji}(-\tau)$, with $i, j = 1,...,6$.

The power distribution, i.e. the distribution of mean square values of amplitudes of particular harmonic components appearing in the random process, and the frequency band width can be analysed by carrying out a spectral analysis **(14,15)**. Probabilistic relations between random components of the vectorial process are also important. These properties are expressed with power spectral density functions (PSDFs), $G_{ij}(f)$, which give us a 6 x 6 matrix for the random tensor being considered:

$$G(f) = \begin{bmatrix} G_{11}(f) & \dots & G_{61}(f) \\ \dots & \dots & \dots \\ G_{61}(f) & \dots & G_{66}(f) \end{bmatrix} \tag{5}$$

One-sided PSDFs, $G_{ij}(f)(i, j = 1,...,6)$, of the stress state components are determined for frequency $f \geq 0$, and are equal to double the values of the two-sided PSDFs, $S_{ij}(f)$:

$$G_{ij}(f) = \begin{cases} 2S_{ij}(f) & \text{for } 0 \leq f < \infty \\ 0 & \text{for } f < 0 \end{cases} \tag{6}$$

where

$G_{ii}(f), S_{ii}(f)$ autospectral density function of the stress $X_i(t)$;

$G_{ij}(f), S_{ij}(f)$ cross-spectral density function between stresses $X_i(t)$ and $X_j(t)$.

The cross PSDFs are complex functions, where

$\mathrm{Re}\lfloor G_{ij}(f)\rfloor$ = coincident spectral density function, i.e. the real part of the complex function $G_{ij}(f)$;

$\mathrm{Im}\lfloor G_{ij}(f)\rfloor$ = quadrature spectral density function, i.e. the imaginary part of the complex function $G_{ij}(f)$.

Since the relationship

$$G_{ij}(f) = G_{ji}(-f) \tag{7}$$

holds, twenty-one PSDFs should be known to describe the frequency structure of the random stress state.

EXPECTED PRINCIPAL STRESS DIRECTIONS

Let us assume to arrange the principal stresses (eigenvalues of the stress tensor) in this order: $\sigma_1 \geq \sigma_2 \geq \sigma_3$; that is, the directions of maximum and minimum principal stresses are called 1-axis and 3-axis, respectively. The eigenvectors of the above tensor represent the principal direction cosines l_n, m_n, n_n, $n = 1, 2, 3$ (Fig.1), and form the following matrix *A*:

$$A = \begin{pmatrix} l_1 & l_2 & l_3 \\ m_1 & m_2 & m_3 \\ n_1 & n_2 & n_3 \end{pmatrix} \tag{8}$$

where only three out of nine elements are independent because of six orthonormality conditions.

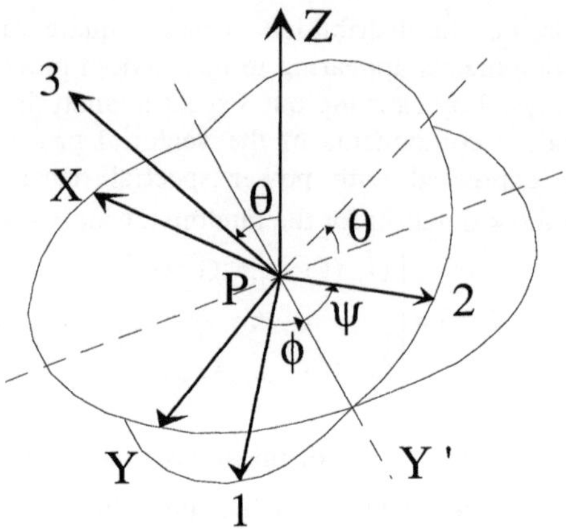

Fig.1 Principal stress directions (1,2,3) described through the Euler angles (ϕ, θ, ψ).

The principal stress directions can also be described through the Euler angles ϕ, θ, ψ, which represent three counterclockwise rotations around Z-axis, Y'-axis and 3-axis, sequentially (Fig.1). Analogously to the case of the direction cosines, we only need three independent parameters to define the principal stress directions. If the above Euler angles are known, matrix *A* of the principal direction cosines can be determined as follows:

$$A = \begin{bmatrix} c_\phi c_\theta c_\psi - s_\phi s_\psi & -c_\phi c_\theta s_\psi - s_\phi c_\psi & c_\phi s_\theta \\ s_\phi c_\theta c_\psi + c_\phi s_\psi & -s_\phi c_\theta s_\psi + c_\phi c_\psi & s_\phi s_\theta \\ -s_\theta c_\psi & s_\theta s_\psi & c_\theta \end{bmatrix} \tag{9}$$

where *s* and *c* correspond to sin and cos, respectively, while the subscripts represent the arguments of such trigonometric functions. On the other hand, the Euler angles of the principal stress directions can be obtained from the components of matrix *A* (see matrix (8)) in a quite simple way, even if some calculation steps are needed **(13)**.
Now consider a multiaxial random stress state, expressed by $X_i(t), i = 1,...,6,$. Every component $\sigma_{ij}(t)$ of such a vectorial process is a random function of time, and therefore we can compute the Euler angles $\phi(t), \theta(t)$ and $\psi(t)$ at each time instant *t*, with $t = t_1, t_2, ..., t_k, ..., t_N$. The calculation of the three Euler angles at the generic time instant t_k from the matrix $A(t_k)$ consists of two stages **(9, 10)**, after which the Euler angle ranges $0 \le \phi(t_k), \psi(t_k) < 2\pi$ and $0 \le \theta(t_k) < \pi$ are reduced to the new

ranges: $0 \le \phi(t_k), \theta(t_k) \le \pi/2$ and $-\pi/2 \le \psi(t_k) \le \pi/2$, in order to average the values of the Euler angles in a correct way with respect to their physical meaning.

If we assume that the expected position $(\hat{\phi}, \hat{\theta}, \hat{\psi})$ of the principal axes under random loading is affected by their generic position $(\phi(t_k), \theta(t_k), \psi(t_k))$ in the same way for any value of t_k, independently of the stress values, the mean principal stress directions $\hat{1}$, $\hat{2}$ and $\hat{3}$ can be obtained from simple arithmetic averages:

$$\hat{\phi} = \frac{1}{N} \sum_{t_1}^{t_N} \phi(t_k), \quad \hat{\theta} = \frac{1}{N} \sum_{t_1}^{t_N} \theta(t_k), \quad \hat{\psi} = \frac{1}{N} \sum_{t_1}^{t_N} \psi(t_k) \tag{10}$$

On the other hand, in order to determine the mean directions of the principal stress axes $\hat{1}$, $\hat{2}$ and $\hat{3}$, it seems logical to carry out the averaging of the Euler angles by employing a suitable weight function, $W(t_k)$, to take into account the main factors influencing the fatigue fracture behaviour **(8-10)**:

$$\begin{aligned} \hat{\phi} &= \frac{1}{W} \sum_{t_1}^{t_N} \phi(t_k) W(t_k), \\ \hat{\theta} &= \frac{1}{W} \sum_{t_1}^{t_N} \theta(t_k) W(t_k), \\ \hat{\psi} &= \frac{1}{W} \sum_{t_1}^{t_N} \psi(t_k) W(t_k), \\ W &= \sum_{t_1}^{t_N} W(t_k). \end{aligned} \tag{11}$$

where W represents the summation of the weights $W(t_k)$, with t_k from t_1 to t_N.

Let us consider the following weight function:

$$W(t_k) = W_1(t_k) = 1, \quad \text{for each } t_k \in [t_1, t_2, ..., t_N], \tag{12}$$

In this case, the summation W is equal to the number *N* of time instants being considered and, consequently, the weighted mean values of the Euler angles, obtained from Eqs (11) for $W_1(t_k)$, coincide with the arithmetic averages, given by Eqs (10).

In order to take into account the effect of the maximum principal stress $\sigma_1(t_k)$ on the expected position of the principal axes, we could adopt the following weight function:

$$W(t_k) = W_2(t_k) = \frac{\sigma_1(t_k) - \sigma_{1,min}}{\sigma_{1,max} - \sigma_{1,min}}, \tag{13}$$

where $\sigma_{1,min}$ and $\sigma_{1,max}$ are the minimum and maximum values, respectively, of the maximum principal stress $\sigma_1(t)$, with $t = t_1, t_2, ..., t_k, ..., t_N$. According to $W_2(t_k)$ defined in Eq.(13), the higher the stress $\sigma_1(t_k)$ is, the more pronounced its effect on the averaging process of the Euler angles becomes. The values of this weight function range between 0 and 1.

Now let us examine another weight function:

$$W(t_k) = W_3(t_k) = \begin{cases} 0 & \text{if } \sigma_1(t_k) < c\,\sigma_{af} \\ \left(\dfrac{\sigma_1(t_k)}{c\,\sigma_{af}}\right)^{m_\sigma} & \text{if } \sigma_1(t_k) \geq c\,\sigma_{af} \end{cases} \qquad 0 < c \leq 1 \tag{14}$$

It only includes into the averaging process, described by Eqs (11), those positions of the principal axes for which the maximum principal stress σ_1 is greater than or equal to the product of the constant coefficient c, with $0 < c \leq 1$, and the fatigue limit stress, σ_{af}, deduced from the S-N curve for uniaxial tension-compression with loading ratio, R, equal to -1 (continuous thick line in Fig.2). The weight of such positions, which is defined in Eqs.(14), exponentially depends on the coefficient $m_\sigma = -\dfrac{1}{m}$, where m is the slope of the S-N curve considered.

Note that the function W_3 is proportional to the reciprocal, $N_{t_k}^{-1}$, of the fatigue life determined from the dashed Wöhler curve (Fig.2) for $\sigma_a = \sigma_1(t_k)$. As a matter of fact, such a curve may be represented by the following expression:

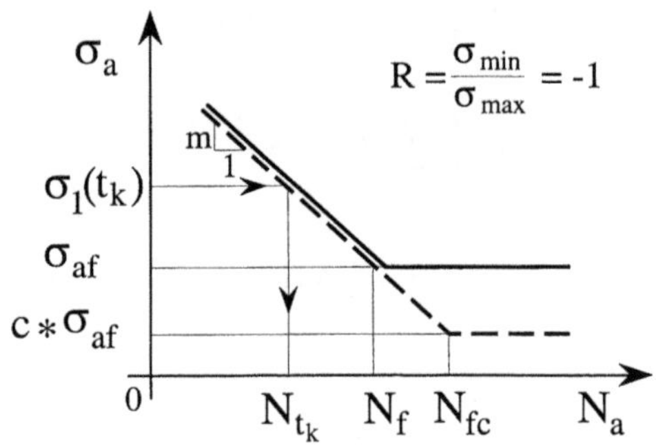

Fig.2 The Wöhler curve for uniaxial tension-compression, with loading ratio $R = -1$.

$$\frac{N_a}{N_{fc}} = \left(\frac{c\,\sigma_{af}}{\sigma_a}\right)^{m_\sigma} \tag{15}$$

with $N_{fc} = N_f / (c)^{m_\sigma}$ = constant for the S-N curve being considered. Since the fatigue life N_a is equal to N_{t_k} for $\sigma_a = \sigma_1(t_k) \geq c\sigma_{af}$, we can obtain from Eq.(15):

$$\frac{N_{fc}}{N_{t_k}} = \left(\frac{\sigma_1(t_k)}{c\,\sigma_{af}}\right)^{m_\sigma} = W_3(t_k)\ . \tag{16}$$

Therefore, by linearly increasing the maximum principal stress $\sigma_1(t_k)$, the S-N fatigue life N_{t_k} exponentially decreases according to the coefficient m_σ., while the proposed weight function $W_3(t_k)$ increases according to the same exponent m_σ.

Finally, the following weight function is analysed:

$$W(t_k) = W_4(t_k) = \begin{cases} 0 & \text{if } \tau_{max}(t_k) < c\,\tau_{af} \\ \left(\dfrac{\tau_{max}(t_k)}{c\,\tau_{af}}\right)^{m_\tau} & \text{if } \tau_{max}(t_k) \geq c\,\tau_{af} \end{cases} \quad 0 < c \leq 1 \qquad (16)$$

where $\tau_{max}(t_k)$ is the maximum shear stress at time instant t_k; τ_{af} is the fatigue limit shear stress determined from the S-N curve for fully reversed torsion (R = -1), while $m_\tau = -\frac{1}{m^*}$ depends on the slope m^* of this S-N curve. It seems interesting to consider such a weight function based on τ_{max}, since the maximum shear stress plays a basic role during the crack initiation stage in a ductile material.

Some of the above weight functions are plotted in Fig.3. In particular, Fig.3(b) shows the constant weight function W_1, whereas Figs.3(c) and 3(d) represent the functions W_2 and W_3, respectively. Finally, the function $\phi^* W_i$ (with $i = 1, 2, 3$), which has to be averaged to determine the expected Euler angle $\hat{\phi}$ according to Eqs (11) to (14), is displayed in Fig.3(e).

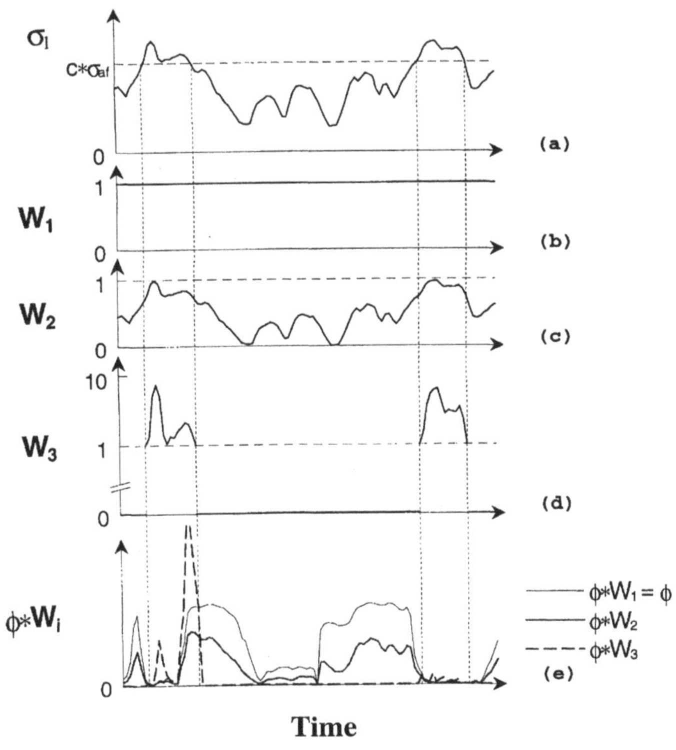

Fig.3 Modifications of the time history of angle $\phi(t)$ by means of weight functions $W_1(t)$, $W_2(t)$ and $W_3(t)$.

EXPERIMENTAL TESTS

The proposed weighted averaging of the three Euler angles is used here to analyse the results obtained from fatigue tests on round specimens made of 10HNAP steel, subjected to a combination of random proportional bending and torsion **(11)**. The steel has a fine-grained ferritic-pearlitic structure, and presents the following properties: tensile strength R_m = 566 MPa, yield point R_e = 418 MPa, Young modulus E = 215 GPa, Poisson ratio ν = 0.29. The characteristic values of the S-N curve for cyclic uniaxial tension-compression with loading ratio *R* equal to -1 are: fatigue limit stress σ_{af} = 252.3 MPa (for $N_f = 1.282 \cdot 10^6$ cycles) and coefficient m_σ = 9.82. Moreover, the following parameters of the S-N curve for fully reversed torsion (*R* = -1) have been determined: fatigue limit shear stress τ_{af} = 182.0 MPa (for $N_f = 5.590 \times 10^6$ cycles) and coefficient m_τ = 8.20.
Stationary and ergodic random loading with zero expected value, normal probability distribution and wide-band frequency spectrum (0-60 *Hz*) has been applied. Fatigue tests for long-life time have been carried out for four combinations of torsional, $M_T(t)$, and bending, $M_B(t)$, moments (Fig.4): 23 specimens for $\alpha = 0$ (pure bending), 21 specimens for $\alpha = \pi/8$, 14 specimens for $\alpha = \pi/4$ and 23 specimens for $\alpha = \pi/2$ (pure torsion), with $\mathrm{tg}\,\alpha = M_T(t)/M_B(t)$. The number of time instants being considered is *N* = 49152. A fragment of the time history of the stresses σ_{xx} and σ_{xy} for $\alpha = \pi/8$ is shown in Fig.5.

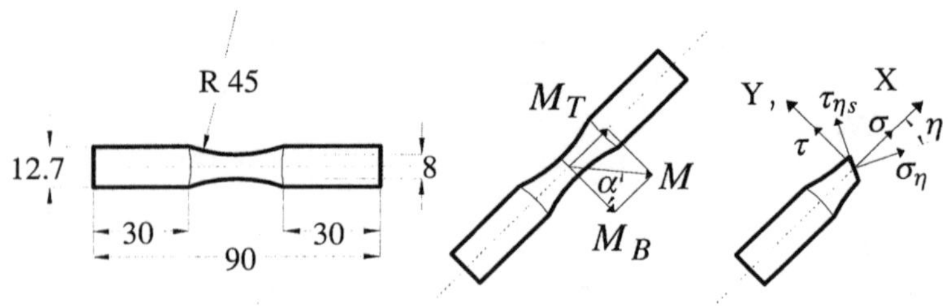

Fig.4 Fatigue tests on round specimens (10HNAP steel): torsion (M_T) and bending (M_B); angle η between normal to the fracture plane and longitudinal X-axis.

From the observation of the experimental fatigue fracture plane (macro), it results that the angle η between the normal vector to the fracture plane and the longitudinal axis of the specimen (Fig.4) presents a mean value, η_{exp}, equal to about $\alpha/2$ for each of the four loading combinations (Table 1). Furthermore the following parameters are shown in Table 1:

- M_{RMS} - root-mean-square values of the total moment $M(t)$;
- $\sigma_{xx,RMS}$, $\sigma_{xy,RMS}$ - root-mean-square values of the normal stress σ_{xx} and shear stress σ_{xy}, respectively.

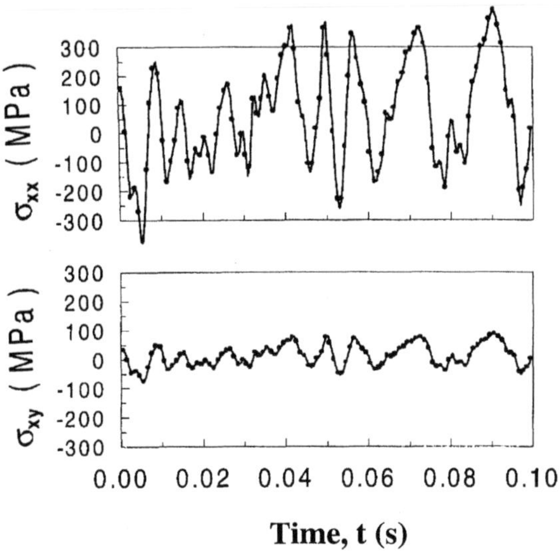

Fig.5 Fragment of time history for the stresses σ_{xx} and σ_{xy}, in the case of $\alpha = \pi/8$.

Table 1 Details of experimental tests under random proportional bending and torsion **(11)**.

α rad	M_{RMS} Nmm	$\sigma_{xx,RMS}$ MPa	$\sigma_{xy,RMS}$ MPa	η_{exp} rad
0	11249	221.29	0	0
$\pi/8$	11091	201.58	41.75	0.062 π
$\pi/4$	10596	147.39	73.70	0.125 π
$\pi/2$	10224	0.39	100.56	0.250 π

For the round specimens of the biaxial random tests examined, the root-mean-square values of σ_{xx} and σ_{xy} are given by:

$$\sigma_{xx,RMS} = \frac{M_{B,RMS}}{W_X} = \frac{\cos\alpha}{W_X} M_{RMS} \tag{18}$$

$$\sigma_{xy,RMS} = \frac{M_{T,RMS}}{2W_X} = \frac{\sin\alpha}{2W_X} M_{RMS} \tag{19}$$

where the section modulus of bending, W_X, is equal to 50.265 mm^3; $M_{B,RMS}$ and $M_{T,RMS}$ are the root-mean-square values of the bending moment $M_B(t)$ and the torsional moment $M_T(t)$, respectively.

The theoretical procedure proposed in the previous section is applied to the above experimental random loadings. Fig.6 shows the fragment of time history of the principal stresses in the case of $\alpha = \pi/8$. It can be observed that the principal stress σ_2 is always equal to zero; moreover, $|\sigma_1|$ is higher than $|\sigma_3|$ for the time instants at which the applied stresses are positive, while $|\sigma_1|$ is lower than $|\sigma_3|$ when the applied stresses are negative.

Table 2 shows the results obtained for the calculated angle η_{cal} between the expected direction $\hat{1}$ of the maximum principal stress and the longitudinal axis of the specimen, for the above four combinations of loadings. In more detail, the Euler angles defining the principal stress directions at time instant t_k, with $k = 1, \ldots, N$, have been averaged by employing the different weight functions W_i, with $i = 1, 2, 3, 4$, discussed in the previous section. Note that the coefficient c for W_3 and W_4 has been assumed equal to 0.5. Then, for each weight function considered, the direction cosines of the expected principal stress axes $(\hat{1}, \hat{2}, \hat{3})$ have been determined from the averaged Euler angles $\hat{\phi}, \hat{\theta}, \hat{\psi}$ by means of matrix (9). Finally, the values $\eta_{cal}(W_i)$, with i from 1 to 4, have been computed through the expression $\eta_{cal} = arc\ cos\,(\hat{l}_1)$, where $\hat{l}_1$ is the direction cosine of the expected principal stress axis $\hat{1}$ with respect to the X-axis.

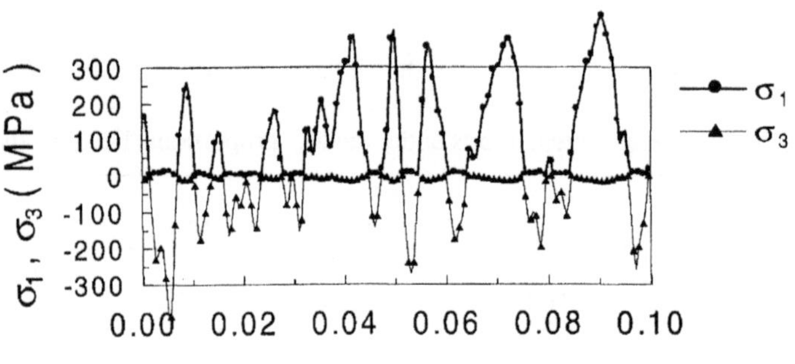

Fig.6 Fragment of time history for the principal stresses σ_1 and σ_3 ($\sigma_2 = 0$), in the case of $\alpha = \pi/8$.

Table 2 Comparison between the normal to the experimental fracture plane and the expected direction of the maximum principal stress, for different weight functions.

α rad	η_{exp} rad	$\eta_{cal}(W_1)$ rad	$\eta_{cal}(W_2)$ rad	$\eta_{cal}(W_3)$ rad	$\eta_{cal}(W_4)$ rad
0	0	0.223 π	0	0	0.167 π
π/8	0.062 π	0.230 π	0.072 π	0.062 π	0.156 π
π/4	0.125 π	0.237 π	0.151 π	0.125 π	0.187 π
π/2	0.250 π	0.250 π	0.250 π	0.250 π	0.250 π

Finally, for each loading case we can compare the theoretical values $\eta_{cal}(W_i)$ with the experimental angle η_{exp} (Table 2). In particular, we can remark that η_{exp} is coincident with $\eta_{cal}(W_3)$ for all the considered values of, α while the simple arithmetic averaging of the Euler angles through the weight function W_1 generally leads to an expected principal direction $\hat{1}$ which does not agree with the normal vector to the experimental fracture plane.

It can be observed that the probability density functions of the principal stresses σ_1 and σ_3 ($\sigma_2=0$) for $\alpha=\pi/8$ (see Fig.7) are strongly asymmetric and differ from the normal probability distribution. Note that the PDF of σ_1 resembles symmetrically to the PDF of σ_3 and such PDFs present nonzero expected values with opposite signs.

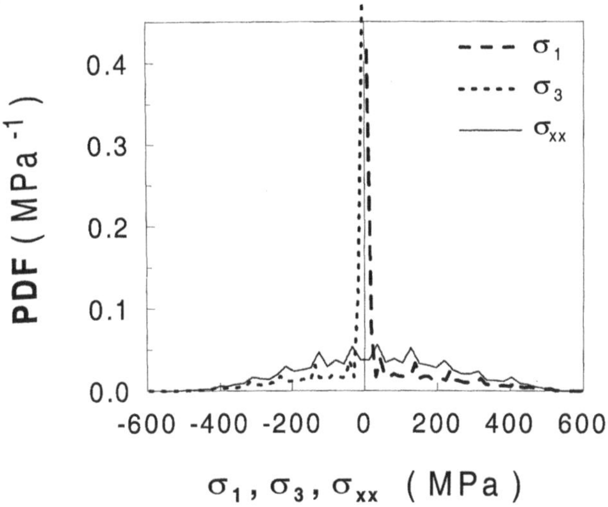

Fig.7 Probability density function (PDF) for the principal stresses σ_1 and σ_3 ($\sigma_2 = 0$) and the applied stress σ_{xx}, in the case of $\alpha = \pi/8$.

CONCLUSIONS

Under multiaxial random loading, the stress tensor and its eigenvectors (i.e., the principal direction cosines) change at each time instant. The instantaneous values of the Euler angles defining the principal stress directions can be averaged by employing some suitable weight functions to take into account the main factors influencing the fatigue fracture process. In this way the expected principal stress directions under multiaxial random stress state are determined.

This theoretical procedure has been applied to an experimental biaxial random stress state. For all the loading combinations examined, it has been observed that, by employing the weight function W_3 analysed above, the calculated angle η_{cal} between the mean direction $\hat{1}$ of the maximum principal stress and the longitudinal axis (X-axis) of the round specimen is coincident with the experimental angle η_{exp} between the normal vector to the fracture plane and the X-axis.

REFERENCES

(1) Irwin G.R. (1960), Fracture mechanics. In Structural Mechanics, Pergamon Press, New York, pp.557-592.

(2) Brown M.W. and Miller K.J. (1979), Initiation and growth of cracks in biaxial fatigue. Fatigue Eng.Mater.Struct., Vol.1, No.2, 231-246.

(3) Socie D. (1987), Multiaxial fatigue assessment. In Low-Cycle Fatigue and Elasto-Plastic Behaviour of Materials, Ed.K.-T.Rie, Elsevier Applied Science, London, pp.465-472.

(4) You B.-R. and Lee S.-B. (1996), A critical review on multiaxial fatigue assessments of metals. Int.J.Fatigue, Vol.18, No.4, 235-244.

(5) Ohnami M., Sakane M. and Hamada N. (1985), Effect of changing principal stress axes on low-cycle fatigue life in various strain wave shapes at elevated temperature. In Multiaxial Fatigue ASTM STP 853, pp.622-634.

(6) McDiarmid D.L. (1987), Fatigue under out-of-phase bending and torsion. Fatigue Eng. Mater. Struct. Vol.9, 457-475.

(7) Macha E. (1988), Generalization of Strain Criteria of Multiaxial Cyclic Fatigue to Random Loadings. Fortschr.-Ber.VDI, Reiche 18, Nr 52, VDI-Verlag, Dusseldorf, p.102.

(8) Macha E. (1989), Simulation investigations of the position of fatigue fracture plane in materials with biaxial loads. Mat.-wiss. u. Werkstofftech., Vol. 20, Heft 4/89, pp.132-136 and Heft 5/89, pp.153-163.

(9) Carpinteri A., Macha E., Brighenti R. and Spagnoli A. (1999), Expected principal stress directions for multiaxial random loading - Part I: Theoretical aspects of the weight function method. Int.J. Fatigue, Vol.21, No. 1, 83-88.

(10) Carpinteri A., Brighenti R., Macha E. and Spagnoli A. (1999), Expected principal stress directions for multiaxial random loading - Part II: Numerical simulation and experimental assessment through the weight function method. Int.J. Fatigue, Vol.21, No. 1, 89-96.

(11) Achtelik H., Będkowski W., Grzelak J. and Macha E. (1994), Fatigue life of 10HNAP steel under synchronous random bending and torsion. Proc. 4th Int. Conf. on Biaxial/ Multiaxial Fatigue, St Germain en Laye (France), Vol.I, 421-434.

(12) Papoulis A. (1985), Probability, Random Variables and Stochastic Processes. McGraw-Hill Inc., New York.

(13) Korn G.A. and Korn T.M. (1968), Mathematical Handbook. Sec.Ed., McGraw-Hill Book Company, New York.

(14) Bendat J.S. and Piersol A.G. (1976), Random Data. Analysis and Measurement Procedures. John Wiley and Sons Inc., New York.

(15) Bendat J.S. and Piersol A.G. (1980), Engineering Applications of Correlation and Spectral Analysis. John Wiley and Sons Inc., New York.

APPLICATION OF BIAXIAL PLASTICITY AND DAMAGE MODELLING TO THE LIFE PREDICTION AND TESTING OF AUTOMOTIVE COMPONENTS

Peter HEYES*, Xiaobin LIN*, Andrzej BUCZYŃSKI** and Mike W. BROWN***

* nCode International Limited, Sheffield, England
** Warsaw University of Technology, Poland
*** Sheffield University, England

ABSTRACT

A method has been developed for making life predictions for engineering components subject to multiaxial loadings using the local strain approach. The method has been incorporated into a computer program, which uses a Mróz-Garud **(1,2)** cyclic plasticity model and the Wang-Brown cycle counting and damage models **(3, 4)**. The plasticity, rainflow cycle counting and damage models have been generalised to deal with any free-surface loading conditions. The program makes calculations based on strain gauge rosette measurements and its application is illustrated by calculations from three typical automotive components. Some interesting methods for visualising the analysis results are explored. In addition to life prediction, the approach also has applications in accelerated durability testing.

KEY WORDS

Biaxial loading, non-proportional loading, damage accumulation, rainflow cycle counting, cyclic plasticity model, life time calculation, industrial applications

INTRODUCTION

Background

There is increasing pressure in the automotive industry to reduce the time taken to bring new designs to production, with typical development times coming down from around 5 years to around 2 years over the last decade. At the same time it is necessary for the vehicles developed to have low weight, high stiffness, good fuel efficiency, good ride and handling and good NVH properties (Noise, Vibration and Harshness) *while maintaining adequate durability*. It is generally recognised that these objectives cannot be reasonably be met by developing the design through the testing and modification of a series of mechanical prototypes, because these methods are too time consuming and expensive. For this reason fatigue life prediction is now becoming an essential part of the development process for many vehicle manufacturers, and methods for reducing the time necessary for the essential fatigue tests are very important.

Fatigue life calculations tend to fall into two categories:

1. Calculations made on the basis of calculated stresses and strain, for instance from Finite Element Analysis (FEA) **(5-7)**.
2. Calculations made on the basis of measured strains, typically from resistance strain gauges.

One of the most common approaches to fatigue analysis used in the automotive industry is the local strain approach. This paper is concerned with the application of multiaxial local strain methods to fatigue calculations of automotive components, based on strain gauge measurements.

SUMMARY OF METHOD

Uniaxial methods for the life prediction using the local strain approach have been in use for some time, having their roots in the work of Basquin **(8)** Manson **(9)** and Coffin **(10, 11)**, incorporating rainflow cycle counting and material memory **(12)** and Miners rule **(13)**. Such methods are available in commercial software products such as FATIMAS **(14)**. Within the well-known limitations of these methods they work quite well for a variety of components where the local loading in the critical area is uniaxial or near uniaxial. This class of components includes many that are subject to complex multiaxial loading environments **(5,6)**. However, there are many other components where a combination of loads and geometric effects generates local loadings, which are proportional or non-proportional multiaxial. It was to deal with these cases that the methods and software discussed in the current paper were developed.

The life prediction process from measured strains can be divided into two steps. The first step is to determine the relationship between the measured strain and all the stress and strain components required for the damage calculation, through application of a cyclic plasticity model. The second step is to carry out cycle and damage accumulation. The current work addresses problems where the strains can be measured with a rosette, i.e. biaxial loading on a free surface. For these problems the process can be summarised by the flow chart in Fig.1.

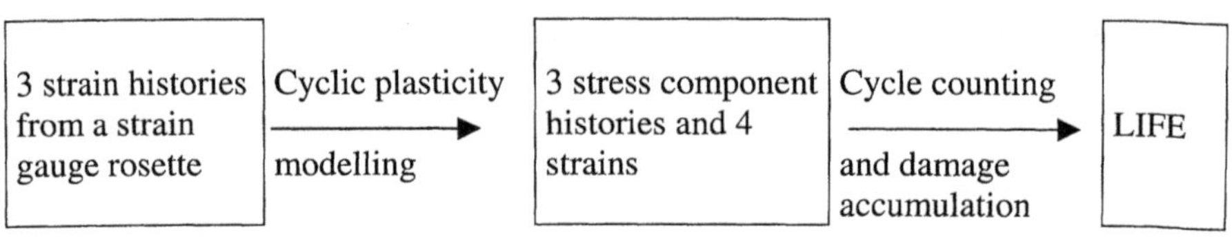

Fig.1 Outline of strain-based life prediction process.

The approach for calculation on the basis of elastic finite element analysis will be similar except that the input strains will be elastic strain components from the surface of the structure, and a notch correction procedure will be required in addition to the cyclic plasticity model to estimate the stresses and elastic-plastic strains **(15)**.

The essential calculations made by the current system are as follows:

1. Take three strain channels from a strain gauge rosette and convert them to the components of strain ε_x, ε_y and ε_{xy} taking into account the transverse

sensitivities of the gauge legs. The gauge co-ordinate system is defined by x parallel to gauge 1 and z as the outward surface normal.

2. Feed these 3 strain channels into the Mróz-Garud cyclic plasticity model, the outputs of which are the remaining non-zero strain component ε_z and the in-plane stresses σ_x, σ_y, and σ_{xy}.
3. Process the resulting 7 components of stress and strain either by the conventional critical plane methods, or ...
4. By multiaxial rainflow counting and then accumulating damage using the Wang-Brown methods.

These calculations are described in more detail in the following sections.

CYCLIC PLASTICITY MODELLING

The stresses and strains required by the damage models can be calculated if the relation between the equivalent plastic strain increment $\Delta\varepsilon^p_{eq}$ and the equivalent stress increment $\Delta\sigma_{eq}$ is known during the application of a given load increment. However, it is known that the current $\Delta\varepsilon^p_{eq} - \Delta\sigma_{eq}$ relation depends on the previous load path and therefore the plasticity model must deal with loading path dependent material constitutive behaviour. Several models are available in the literature **(1, 2, 16, 17)** of which the model proposed by Mróz **(1)** and recently modified by Garud **(2)** are the most popular. Mróz **(1)** has proposed that the uniaxial stress-strain material curve is represented by a set of plasticity surfaces in three-dimensional stress space. In the case of a two dimensional stress state, the plasticity surfaces reduce to ellipses on the plane of principal stresses described by

$$\sigma_{eq} = \sqrt{\sigma_1^2 - \sigma_1 \cdot \sigma_2 + \sigma_2^2} \tag{1}$$

and illustrated in Fig.2.

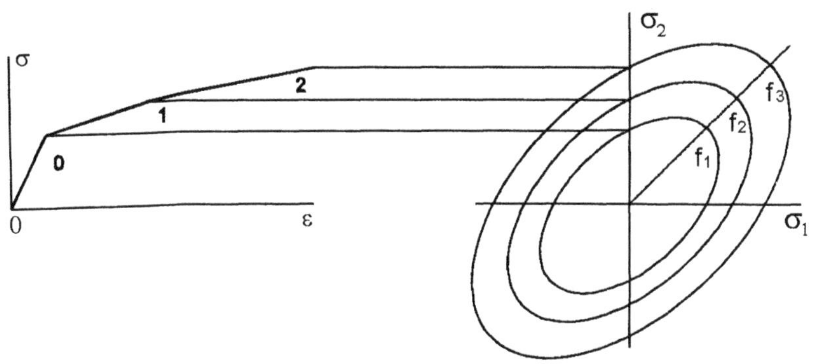

Fig.2 Linearisation of the material $\sigma - \varepsilon$ curve and corresponding plasticity surfaces.

The load path dependent memory effects are modelled by prescribing a translation rule for the ellipses moving with respect to each other over distances given by the stress increments. It is also assumed that the ellipses move inside each other and they do not intersect. If the ellipses come in contact with one another they move together as a rigid body.

The translation rule proposed by Garud **(2)** avoids the intersection of the ellipses that could occur in some cases in the original Mróz **(1)** model. The Garud translation rule is illustrated in Fig.3 and can be described by a model consisting, for simplicity, of only two plastic surfaces (ellipses).

In order to predict material response due to the stress increment $d\sigma$, the following steps should be made:

1. Extend the stress increment $d\sigma$ to intersect the first external non-active plastic surface f_2 at point B_2
2. Connect point B_2 and the centre O_2 of the intersected plastic surface f_2
3. Find point B_1 on the active plastic surface f_1 by drawing a line parallel to the line O_2B_2 through the centre O_1 of the surface f_1
4. Connect the conjugate points B_1 and B_2 by the line B_1B_2
5. Translate the ellipse f_1 in the direction of B_1B_2 from point O_1 to O_1' until the end of the vector $d\sigma$ lands on the moving ellipse f'_1.

The translation rule assures that the ellipses are tangential with the common point B_1B_2 without intersecting each other. Two or more tangential ellipses translate as a rigid body and the largest moving ellipse (Fig.3) indicates the proper constitutive relation (linear segment) to be used for a given stress increment.

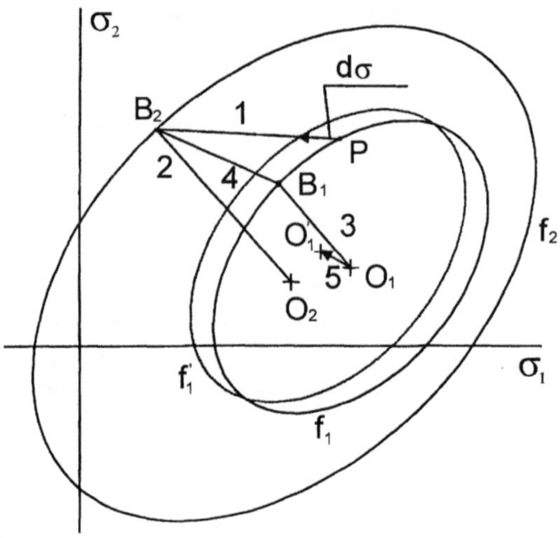

Fig.3 Geometrical interpretation of the Mróz-Garud incremental plasticity model.

MULTIAXIAL RAINFLOW COUNTING

Wang and Brown **(18)** proposed a multiaxial cycle counting method on the basis of strain hardening behaviour under non-proportional variable amplitude loading. Relative stresses and strains were introduced so that a pair of turning points defines the start and end points of a reversal, where the equivalent relative strain rises monotonically to a peak value. Since plastic deformation generates the driving force for small fatigue cracks, hysteresis hardening provides a physical parameter for cycle counting, analogous to rainflow counting in the uniaxial case.

Each reversal commences with elastic unloading, which is followed by reloading and plastic strain hardening up to the next turning point. The most significant turning point occurs at the highest value of equivalent strain. This is illustrated at the time 0 in Fig.4, which shows a repeating block of a combined tension/torsion non-proportional load history. The equivalent strain is defined as the von Mises strain.

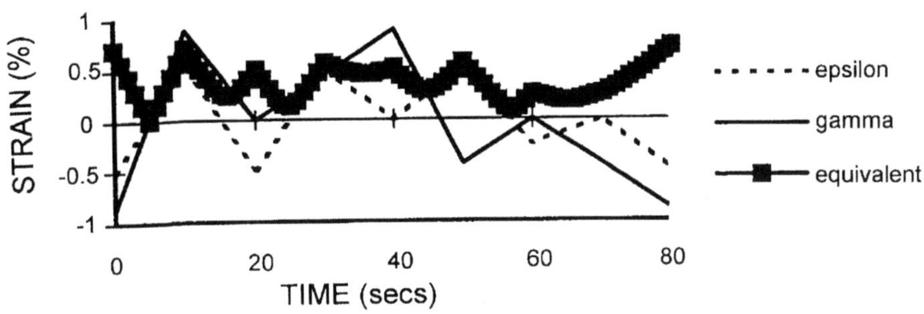

Fig.4 A variable amplitude non-proportional strain history, showing applied tensile (epsilon) and torsional (gamma) strains with the absolute equivalent strain.

The cycle counting method is illustrated by the following example. Starting from the most significant turning point, a graph is drawn for the loading block of relative equivalent strain, where relative strain $\varepsilon^*_{ij} = \varepsilon_{ij} - \varepsilon^A_{ij}$ represents the change of strain since time A. Fig.5 shows the relative equivalent strain, with respect to times 0, 10 and 20 seconds. Using the relative strain, a reversal can be defined starting from 0, up to the maximum value 10 seconds. To obtain the second reversal the relative strain is re-plotted starting from the turning point where unloading commences (at 10 seconds), and the portions of the strain hardening curve for the reversal are selected by a traditional rainflow procedure **(3)**. The region of unloading within that reversal is counted in the next step.

Using the next turning point, relative strain is re-plotted with respect to 20 seconds for the subsequent continuous fragment of strain history, yielding the third reversal in Fig.5. This procedure is repeated for each turning point in chronological order, until every fragment of strain history has been counted.

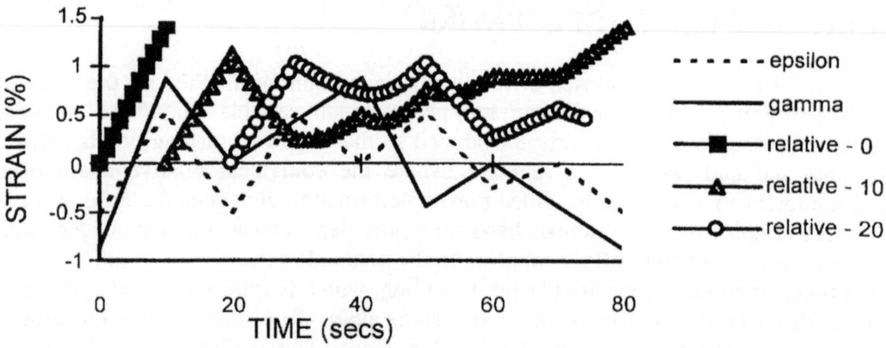

Fig.5 The variable amplitude history, showing relative equivalent strains plotted with respect to times 0, 10 and 20 seconds respectively.

FATIGUE DAMAGE CALCULATION

The counting method described above is independent of fatigue damage parameters, being based on hysteresis deformation behaviour. Being unrelated to material properties, it can be integrated with any multiaxial fatigue damage model. If the counted reversals are non-proportional, a fatigue damage parameter that accounts for non-proportional straining effects is required. The path-independent damage parameter proposed by Wang and Brown **(19)** has been shown to provide good correlation for several materials under proportional and non-proportional loading,

$$\hat{\varepsilon} \equiv \frac{\gamma_{mx} + S \cdot \delta\varepsilon_n}{1 + \nu' + S(1-\nu')} = \frac{\sigma_f' - 2\sigma_{n,mean}}{E} \cdot (2N_f)^b + \varepsilon_f' \cdot (2N_f)^c \qquad (2)$$

where γ_{max} is the maximum shear strain amplitude on a critical plane (proportional or non-proportional), $\delta\varepsilon_n$ is the normal strain excursion between the two turning points of the maximum shear strain (that is the range of normal strain experienced on the maximum shear plane over the interval from start to end of the reversal), and $\sigma_{n,mean}$ is the mean stress normal to the maximum shear plane. The term S is a material constant determined from a multiaxial test (typically between 1 and 2 for Case A and around 0 for Case B) and ν' is the effective Poisson's ratio. The right hand side of the equation is the same as the uniaxial strain life equation, with a Morrow mean stress correction **(20)**. Mean stress is measured as the average of the maximum and minimum stress values over the reversal. The total damage induced by a loading history is calculated using Miner's rule **(13)**.

Fig.6 is a plot of predicted life against experimental life for a variety of proportional and non-proportional tests on laboratory specimens, from Ref. **(4)**.

The other multiaxial damage parameters considered in this paper are the more conventional critical plane parameters. In these methods, stresses and strains are resolved onto a particular plane, inclined at an angle θ = 90 degrees (Case A) and/or θ = 45 degrees (Case B) to the free surface. Cycle counting (uniaxial) and damage parameter calculation is carried out on the critical plane and the damage accumulated.

The orientation ϕ of the projection of the normal to the damage plane is increased by 10-degree increments. The plane with the largest accumulated damage is said to be the critical plane. The models are:

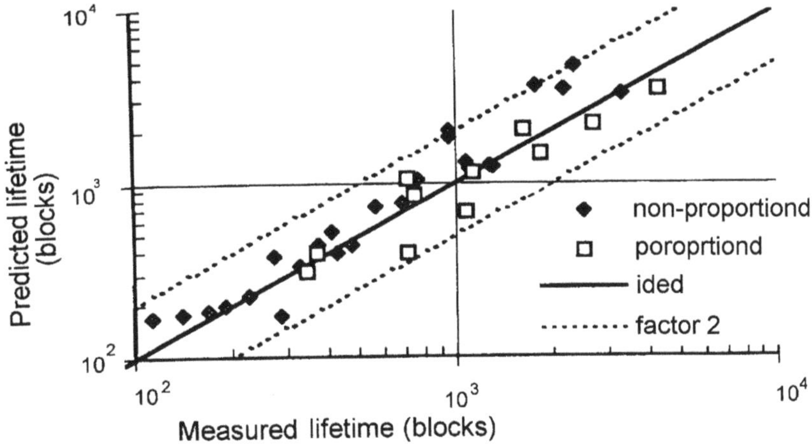

Fig.6 Comparison of experimental and predicted results for the Wang-Brown method.

1. Normal strain (θ=90 degrees only):

$$\frac{\Delta\varepsilon_n}{2} = \frac{\sigma_f'}{E}\cdot(2N_f)^b + \varepsilon_f'\cdot(2N_f)^c \tag{3}$$

ε_n is the strain amplitude normal to the critical plane. Otherwise this is the usual Coffin-Manson-Basquin equation.

2. Shear:

$$\frac{\Delta\gamma}{2} = \frac{(1+\nu_e)\cdot\sigma_f'}{E}\cdot(2N_f)^b + (1+\nu_p)\cdot\varepsilon_f'\cdot(2N_f)^c \tag{4}$$

3. Smith-Watson-Topper/Bannantine ((θ=90 degrees only) **(21)**:

$$\frac{\Delta\varepsilon_n}{2}\cdot\sigma_{n,max} = \frac{\sigma_f'^2}{E}\cdot(2N_f)^{2b} + \sigma_f'\cdot\varepsilon_f'\cdot(2N_f)^{b+c} \tag{5}$$

$\sigma_{n,max}$ is the maximum normal strain on the critical plane, which occurs during each rainflow cycle. Otherwise this is the Smith-Topper-Watson method **(22)**.

4. Fatemi-Socie **(23)**:

$$\frac{\Delta\gamma}{2}\cdot\left(1+n\cdot\frac{\sigma_{n,max}}{\sigma_y}\right) = \frac{(1+\nu_e)}{E}\cdot\sigma_f'\cdot(2N_f)^b + \frac{n\cdot(1+\nu_e)\cdot\sigma_f'^2}{2E\sigma_y}\cdot(2N_f)^{2b} +$$
$$(1+\nu_p)\cdot\varepsilon_f'\cdot(2N_f)^c + \frac{n\cdot(1+\nu_p)\cdot\varepsilon_f'\cdot\sigma_f'}{2\sigma_y}\cdot(2N_f)^{b+c} \tag{6}$$

APPLICATIONS

The new program described in this paper may be used to predict the life components and structures, to provide some help in design optimisation, and to assist in the design of efficient tests through fatigue editing. The main drawback of the program is that it is necessary to place a strain gauge rosette on the critical location or crack initiation site in order to get a reliable prediction. This is due to the fact that there is no simple transfer function between the measured strains at a non-critical location and the strains at a nearby critical location when the loadings non-proportional. For this reason, multiaxial fatigue life prediction is likely to be more useful when used in conjunction with Finite Element Analysis.
In this section, three applications of the multiaxial fatigue program will be described to illustrate the method.

Example 1: Fatigue calculations on an automotive wheel.

This example is based on strain gauge measurements from an automotive wheel. The wheel was tested on a rig simulating the rotation of the wheel under load, and the strain gauge measurements were taken using a 45-degree rosette placed in the most critical location. This location was identified by previous stress analysis. A small section of the three strain gauge channels illustrated in Fig.7.

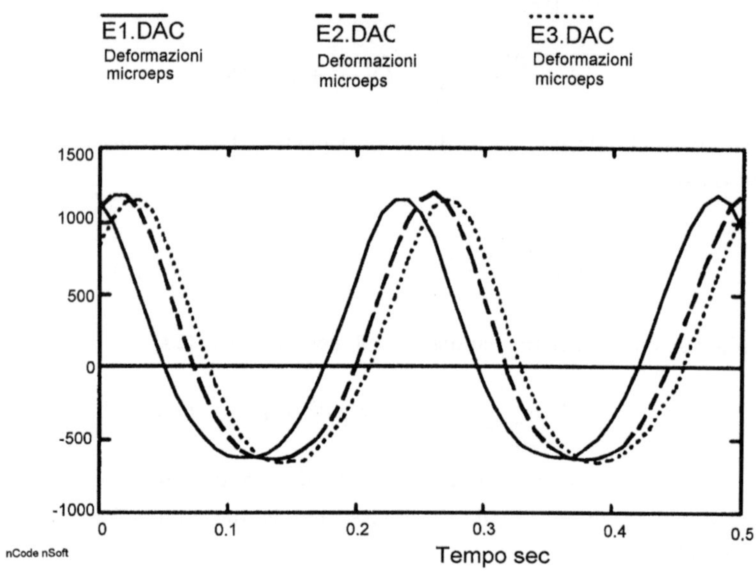

Fig.7 Section of strain measurements from wheel critical location.

The three channels of strain are more or less sinusoidal with a phase difference between them, indicating that the stress state is multiaxial and non-proportional. The two cycles shown above are taken from a longer history. The stress-strain state variations can be better understood qualitatively by looking at the behaviour of the principal stresses as illustrated in Fig.8, where it can be seen that the two principal stresses are almost in phase, but with slightly different amplitudes and means. Notice that the absolute maximum principal stress (the principal stress with the largest magnitude) "lips" between the maximum and the minimum principals. Whenever it does this of course, the angle also flips through 90 degrees indicating that the state of stress has reached a condition where E.MAX and E.MIN are equal and opposite, i.e. pure shear. In fact the stress state behaviour can be summed up by saying that when the stress is at a maximum or minimum, the stress state is almost equibiaxial, and in between times it passes through a condition of pure shear where the stresses are lower. Between the two limits the principal stress axis rotates through 90 degrees, so that there is a complete rotation (180 degrees) per cycle.

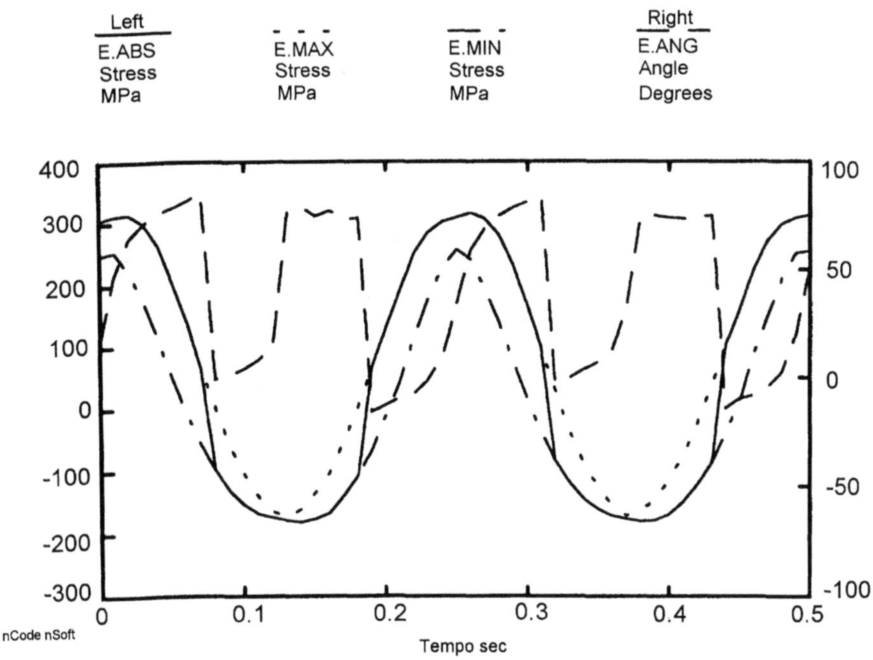

Fig.8 Plot of absolute maximum stress (E.ABS), maximum principal stress (E.MAX), minimum principal stress (E.MIN) and the angle of the absolute maximum principal stress to gauge 1 against time.

Another way of looking at this data is to plot the biaxiality ratio (the ratio of the smaller in magnitude to the larger principal stress) and the orientation against the largest principal stress. This is shown in Fig.9 for 1 cycle.

The results of analysis of the full loading history are given in Table 1, with the results of some other methods for comparison.

These results are all very similar, with the notable exception of the normal strain method, which is relatively non-conservative.

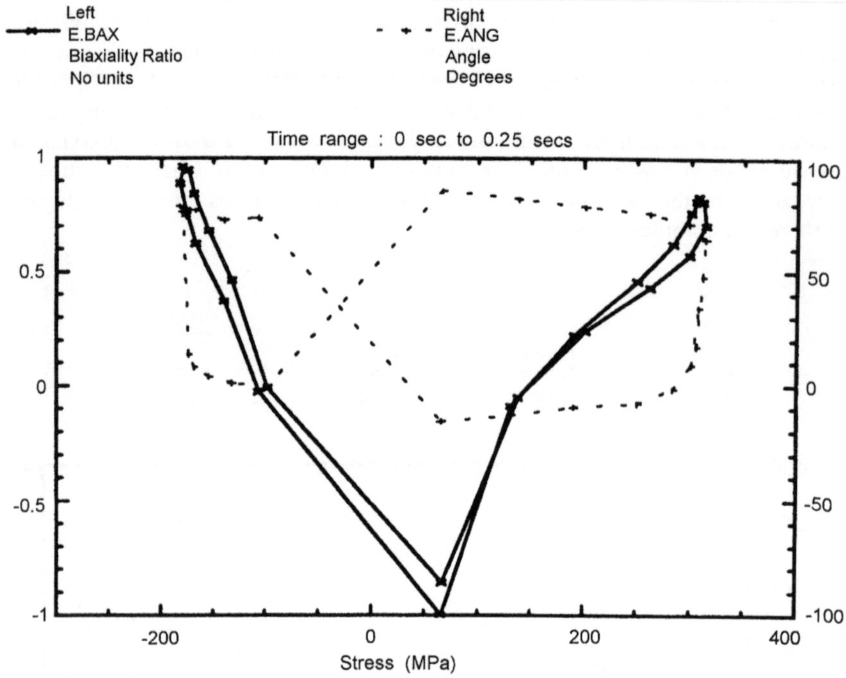

Fig.9 Plot of biaxiality ratio and angle against absolute maximum principal stress.

Table 1 Results of life calculations using various methods.

Method	Life (Rotation)
Wang-Brown (S=1 Case A, 0 for Case B)	1.03 x 10^6
Wang-Brown + Mean (S=1.0)	7.07 x 10
Normal Strain	5.54 x 10
Smith-Topper-Watson (Bannantine)	9.43 x 10
Shear Strain	5.95 x 10
Fatemi-Socie (n=0.6)	5.17 x 10
Experiment	>200 000 (Test stopped with no failure)

It is normal to view the accumulated rainflow cycles in an analysis in the form of a 3D-histogram plot. Normally the axes are range-mean or max-min or from-to. With the Wang-Brown method, each cycle is characterised by shear strain range, normal strain range, mean normal stress, two angles θ and ϕ. The distribution of reversal numbers and damage may reasonably be visualised in relation to any two of these parameters. Fig.10 shows the distribution of reversals in relation to shear strain range and normal strain range.

Another potentially useful way of viewing the results is in the form of a polar plot of damage, as shown in Fig.11. In this case there are only Case B damaging cycles, so there is only one curve on the plot.

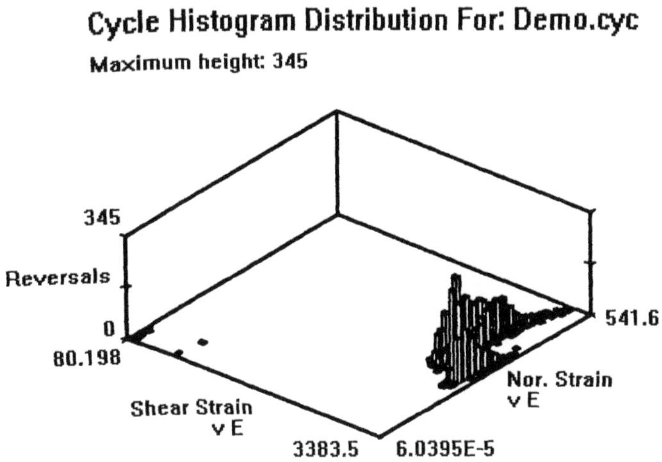

Fig.10 One representation of rainflow counting.

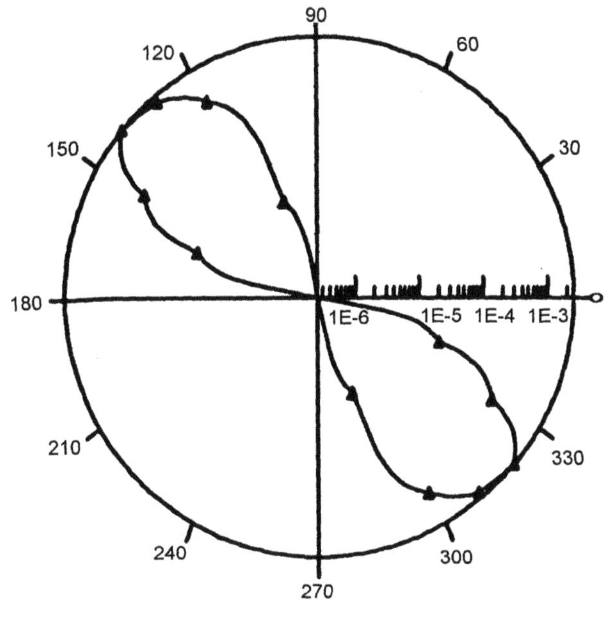

Fig.11 Polar plot of log (damage) against orientation ϕ.

Example 2: Fatigue life prediction of a steering rack mounting bracket.

This example concerns a sign-off test, which is carried out on a front suspension cross-member and steering rack assembly. The steering rack is mounted to the cross-member by means of two small brackets, which are fatigue-sensitive areas. The assembly is tested by fixing the cross-member and applying constant amplitude, unidirectional load to the steering rack. The resulting stress state variations are illustrated in Fig.12.

These plots show that the single-axis loads applied generate locally a near-uniaxial, near proportional stress state. The results of analysis with the new program, assuming 99.9% certainty of survival (based on the calculated standard errors on the material parameters) are as given in Table 2

Table 2 Fatigue life predictions on steering rack mounting bracket.

Method	**Life (Repeats)**
Wang-Brown (S=1,0)	24
Wang-Brown + Mean (S=1.0)	23
Normal Strain	33
Smith-Topper-Watson (Bannantine)	22
Shear Strain	23
Fatemi-Socie (n=0.6)	23

Example 3: Fatigue editing on an off-road vehicle rear axle.

This example illustrates how the new program can be used to reduce testing times through multiaxial, multi-channel fatigue editing of simulation test rig drive files. The component in question is a rear axle from an off-road vehicle. The axle is instrumented with three strain gauge rosettes close to critical location, and analysis of the measurements from all three strain gauge rosettes show that the loading is non-proportional. It is desired to reduce the testing time by editing the rig drive signals while retaining all the sections of the original sections that cause significant damage. This was carried out using a simple editing procedure.

One of the outputs of the life prediction program is a time correlated damage file, essentially a time history of damage. This file is a record of the damage accumulated, with damage for each reversal being distributed between the opening and closing point of each reversal. The signal is divided up into a number of equally spaced windows, and the least damaging are discarded until a certain percentage of damage is retained, in this example 95% was retained. The time slices to be retained are compared across the three channels with a logical OR operation and all resulting windows are retained. The resulting edit vector can be used to edit the rig drive signals, inserting suitable joining functions for continuity. It is also important to edit the strain signals and recalculate the life in order to check the likely effect of editing on the fatigue damage. This is important because of the possibility of reversals starting in one window and ending in another, and also because in multiaxial loading the loading path is important as well as the peaks. In some cases, editing the signals may actually cause a slight increase in fatigue damage.

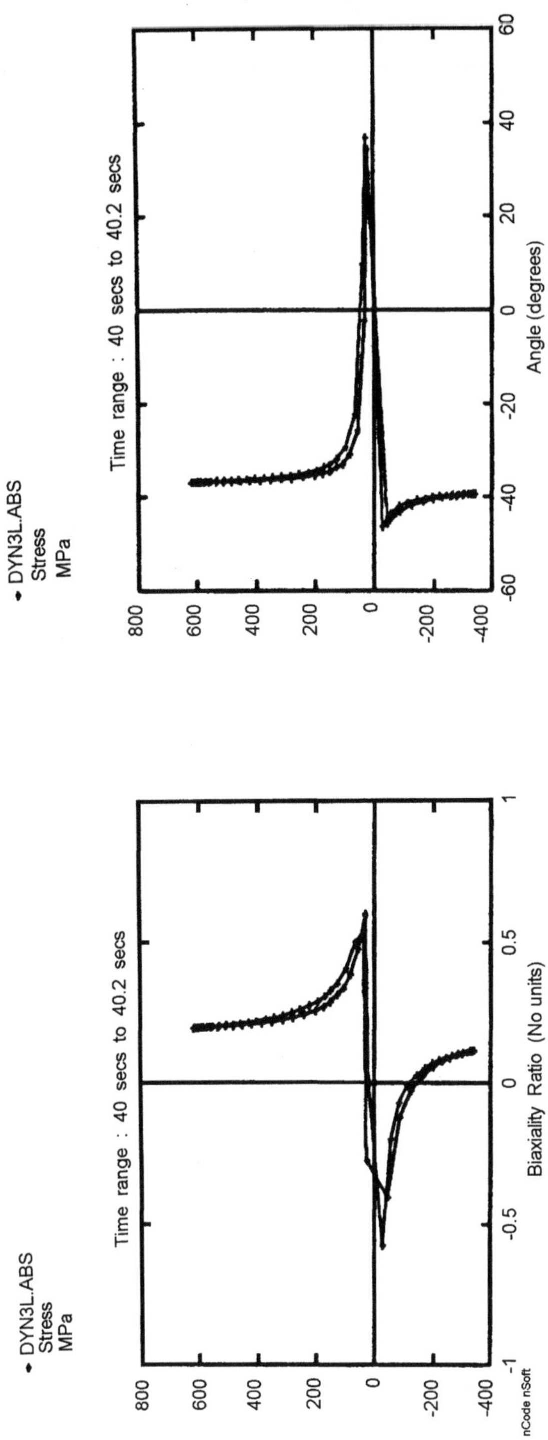

Fig.12 Plots of principal stress ratio (left) and principal axis orientation (right) plotted against absolute maximum principal stress.

This is why methods such as described in Ref. **(24)** are not really suitable for non-proportional loadings, because they do not retain the important sections of the loading path. The results of the analysis described are summarised in Table 3 and the original and edited strain signals for gauge 1 are illustrated in Fig.13.

Table 3 Summary of fatigue editing results.

Gauge Number	Original Signal Predicted Life (repeats)	Edited Signal Predicted Life (repeats)	Actual Damage Retained	Data Reduction Factor
1	13866	14079	98.5	8.25
2	29770	30188	98.6	8.25
3	53200	53800	98.8	8.25

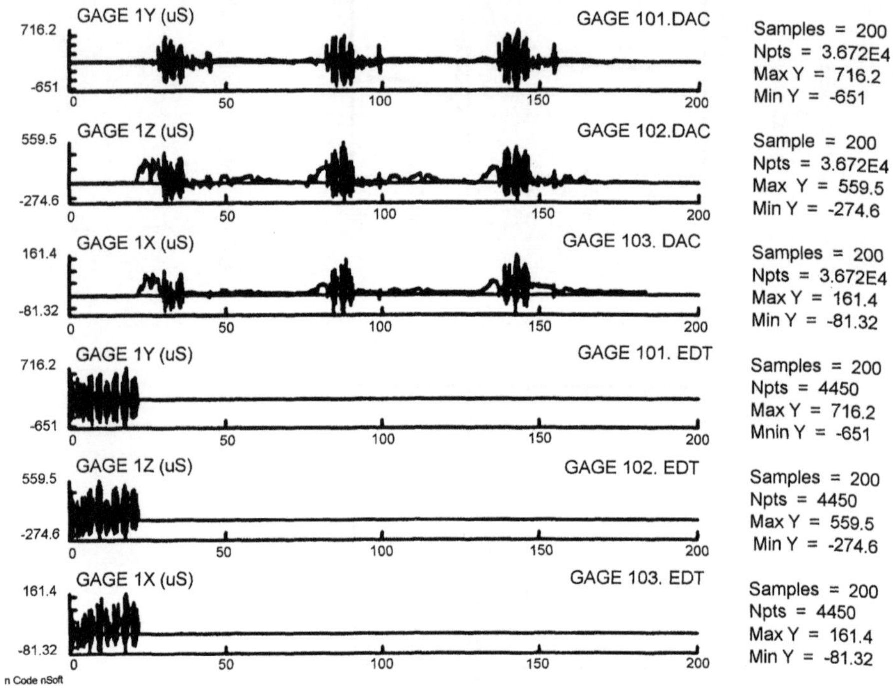

Fig.13 Strain signals before and after editing.

The advantages of this method are that it retains both the most important sections of loading path, and the essential frequency content of the loading, essential for components with dynamic behaviour.

CONCLUDING REMARKS

1. A new computer program has been written which incorporated a Mróz-Garud **(1, 2)** cyclic plasticity model, together with a generalised form of the Wang-Brown **(3, 4)** multiaxial rainflow and damage accumulation procedures. The program requires measured strains from strain rosette as input, together with uniaxial cyclic material properties. The software has a graphical user interface and various post-processing options.
2. The software has been applied to strain measurements from three different components, including one with an interesting Case B biaxial non-proportional loading.
3. The software can handle general non-proportional biaxial loadings.
4. A variety of multiaxial damage parameters have been applied and compared.
5. The approach described here would benefit from more validation within an industrial context, and will also benefit from being extended to interface to finite element analysis. This is planned for the near future.

REFERENCES

(1) Mróz Z., (1967), On the description of anisotropic work hardening, Journal of Mechanics and Physics of Solids, vol. 15, pp. 163-175.

(2) Garud Y.S., (1981), A new approach to the evaluation of fatigue under multiaxial loading, Trans. ASME, J.Eng Mater.Techn., vol. 103, pp. 118-125.

(3) Wang C.H. and Brown M.W., (1996), Life prediction techniques for variable amplitude multiaxial fatigue – Part 1: Theories, Trans. ASME J. Eng. Mater. Techn., vol. 118, pp. 367-370.

(4) Wang C.H. and Brown M.W., (1996), Life prediction techniques for variable amplitude multiaxial fatigue – Part 2: Comparison with experimental results, J. Eng. Mater. Techn, vol. 118, pp. 371-374.

(5) Heyes P.J., Milsted M.G. and Dakin J., (1996), Multiaxial fatigue Assessment of automotive chassis components on the basis of finite-element models, Multiaxial Fatigue and Design, ESIS 21 (Edited by A. Pineau, G. Cailletaud and T.C. Lindley) MEP London, pp. 461-475.

(6) Heyes P., Dakin J. and S.T. John C., (1995), The assessment and use of linear static FE stress analysis for durability calculations, Proc. Ninth Int. Conf. on Vehicle Structural Mechanics and CAE, pp. 189-199.

(7) Heyes P. and Fermer M., (1996), A program for the fatigue analysis of automotive spot-welds based on finite element calculations, Proc. Symp. International Automotive Technology, SAE Technical Paper 962507.

(8) Basquin O.H., (1910), The exponential law of endurance tests, Proc. American Society for Testing Materials, vol. 10, pp. 625-630.

(9) Manson S.S., (1953), Behaviour of materials under conditions of thermal stress, Heat Transfer Symposium, Univcrsity of Michigan Engineering Research Institute, pp. 9-75.

(10) Coffin L.F., (1954), The problem of thermal stress fatigue in austenitic steels at elevated temperatures, ASTM STP No. 165, p. 31.

(11) Coffin L.F., (1954), A study of the effects of cyclic thermal stresses on a ductile metal, Trans. ASTM, vol. 76, pp. 931-950.

(12) Matsuishi M. and Endo T., (1968), Fatigue of metals subjected to varying stress, Presented to Kyushu District Meeting, JSME.

(13) Miner M.A., (1945), Cumulative damage in fatigue, Journal of Applied Mechanics, vol. 12, pp. A159-A164.

(14) nCODE International Ltd., (1997), nSoft-E FATIMAS Software Manual.

(15) Buczyński A. and Glinka G., (1997), Elastic-plastic stress-strain analysis of notches under non-proportional cyclic loading paths, Proc.5th Int. Conf. on Biaxial/Multiaxial Fatigue and Fracture, Eds E.Macha and Z.Mroz, TU Opole, Poland, vol. I, pp.461-480

(16) Chu C.C., (1989), A three-dimensional model of anisotropic hardening in metals and its application to the analysis of sheet metal formability, J. Mech. Physics, vol. 22, No. 3, pp. 197-212.

(17) Armstrong P.J. and Frederic C.O., (1966), A mathematical representation of the multiaxial Bauschinger effect, CEGB Report RD/B/M731, Berkely Nuclear Laboratories.

(18) Wang C.H. and Brown M.W., (1993), Inelastic deformation and fatigue under complex loading, Proc. 12th Int. Conf. on Structural Mechanics in Reactor Technology, vol. L, pp. 159-170.

(19) Wang C.H. and Brown M.W., (1993), A path-independent parameter for fatigue under proportional and non-proportional loading, Fatigue Fract. Eng. Mater. Struct., vol. 16, pp. 1285-1298.

(20) Brown M.W., Suker D.K. and Wang C.H., (1996), An analysis of mean stress in multiaxial random fatigue, Fatigue Fract. Eng. Mater. Struct., vol. 19, No. 2/3, pp. 323-333.

(21) Bannantine J.A., (1989), A variable amplitude multiaxial fatigue life prediction method, PhD thesis, University of Illinois at Urbana-Champaign.

(22) Smith K.N., Watson P. and Topper T.H., (1970), A stress-strain function for the fatigue of metals, Journal of Materials, vol. 5, No. 4, pp. 767-778.

(23) Fatemi A. and Socie D.F., (1988), A critical plane approach to multiaxial fatigue damage including out-of-phase loading, Fatigue Fract. Eng. Mater. Struct., vol. 11, No. 3, pp.149-165.

(24) Dressler K., Köttegen V.B. and Kötzle H., (1995), Tools for fatigue evaluation of non-proportional loading, Proc. of Fatigue Design 1995 (Eds G.Marquis and J. Solin), vol. 1, pp. 261-277.

Acknowledgements

The authors are grateful for financial support from Ford Motor Company and the Department of Trade and Industry. We would like also to thank Prof. Grzegorz Glinka for his support and Fiat Auto, The Ford Motor Company and Jaguar Cars for providing information about the applications.

OVERVIEW OF THE STATE OF THE ART ON MULTIAXIAL FATIGUE OF WELDS

Cetin Morris SONSINO

Fraunhofer-Institute for Strength of Structures under Operational Loading (LBF), Darmstadt / Germany

ABSTRACT

Flange-tube joints from fine grained steel StE 460 with unmachined welds were investigated under biaxial constant and variable amplitude loading (bending and torsion) in the range of 10^3 to $5 \cdot 10^6$ cycles to crack initiation and break-through, respectively. In order not to interfere with residual stresses they were relieved by a heat treatment. In-phase loading can be treated fairly well using the conventional hypotheses (von Mises or Tresca) on the basis of nominal, structural or local strains or stresses. But the influence of out-of-phase loading on fatigue life is severely overestimated if conventional hypotheses are used. However, the introduced hypothesis of the effective equivalent stress leads to fairly good predictions. Therefore, the knowledge of local strains or stresses is necessary. They are determined by boundary-element analyses in dependency of weld geometry. This hypothesis consider the fatigue-life reducing influence of out-of-phase loading by taking into account the interaction of local shear stresses acting in different surface planes of the material. Further more, size effects resulting from weld geometry and loading mode were included. Damage accumulation under a Gaussian spectrum of amplitudes can be assessed for in- and out-of-phase combined bending and torsion using an allowable damage sum of 0.35.

KEY WORDS

Welded joints, combined torsion and bending, constant and variable amplitude loading, in- and out-of-phase, nominal and local stresses, equivalent stress, damage accumulation

NOTATION

Stresses

σ bending or normal stress

τ shear stress

$æ^*$ normalised stress gradient

Strains

ε axial, bending or normal strain

γ shear strain

Indexes

a amplitude
a, b, t axial, bending, torsion
arith arithmetic
eq equivalent
E endurance
n nominal, normal
m mean
x, y, z coordinates

Other symbols

D damage sum
E Young modulus
G sliding modulus, ratio of stress concentration factors or of normalised stress gradients
K_t stress concentration factor
N number of cycles
P_s probability of survival
R stress ratio
T scatter
f frequency
f_G size effect factor
k slope of S-N curve
$\tilde{m}$ slope of prolonged S-N curve
t time, depth
φ angle of a reference plane
δ phase angle
μ Poisson ratio
ω angular frequency ($2 \pi f$)

INTRODUCTION

Multiaxial random fatigue has been ignored by engineers for a long times, despite the fact that fatigue critical areas like weld toes of many structures, e.g. Figs. 1, 2 and 3, are subjected to multiaxial states of stress/strain. The latter do not result only from local constraints (stress concentrations) but can also be caused by multiaxial external loading like combined bending and torsion. The most complex local multiaxial stress/strain states are those with varying directions of principal stress/strain directions under random loading.

The designer is confronted with the following problems in the assessment of multiaxial stress/strain states:

- Which kind of stresses / strains (nominal, hot-spot, structural, local) should be used?
- Which hypothesis should be used for the transformation of the multiaxial state into an equivalent one (von Mises or a modification)?
- Can design S-N curves obtained under uniaxial loading be applied for the assessment of multiaxial loading?

– Which damage accumulation hypothesis (Palmgren -Miner or a modification) and which allowable damage sum ($\Sigma\, (n/N)_i \leq D_{al}$, $D_{al} = 1.0$ or smaller) should be used in the case of random multiaxial loading?

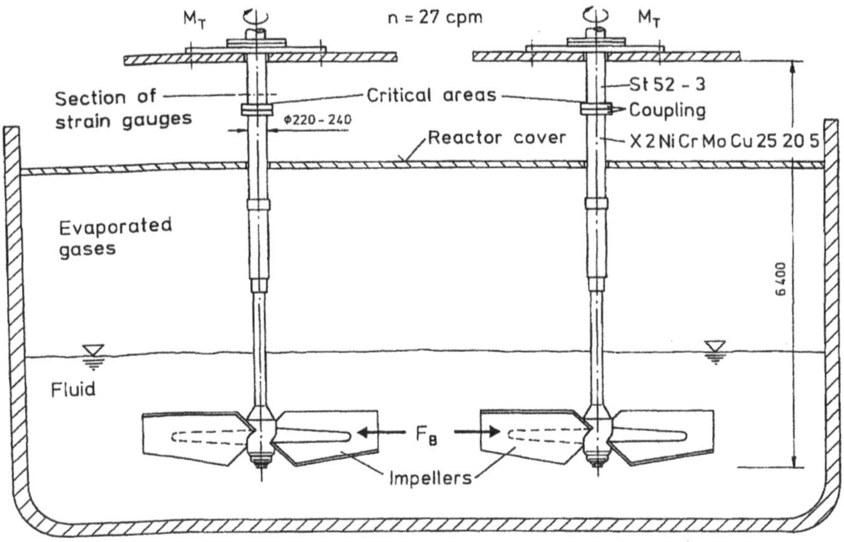

Fig.1 Stirrers of a fertiliser plant

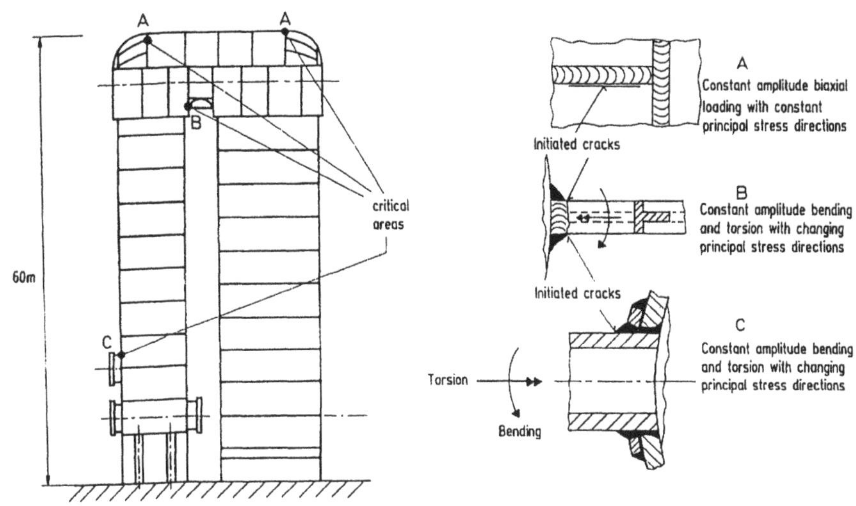

Fig.2 Hot-blast furnace and critical areas

The paper will briefly outline the state of the art assessment of multiaxial fatigue of welds including design codes and will demonstrate some examples of the problems and show possible solutions.

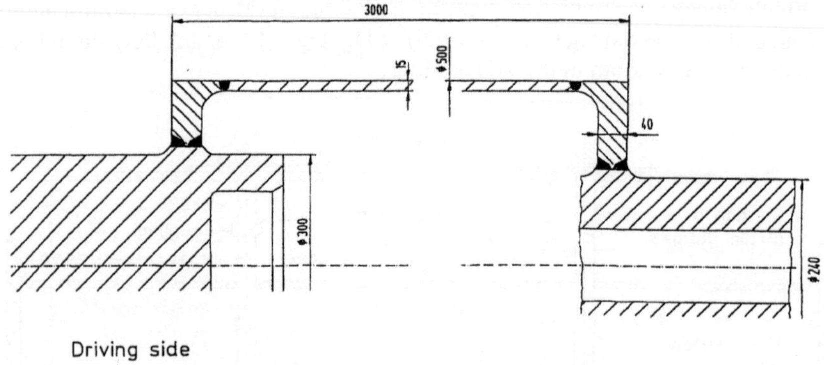

Fig.3 Fatigue critical areas of a welded stirrer

SHORT OVERVIEW OF DIFFERENT FATIGUE EVALUATION CONCEPTS

Definition of stress categories (nominal, hot-spot, structural, local)

There exist different concepts for the evaluation of the fatigue strength of welded structures. The most common concept is the nominal one. Generally, all design codes for welded structures are based on this concept **(1-7).** Provided that a nominal stress can be defined, it is assigned to the design curve of the particular geometry (design category), Fig.4. These design curves are mostly obtained from uniaxial loading; the failure criterion is total failure. The different safety concepts (probability of failure) are not discussed in this paper.

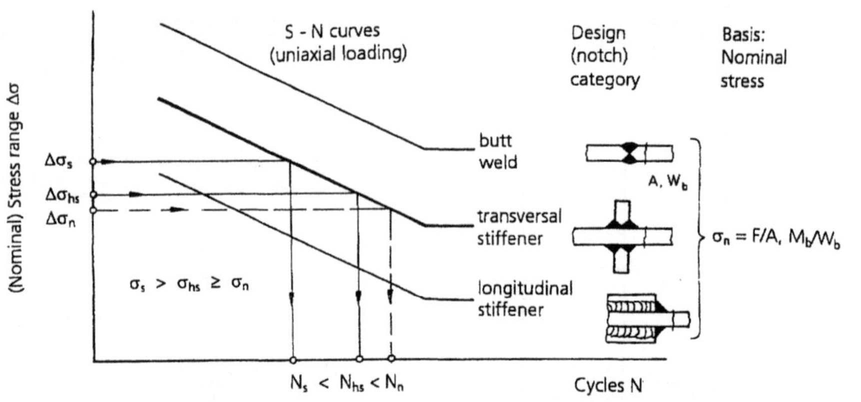

Fig.4 Fatigue evaluation concept in design codes using the nominal system

The design curves presented on the basis of nominal stresses are also used for two other concepts, namely the hot-spot **(7, 8)** and the structural stress concept **(9)**, Fig.5. The hot-spot stress is a fictional value derived by linear extrapolation of a calculated or measured stress-distribution to the weld toe. For structures where a nominal stress can be calculated by the use of basic mechanical equations like $\sigma_n = F/A$ or M_b / W_b the hot-spot stress is equal to the nominal stress. However, the structural stress close to the weld toe is significantly higher than the nominal or hot-spot ones, because it is also influenced by the geometrical transition due to the weld and the local notch effect, Fig.5, left. If the structural stress, e.g. measured or calculated at a distance of 1 to 2 mm of the weld toe **(9)**, is assigned to the particular design category, the fatigue life will be smaller than for the hot-spot or nominal stress, Fig.4. (The conservative result can be corrected, if the S-N curve is not presented in dependency of the nominal stress but of the structural stress at a defined distance to the weld toe.)

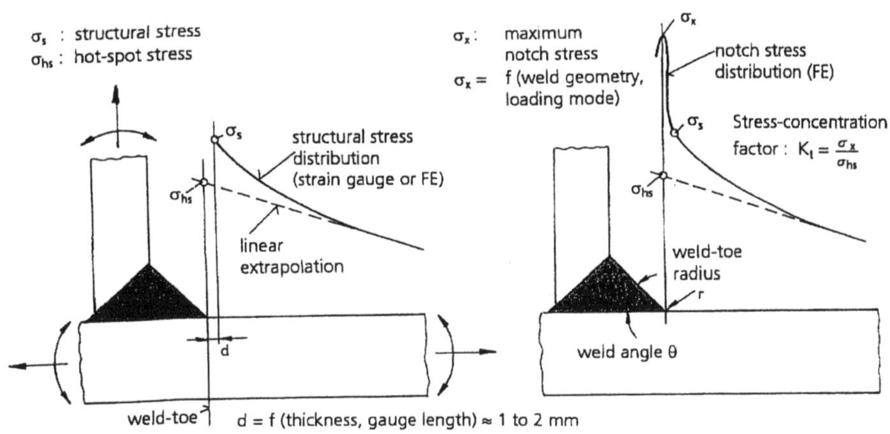

Fig.5 Definition of stresses in a weld

Recently, different variants of local concepts were also applied to welds **(12-20)**. The basic idea is to treat the weld as a notch and calculate the local stress distribution in the weld toe, Fig.5, right. The notch (principal-) stress distribution is nothing other than the continuation of the structural stress distribution. The ratio between the maximum local (principal-) stress and the nominal or hot-spot stress corresponds to the theoretical stress-concentration factor K_t. The maximum local stress in the weld toe is the decisive value dominating the fatigue life.

For the case of uniaxial loading and the use of nominal, hot-spot or structural stresses the knowledge of the local stress is not mandatory as long as the design curve used is known for the assessed weld detail in the nominal system. But for combined multiaxial loading knowledge of the particular maximum local stress components and of the nominal design curve is not sufficient, because the nominal stress components, e.g. for bending σ_n and for torsion τ_n, do not describe the local situation.

Figure 6 shows the calculated stress-distributions for bending and torsion of a welded flange-tube connection **(18, 19)**. The stress raising effect of the notch under bending more pronounced than for torsion.

TRANSFORMATION OF DESIGN CURVES FROM THE NOMINAL INTO THE LOCAL SYSTEM

For the evaluation of the fatigue strength of welded structures the nominal design curve must be transformed into a local one. The nominal design curve is shifted into the local curve, Fig.7, by determination of the theoretical stress-concentration factor K_{tb} for bending and K_{tt} for torsion, e.g. by finite or boundary element calculations or from analytical solutions **(10-13)**, and by the application of the von Mises equivalent stress criterion valid for ductile material state.

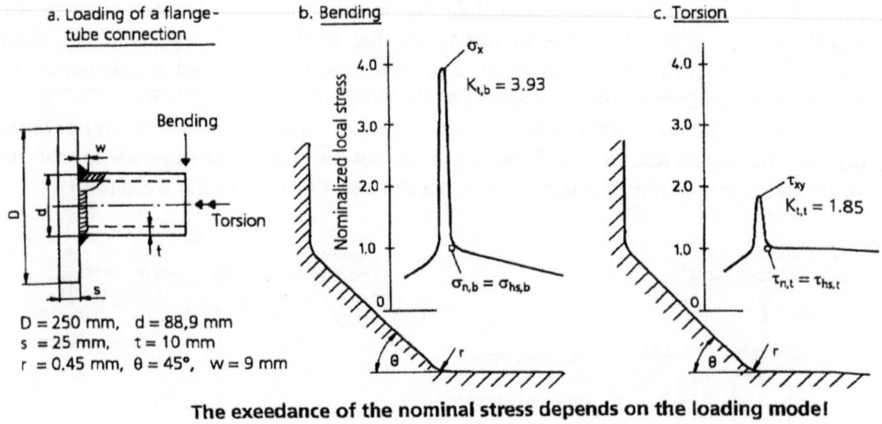

Fig.6 Loading mode and stress concentration factors of a weld

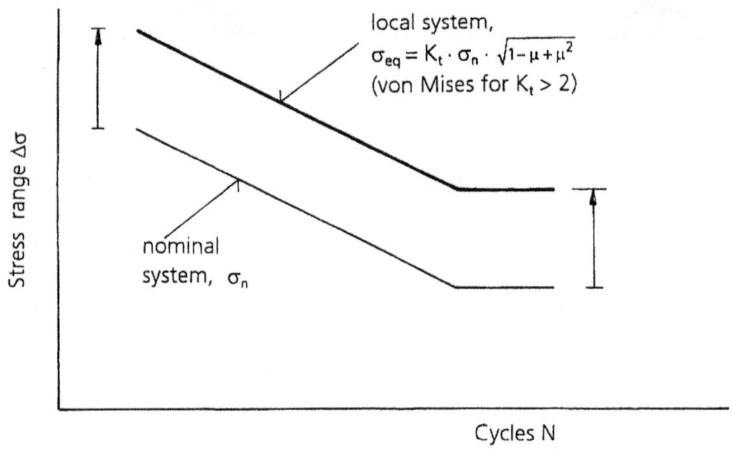

Fig.7 Transformation of an S-N curve from the nominal into the local system

Under uniaxial loading (bending or axial) in the weld toe a biaxial stress state with the local stress components σ_x and σ_y is produced. The local equivalent stress according to the von Mises hypothesis that is suggested in several design codes for ductile materials reads with

$$\sigma_x = K_{tb} \cdot \sigma_n \text{ and} \tag{1}$$

$$\sigma_y = \mu \cdot \sigma_x \text{ for } K_{tb} > 2 \text{ (21)} \tag{2}$$

$$\sigma_{eq} = \sqrt{\sigma_x{}^2 + \sigma_y{}^2 - \sigma_x \sigma_y} = K_{tb} \cdot \sigma_n \cdot \sqrt{1-\mu+\mu^2} \tag{3}$$

In the case of uniaxial torque, the local equivalent stress is

$$\tau_{xy} = K_{tt} \cdot \tau_n \tag{4}$$

$$\sigma_{eq} = \sqrt{3}\, K_{tt} \cdot \tau_{xy} \ . \tag{5}$$

If a shear stress is imposed on the bending due to an additional torsion, the equivalent stress is then extended to

$$\sigma_{eq} = \sqrt{\sigma_x^2 + \sigma_y^2 - \sigma_x \cdot \sigma_y + 3 \cdot \tau_{xy}^2} \tag{6}$$

Figure 8 demonstrates an example for the assessment of a multiaxial stress state. The incorrect use of combined nominal stress components results in an underestimation of fatigue life, because the effect of the shear stress in relation to the normal stress is overestimated. Therefore, the correct evaluation of a multiaxial stress state requires the use of physically decisive maximum local stress components.

a. Nominal system

$\Delta\sigma_n = 100$ MPa

$\Delta\tau_n = 58$ MPa

$\Delta\sigma_{n,eq} = \sqrt{\Delta\sigma_n^2 + 3\Delta\tau_n^2}$ (von Mises)

$\Delta\sigma_{n,eq} = 141$ MPa

b. Local system ($K_{tb} = 3.93$; $K_{tt} = 1.85$)

$\Delta\sigma_x = K_{tb} \cdot \Delta\sigma_n = 393$ MPa

$\Delta\sigma_y = \mu \cdot \Delta\sigma_x = 118$ MPa

$\Delta\tau_{xy} = K_{tt} \cdot \Delta\tau_n = 107$ MPa

$\Delta\sigma_{loc,eq} = \sqrt{\Delta\sigma_x^2 + \Delta\sigma_y^2 - \Delta\sigma_x \cdot \Delta\sigma_y + 3\Delta\tau_{xy}^2}$

$\Delta\sigma_{loc,eq} = 395$ MPa

c. Evaluation

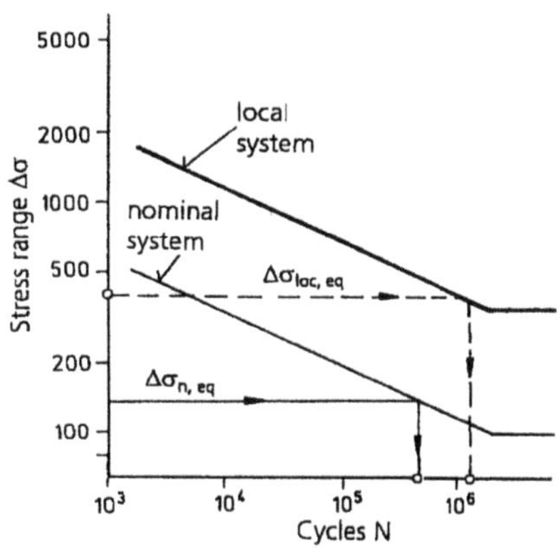

Fig.8 Example of the assessment of a multiaxial stress state

However, the physically correct use of local maximum stress does not always result in a correct equivalent stress if the von Mises hypothesis is applied. This hypothesis is only valid for proportional (in phase) multiaxial loading without changing principal stress directions **(18, 19)**. The next section will indicate how to overcome the limitations of the von Mises criterion.

ASSESSMENT OF MULTIAXIAL STRESS STATES

Constant amplitude loading

Experimental

In design codes the application of the von Mises hypothesis is usually recommended on the basis of nominal, hot-spot or structural stresses depending on the stress calculation method, but not on the basis of local notch stresses (or strains). Experiments were carried out with as welded and stress relieved, unmachined flange-tube connections from the structural fine-grained steel StE 460, Fig.9. The failure criterion was the break-through of the tube, Fig.10. The following results were obtained:

a. Specimen and clamping

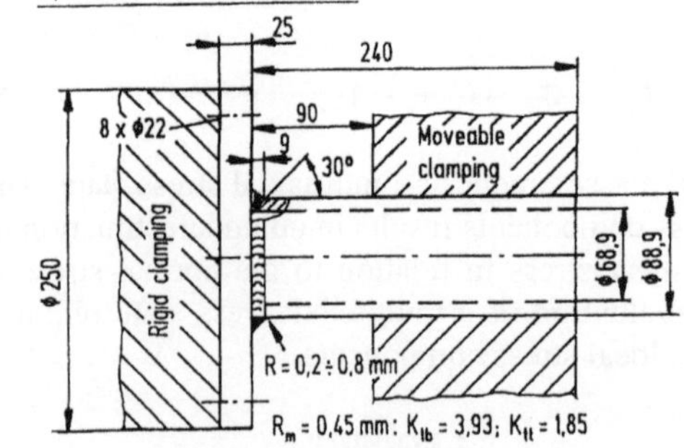

Material: StE 460, R_m = 670 MPa, $R_{p0,2}$ = 520 MPa

Welding: MAG, stress-relief annealed

Hardness: Base material 200 HV 1, weld 320 HV 1

b. Microstructure

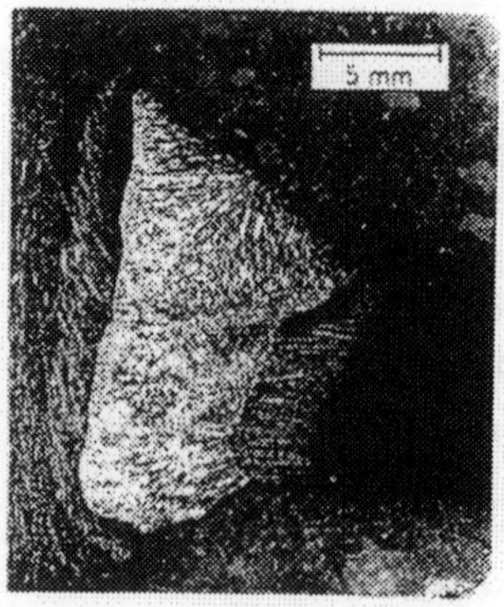

Fig.9 Unmachined flange-tube-connection

a. Crack along tube wall

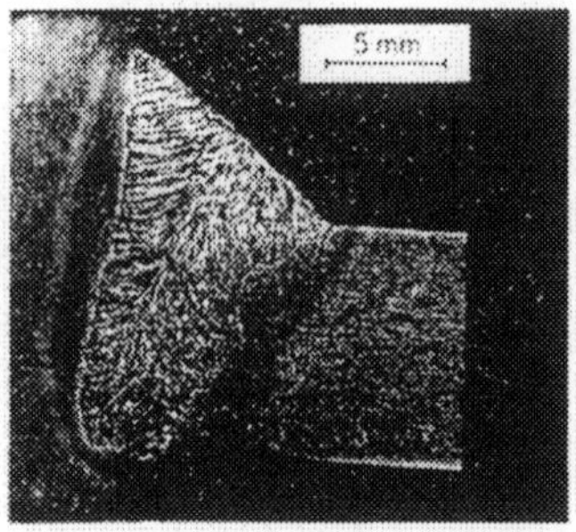

b. Fracture surface

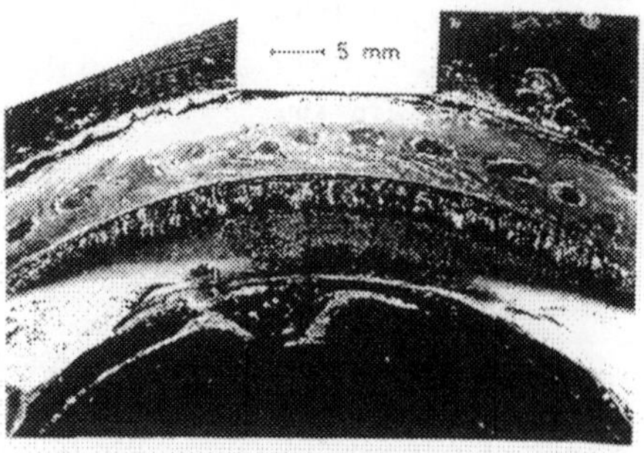

Fig.10 Fracture of a welded flange-tube-connection

When a combined in-phase bending and torsion (nominal shear stress was selected to be 58 % of nominal bending stress) with constant principal stress directions is applied, the fatigue life can be assessed using any of the mentioned stress concepts, Fig.11. But for out-of-phase loading, which simulates the changing of principal stress directions, these concepts fail when the von Mises hypothesis is applied; this hypothesis severely overestimates the fatigue life, Fig.11.

In order to overcome the deficiencies of this conventional hypothesis, a local stress based modification of the von Mises, hypothesis of effective equivalent stress (EESH) was developed **(18, 19)**.

Effective equivalent stress hypothesis (EESH)

The local components of the stress tensor for the most common biaxial stress state

$$[\sigma] = \begin{bmatrix} \sigma_x & \tau_{xy} \\ \tau_{xy} & \sigma_y \end{bmatrix} \tag{7}$$

generated by sinusoidal combined bending and torsion form in the weld toe (notch ground surface) read in general as:

$$\sigma_x(t) = \sigma_{xm} + \sigma_{xa} \sin \omega_x t \,, \tag{8}$$

$$\sigma_y(t) = \sigma_{ym} + \sigma_{ya} \sin(\omega_y t - \delta_y) \,, \tag{9}$$

$$\tau_{xy}(t) = \tau_{xym} + \tau_{xya} \sin(\omega_{xy} t - \delta_{xy}) \,. \tag{10}$$

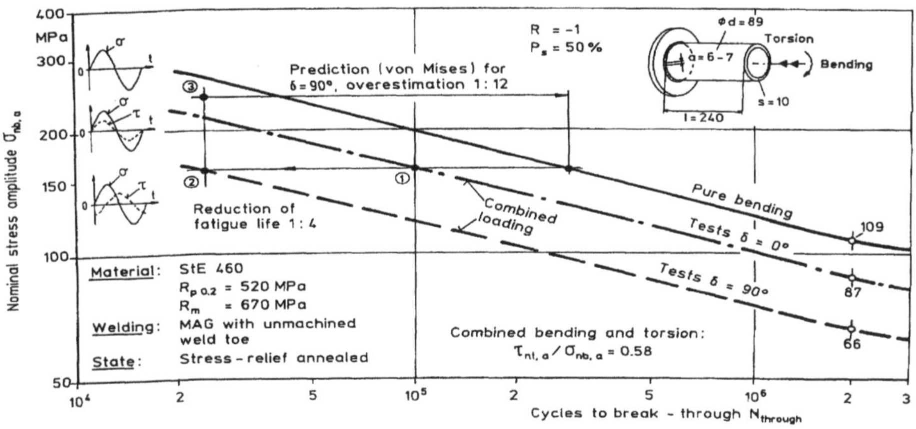

Fig.11 Fatigue strength of welded unmachined flange-tube-connections under multiaxial loading.

In the case of fully reversed stress (R = -1) the local mean stresses are σ_{xm}, σ_{ym} and $\tau_{xym} = 0$. The coordinate stresses σ_x and σ_y depend on each other, see Eqs (1) and (2), and are synchronous, i.e. $\delta_y = 0$. Only the shear stress is imposed with a phase displacement $\delta = \delta_{xy}$. The frequency of the tests discussed in this paper was kept the same for all stress components: $\omega = \omega_x = \omega_y = \omega_{xy}$. With this data the stresses acting in various reference planes φ of a surface element, Fig.12, can be calculated:

$$\sigma_n(\varphi) = \sigma_x \cos^2 \varphi + \sigma_y \sin^2 \varphi + 2\,\tau_{xy} \cos \varphi \sin \varphi \quad \text{and} \tag{11}$$

$$\tau_n(\varphi) = \tau_{xy}(\cos^2 \varphi - \sin^2 \varphi) - (\sigma_x - \sigma_y) \cos \varphi \sin \varphi. \tag{12}$$

The EESH assumes that failure of ductile materials under multiaxial stress states is initiated by shear stresses $\tau_n(\varphi)$. The interaction of shear stresses in various reference planes φ, in particular in the case of time-variable principal stress directions, which initiate corresponding dislocations, is taken into account by generating an effective shear stress

$$\tau_{arith} = \frac{1}{\pi} \int_0^{\pi} \tau_n(\varphi)\, d\varphi \tag{13}$$

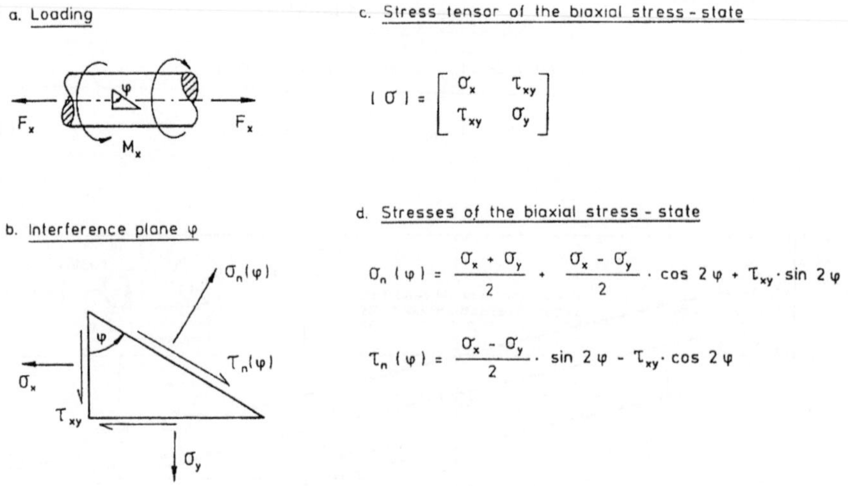

Fig.12 Reference plane stresses.

The effective shear stress is then used for determining the effective equivalent stress:

$$\sigma_{eq}(\delta)=\sigma_{eq}(\delta=0°)\frac{\tau_{arith}(\delta)}{\tau_{arith}(\delta=0°)}\sqrt{G\exp\left[1-\left(\frac{\delta-90°}{90°}\right)^2\right]}\ , \tag{14}$$

with

$$\sigma_{eq}(\delta=0°) = \sqrt{\sigma_x^2+\sigma_y^2-\sigma_x\sigma_y+f_G^2 3\tau_{xy}^2}\ , \tag{15}$$

$$f_G=\frac{\sigma_{eq,DEH}(\text{pure axial or bending load})}{\sigma_{eq,DEH}(\text{pure torsion})}=\frac{\sqrt{\sigma_x^2+\sigma_y^2-\sigma_x\sigma_y}}{\sqrt{3}\cdot\tau_{xy}}\ , \tag{16}$$

and

$$G=\frac{1+K_{ta}}{1+K_{tt}} \text{ or } \frac{1+K_{tb}}{1+K_{tt}} \text{ or } \frac{1+æ_a^*}{1+æ_t^*} \text{ or } \frac{1+æ_b^*}{1+æ_t^*}\ . \tag{17}$$

Here, f_G is the size effect factor which is determined by comparing the S-N curve for pure axial or bending stress with that for pure torsion on the basis of local supportable stresses. This factor reflects the influence of the maximum stressed material volume on the supportable local stress. The root in Eq. (14) considers the influence of the material volume affected by the rotating principal stress and principal strain axes on the magnitude of the effective equivalent strain in the case of a phase displacement, according to a model developed for semiductile materials. The ratio G in the Eqs (14) and (17) can be computed with the corresponding stress concentration factors. In the case of complex geometries, however, the corresponding stress gradients can, for instance, be found by a finite element computation and the referenced normalised stress gradients

$$\text{æ}^* = \frac{1}{\sigma_{x,max}} \cdot \frac{d\sigma_x}{dx} \tag{18}$$

can thus be determined. The basis for the effective equivalent stress is given by the appropriate S-N curve for pure bending load. The local coordinate stresses are calculated from the nominal stress and the stress concentration factor according to the Eqs.(1) and (2) and used for calculating the local equivalent stress according to the Eq.(3).

For evaluating the multiaxial stress state, the effective equivalent stress calculated according to Eq.(14) is then allocated to the reference S-N curve thus derived.

EVALUATION OF RESULTS FOR WELDED FLANGE-TUBE CONNECTIONS

Figure 13 shows the reference S-N curve with the scatter band between P_S = 10 and 90% for the unmachined flange-tube connections. This curve reflects the interdependence of the local equivalent stress amplitude according to Eq.(3) and the cycles to the break-through for the tests carried out under pure bending. Most of the results from tests performed under pure bending, pure torsion and with combined in-phase and out-of-phase loading lie within the scatter band of the reference S-N curve if they were derived according to the EESH.

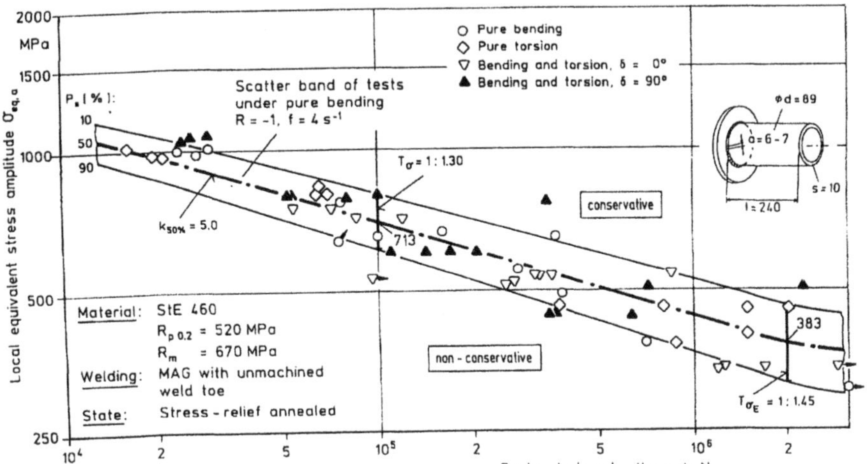

Fig.13 Evaluation of multiaxial stress states with the hypothesis of effective equivalent stress (EESH) for welded, unmachined flange-tube connections

This hypothesis was also successfully applied to other welded joints, such as machined flange-tube connections having lower stress concentrations than unmachined ones, to unmachined and machined tube-tube connections, all subjected to combined constant amplitude in- and out-of-phase bending and torsion **(18, 19)**.

The question that arises is, if the EESH is also applicable to variable in- and out-of-phase amplitude loading. To answer this question, first the cumulative fatigue behaviour **(22)** of the welded connections under uniaxial bending and torsion must be known.

Cumulative fatigue behaviour under uniaxial bending and uniaxial torsion

Damage accumulation

The fatigue life of welded connections is generally calculated according to the Palmgren-Miner cumulative rule

$$\sum \left(\frac{n}{N}\right)_i \leq D \ . \tag{19}$$

According to Haibach **(23)** after the knee point (this can be assumed for welded steels at $2 \cdot 10^6$ cycles or more) the inclination $\overline{m}$ of the S-N curve becomes shallower than its inclination k before the knee point:

$$\overline{m} = 2\,k - 1. \tag{20}$$

In most design codes the value D = 1.0 is prescribed for the damage sum. Based on different experiences under uniaxial loading the value of D = 0.5 seems to be more realistic **(24, 25)**. However, information on realistic damage sums under combined multiaxial random loading was not available until now. It is not even known whether damage sums obtained under uniaxial random loading can be used for the assessment of multiaxial random loading. Therefore, a systematic investigation was carried out **(26)** beginning with the determination of damage sums for uniaxial random bending and uniaxial random torsion applied to welded unmachined, stress-relieved flange-tube specimens, Fig.9, and then extended to multiaxial random loading. (Different research projects on this topic **(26, 27)** are currently running in Germany.)

Results of uniaxial spectrum loading

The tests were carried out under a Gaussian spectrum of amplitudes with a sequence length of $L_S = 5 \cdot 10^4$ cycles, Fig.14 **(26)**. The failure criterion was again the break-through of the tube. Figures 15 and 16 show the fatigue-life curves obtained under uniaxial variable amplitudes bending and uniaxial variable amplitudes torsion; the nominal stress amplitude presented is the maximum value of the spectrum. Figures 15 and 16 also contain the results determined under constant amplitude loading.

a. Amplitude distribution

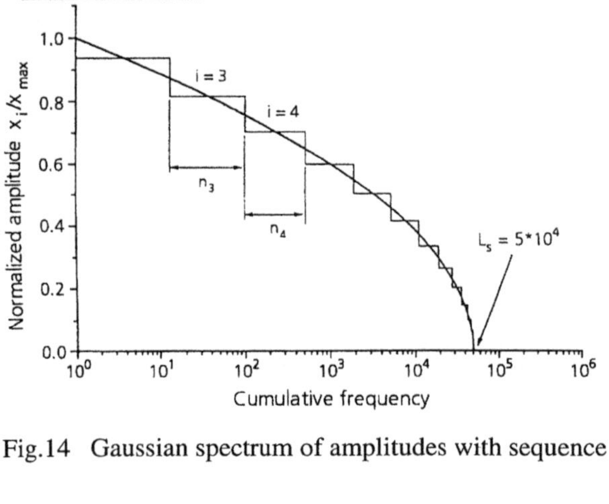

b. Stepping of the spectrum

i	x_i	n_i
1	1.000	1
2	0.938	12
3	0.817	87
4	0.704	418
5	0.600	1377
6	0.504	3256
7	0.417	5768
8	0.337	7964
9	0.267	8859
10	0.204	8156
11	0.150	6325
12	0.104	4158
13	0.067	2285
14	0.037	997
15	0.017	299
16	0.004	38
	$\Sigma n_i = H_0 = 50000$	

Fig.14 Gaussian spectrum of amplitudes with sequence length $L_s = 5 \cdot 10^4$ cycles.

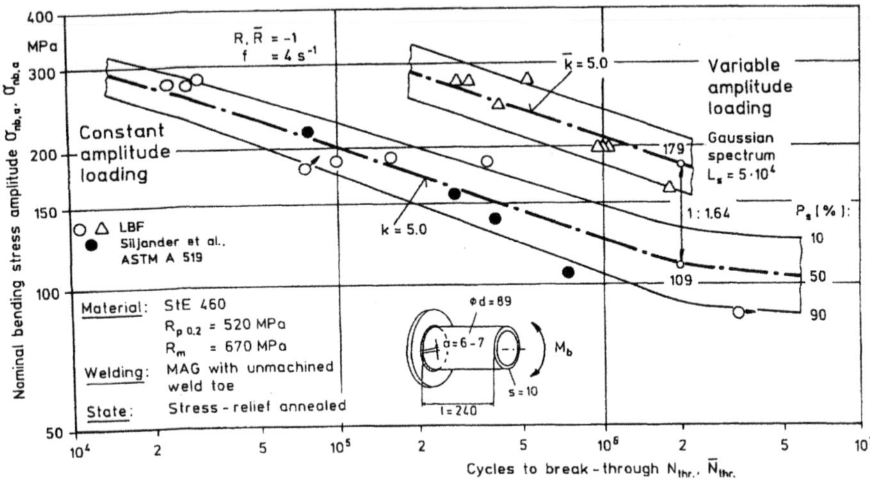

Fig.15 Fatigue strength of welded unmachined flange-tube-connections under uniaxial constant and variable amplitude bending.

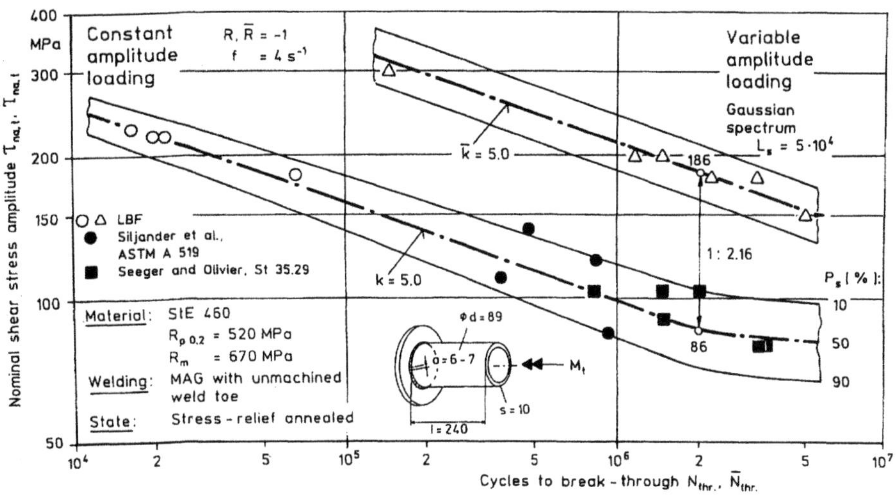

Fig.16 Fatigue strength of welded unmachined flange-tube-connections under uniaxial constant and variable amplitude torsion.

For a given fatigue life, e.g. at N = 2 · 10⁶ cycles, the ratio between the endurable maximum stress amplitude of the spectrum and the constant stress amplitude is higher for torsion than for bending, factor 2.16 vs 1.64. This indicates different damage sums for the particular loading. The determination of the real damage sums

$$D_{real} = \frac{\overline{N}_{experiment,Ps=50\%}}{\overline{N}_{calculated(D=1.0)}} \tag{21}$$

according to the Eqs.(19) and (20) results for bending $D_{real,b} = 0.08$ and for torsion $D_{real,t} = 0.38$, Table 1a.
The different damage sums for uniaxial variable amplitudes bending and uniaxial variable amplitudes torsion justify the necessity for investigations under combined multiaxial in- and out-of-phase variable amplitudes loading in order to obtain the appropriate D_{real}.

Cumulative fatigue behaviour under multiaxial variable amplitude loading

The tests under combined bending and torsion of flange-tube specimens will be carried out under the Gaussian spectra of amplitudes with a sequence length of $L_S = 5 \cdot 10^4$ cycles. The bending and-torque-time histories will be applied in- and out-of-phase. The nominal shear stress will be according to the von Mises hypothesis ($\tau_n/\sigma_n = 1/\sqrt{3}$) 58 % of the nominal bending stress.
Figure 17 presents the test results obtained from constant amplitude in- and out-of-phase multiaxial loading as a basis for the damage accumulation calculation. (Fig.11 contains the mean curves with $P_s = 50$ %.)

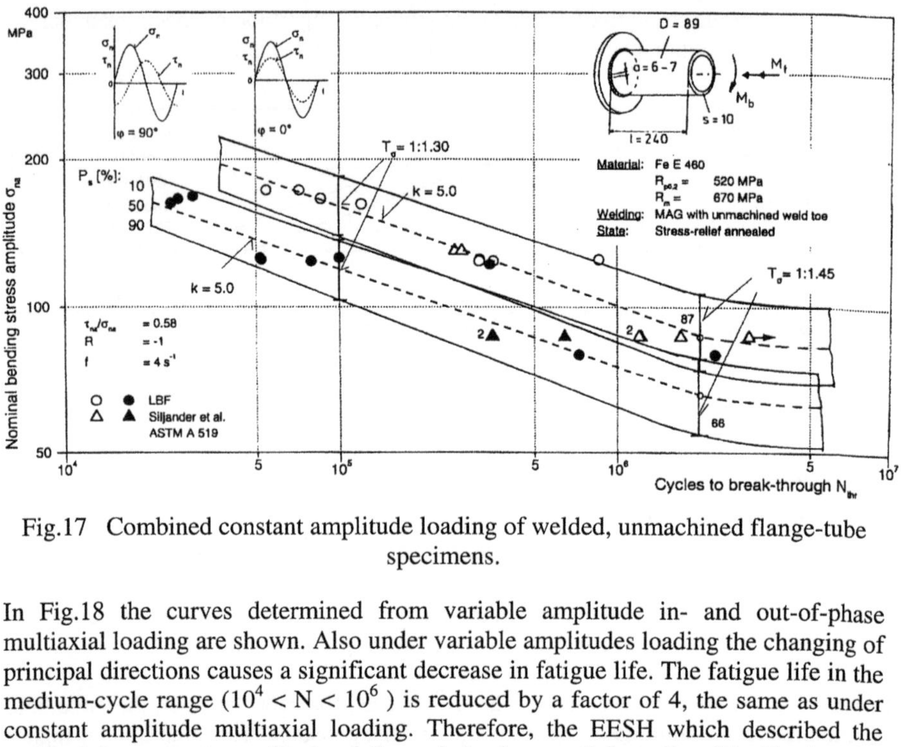

Fig.17 Combined constant amplitude loading of welded, unmachined flange-tube specimens.

In Fig.18 the curves determined from variable amplitude in- and out-of-phase multiaxial loading are shown. Also under variable amplitudes loading the changing of principal directions causes a significant decrease in fatigue life. The fatigue life in the medium-cycle range ($10^4 < N < 10^6$) is reduced by a factor of 4, the same as under constant amplitude multiaxial loading. Therefore, the EESH which described the multiaxial constant amplitude fatigue behaviour satisfactorily, Fig.13, is also applicable to in- and out-of-phase multiaxial variable amplitude loading because of the same slopes of the fatigue life curves and the same fatigue life reduction factor.
Figures 19 and 20 compare the appertaining constant and variable amplitude amplitude curves with $P_s = 50$ % for in- and out-of-phase loading.
For the fatigue life of $N = 2 \cdot 10^6$ cycles the multiaxial constant amplitudes are exceeded under in- as well as out-of phase multiaxial variable amplitude loading by

nearly the same factor 2.11 to 2.12. The determined real damage sums according to the equations are $D_{real} = 0.35$ and 0.38, Table 1b.

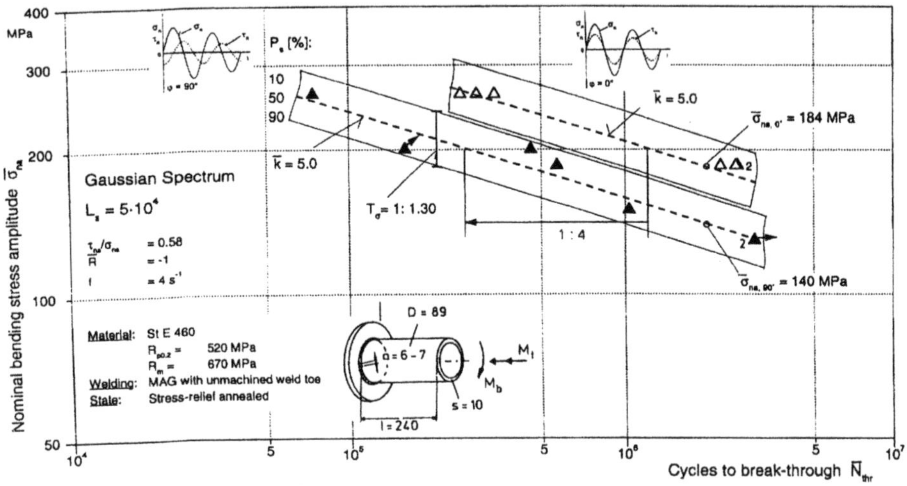

Fig.18 Combined variable amplitude loading of welded, unmachined flange-tube specimens

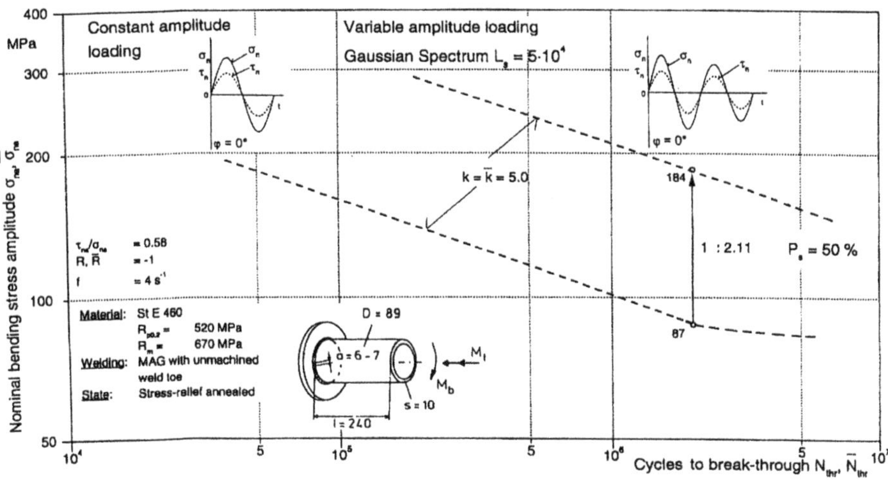

Fig.19 Comparison of in-phase combined constant and variable amplitude loading of welded, unmachined flange-tube specimens.

Figure 21 compares the obtained real damage sums for the investigated loading cases. However, the determined real damage sums for multiaxial variable amplitude loading should not be generalised before testing other combinations of the bending and shear stresses.

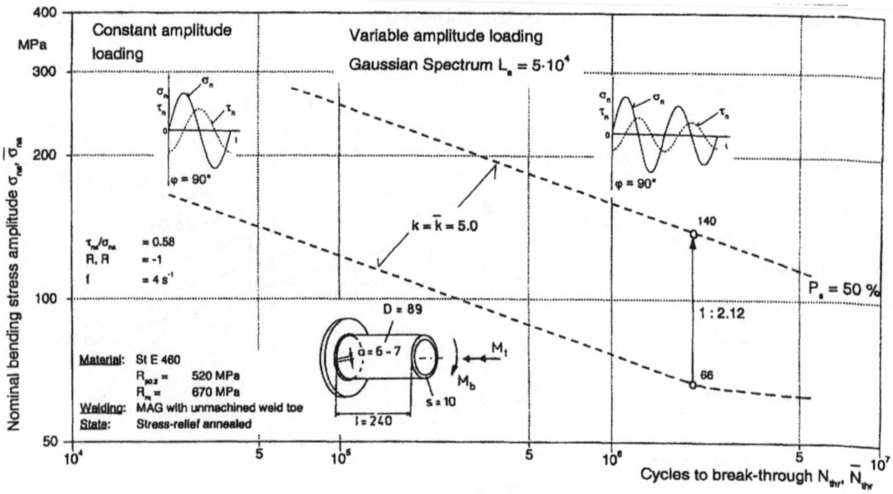

Fig.20 Comparison of out-of-phase combined constant and variable amplitude loading of welded, unmachined flange-tube specimens.

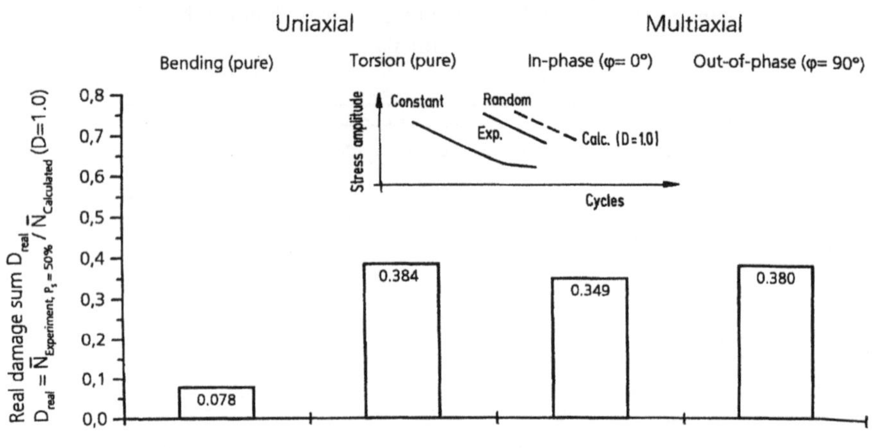

Fig.21 Real damage sums for uniaxial and multiaxial variable amplitudes loading of welded, unmachined, flange-tube specimens.

ASSESSMENT OF CUMULATIVE FATIGUE LIFE ACCORDING TO THE EESH

Figure 22 traces briefly how to apply the Effective Equivalent Stress Hypothesis (EESH) for the assessment of the cumulative fatigue life under multiaxial variable amplitudes loading, when the real damage sum D_{real} is known:

- The local stress time-histories $\sigma_x(t)$, $\sigma_y(t) = \mu \cdot \sigma_x(t)$ and $\tau_{xy}(t)$ corresponding spectra of amplitudes with the maximum values $\bar{\sigma}_x$, $\bar{\sigma}_y = \mu \cdot \bar{\sigma}_x$ and $\bar{\tau}_{xy}$ are calculated using the nominal bending and shear stresses and the stress concentration factors for bending K_{tb} and torsion K_{tt}, respectively.

Table 1 Real damage sums for variable amplitude loading of welded, unmachined flange-tube specimens under pure bending, pure torsion and combined in- and out-of-phase loading.

a. Pure bending and pure torsion

Loading	Knee-point of the S-N curve $N_E (P_s = 50\ \%)$	Endurance limit $(P_s = 50\ \%)$ $\overline{\sigma}_E ; \overline{\tau}_E$ in MPa	Slope of the S-N curve $k_{50\%}$	$k' = 2k-1$	$D_{Spectrum}$ $L_{Spectrum} = 5 \cdot 10^4$	Calculated life $\overline{N}_{cal}$ (D=1.0) for $\overline{\sigma}_{na} ; \overline{\tau}_{na} = 200$ MPa	Experimental life $\overline{N}_{exp}$ for $\overline{\sigma}_{na} ; \overline{\tau}_{na} = 200$ MPa	Real damage sum $D_{real} = \overline{N}_{exp} / \overline{N}_{cal}$
Bending	2 000 000	109	5.0	9.0	0.0034	14 618 060	1 148 500	0.078
Torsion	2 000 000	86	5.0	9.0	0.0138	3 621 334	1 391 400	0.384

b. Combined bending and torsion

Loading	Knee-point of the S-N curve $N_E (P_s = 50\ \%)$	Endurance Limit $(P_s = 50\ \%)$ $\overline{\sigma}_E$ MPa	Slope of the S-N curve $k_{50\%}$	$k' = 2k-1$	$D_{Spectrum}$ $L_{Spectrum} = 5 \cdot 10^4$	Calculated life $\overline{N}_{cal}$ (D=1.0) for $\overline{\sigma}_{na} = 200$ MPa	Experimental life $\overline{N}_{exp}$ for $\overline{\sigma}_{na} = 200$ MPa	Real damage sum $D_{real} = \overline{N}_{exp} / \overline{N}_{cal}$
In-phase $\varphi = 0°$	2 000 000	87	5.0	9.0	0.0129	3 871 012	1 350 000	0.349
Out-of-phase $\varphi = 90°$	2 000 000	66	5.0	9.0	0.0562	890 320	338 000	0.380

$$D_{Spectrum} = \Sigma \left(\frac{n}{N} \right)_i \qquad D_{real} = \frac{D_{Spectrum}}{L_{Spectrum}} \cdot \overline{N}_{exp} = \overline{N}_{exp} / \overline{N}_{cal} \qquad \overline{N}_{cal} = \frac{L_{Spectrum}}{D_{Spectrum}}$$

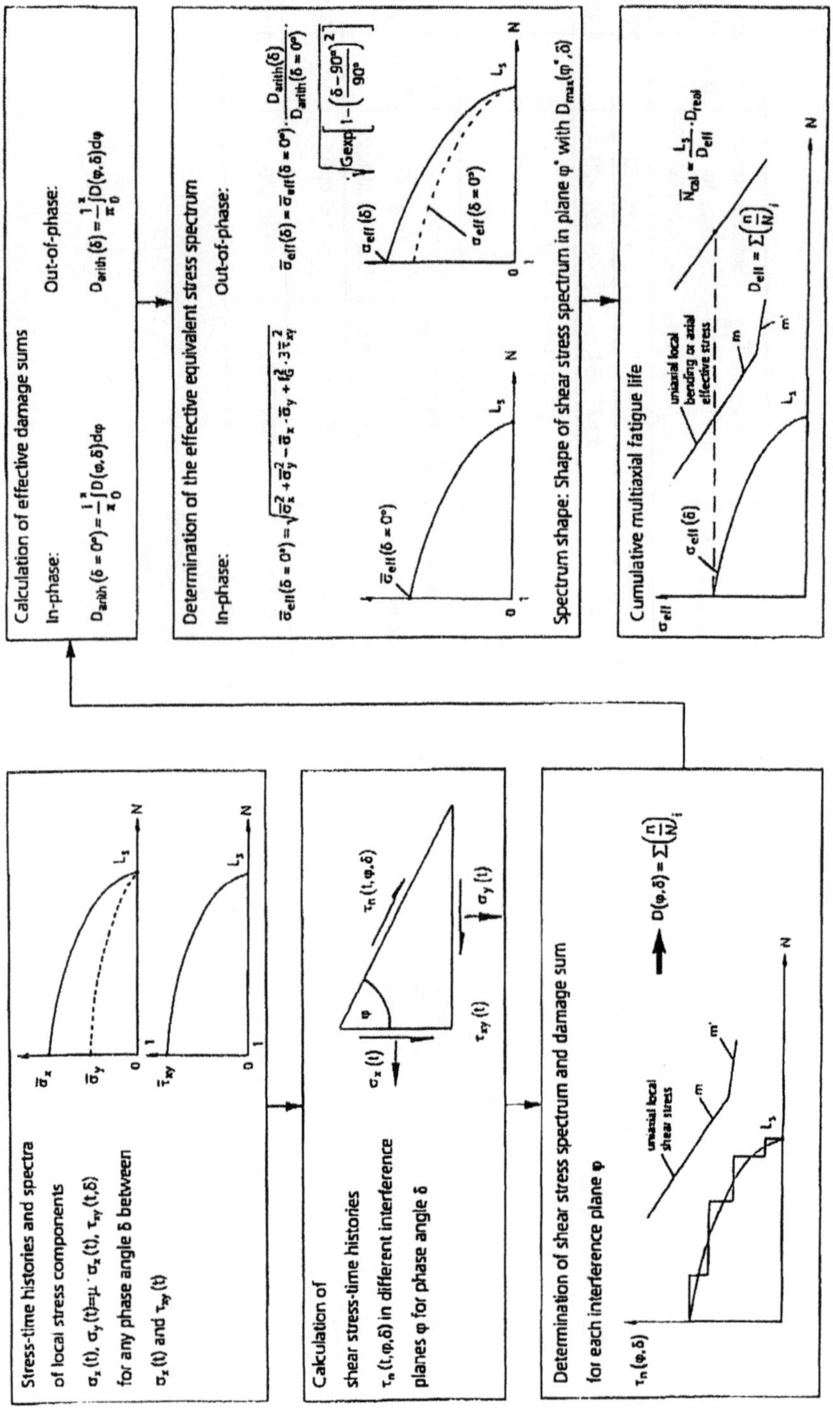

Fig.22 Application of the effective equivalent stress hypothesis to multiaxial variable amplitudes loading.

- From the local stress-time histories the shear stress time histories in different reference planes φ for the given phase angle δ between $\sigma_x(t)$ and $\tau_{xy}(t)$ are determined according to Eq. (12).
- For a given phase angle δ the shear stress spectrum of amplitudes in each plane φ is determined. By assigning the spectra of amplitudes to the uniaxial local shear stress-cycle curve the damage sums D (φ, δ) are calculated for each plane. Damage occurs in the plane φ* where damage is maximum: D_{max} (φ*, δ).
- As steel welds are ductile the interaction of shear stresses acting in different planes φ must be considered by calculation of the effective (arithmetic average) damage sum for in-phase loading $\delta = 0°$and for a given phase angle δ analogous to Eq. (13).
- The effective equivalent stress spectrum of amplitudes is determined, while for the shape of this effective spectrum the shape of the shear stress spectrum of amplitudes with $D_{max}(\varphi*, \delta)$ is taken. The maximum value of the effective equivalent spectrum of amplitudes is calculated using the maximum values of the input spectra according to Eqs.(14) to (17).
- Finally, the cumulative fatigue life under multiaxial variable amplitudes loading is calculated by assigning the effective equivalent spectrum of amplitudes to the local S-N-curve obtained from uniaxial bending. The calculated fatigue life

$$\overline{N}_{cal} = \frac{L_s}{D_{eff}} \cdot D_{real} \tag{22}$$

  must be connected with the real damage sum D_{real} that is obtained from presented multiaxial variable amplitude tests.

Present recommendations

Although not all the information knowledge from current on multiaxial random fatigue **(26, 27)** is available, i.e. the obtained real damage sums for multiaxial random loading cannot yet be generalised, but some recommendations can be given for the dimensioning of welded structures against such complex loading situations:

- It is necessary to determine the local stress-time histories of the stress components. As the local values cannot be measured, it is sufficient if structural stresses σ_{sx}, σ_{sy}, τ_{sxy} are measured and then transformed by stress concentration factors K_{tb}, K_{tt} into local values σ_x, σ_y, τ_{xy}. If stress concentration factors are not available, a boundary or finite element modell can be helpful.
- In structures random torsion and bending stresses can differ very much with regard to frequencies and spectrum shapes. Fig.23 shows such bending- and torque-time histories measured at the shaft of a fertiliser stirrer, Fig.1 **(28)**. The bending stresses are fluctuating (R = -1), while the torsion is a start-stop event with small fluctuations about a positive mean stress.
- The particular spectra of amplitudes are presented on Fig.24. In the weld toe shear stresses could be calculated for different reference planes φ according to Eq (12), and effective damage sums could be determined. As for this complex load situation the knowledge for assessment is still not available but a practical way to solve the problem with conservative assumptions can be proposed:
- The influence of different mean bending and shear stresses, Fig.24, is corrected by application of the mean stress sensitivity of the steel used. The shear stress is transformed to the stress ratio R = -1 of the bending stress. The shape and size of the effective equivalent stress spectrum of amplitudes is determined by the

spectrum of amplitudes shape of the most intensive local stress component. In this case it is the bending stress which has a fuller shape in comparison to the shear stress spectrum of amplitudes.

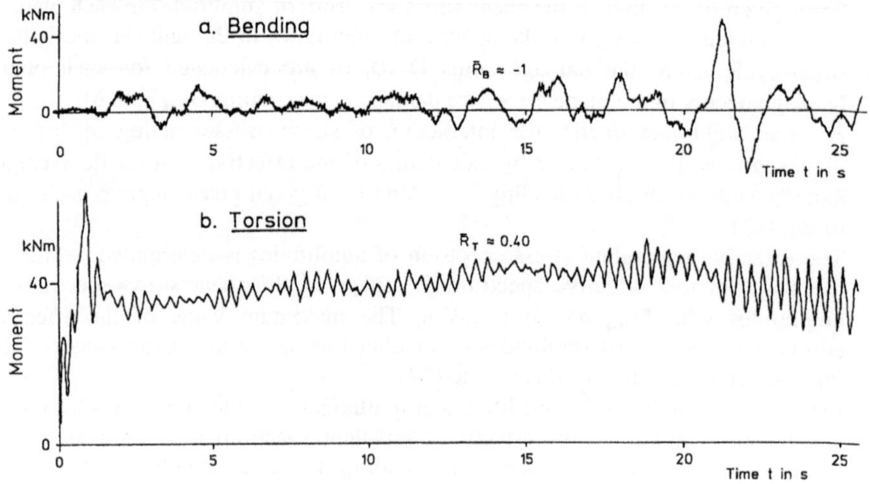

Fig.23 Multiaxial random-load-time-histories.

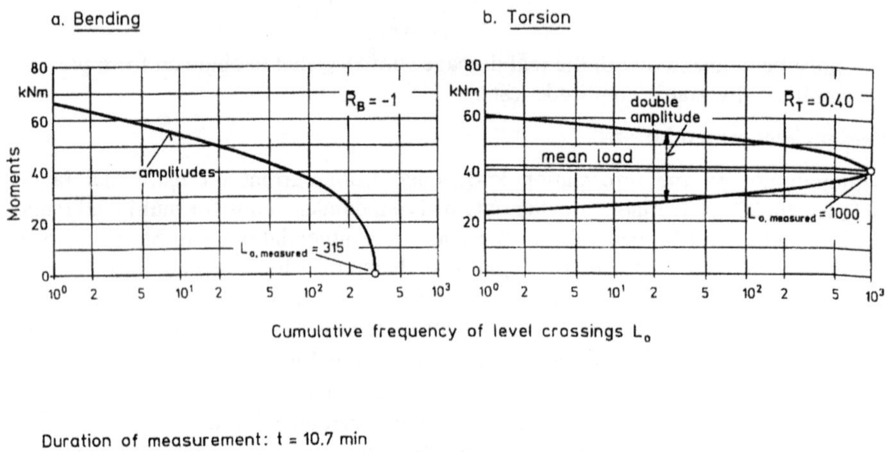

Fig.24 Cumulative frequency distributions.

- This means that if the torsion fluctuates with the same frequency of bending moment and additionally has the shape of the bending spectrum of amplitudes; i.e. the shear stress would act in a more damaging (intensive) way as in reality.
- The maximum effective equivalent stress is obtained by application of the EESH. A constant amplitude in- and out-of-phase simulation with the maximum bending and shear stresses of the corresponding spectra of amplitudes is carried out with the most critical phase angle of $\delta = 90°$ for ductile steels.
- The effective spectrum of amplitudes is then assigned to the local S-N curve of the welded design detail, Fig.25.
- The cumulative fatigue life is calculated assuming an allowable damage sum of $D_{al} = 0.50$.

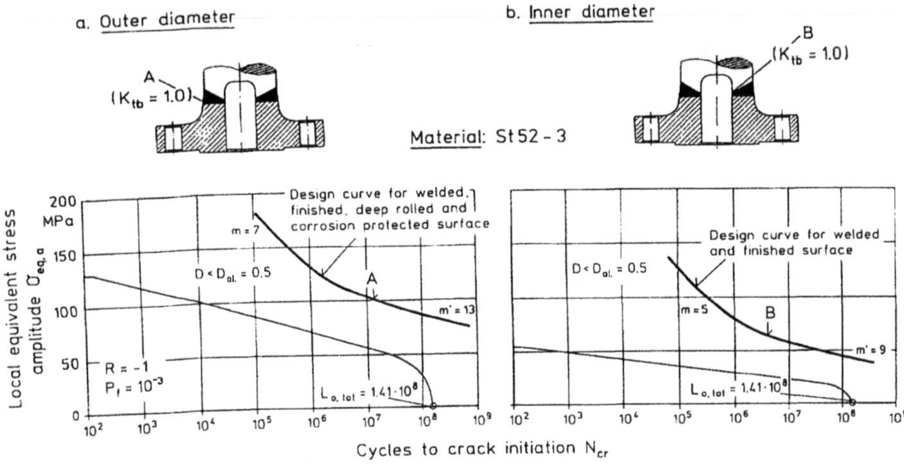

Fig.25 Fatigue life evaluation of the final design of the stirrer coupling.

The stirrers discussed here, Fig.1, were redesigned, Fig.25, fifteen years ago after several structural failures applying very conservative assumptions **(28)**. (After the redesign of the shafts no failures occurred.) The use of the present knowledge on multiaxial fatigue results in a similar effective equivalent stress spectrum of amplitudes. However, the extent of conservatism introduced can be evaluated only when more detailed results on multiaxial random fatigue are available.

CONCLUSIONS AND OUTLOOK

Principally, the evaluation of constant, variable amplitude or random multiaxial fatigue of welds requires the knowledge of the local stress state in the weld toe (notch). It can be determined by boundary or finite element modelling of the weld detail. The observed fatigue life reduction of ductile steel welds under constant and variable amplitude multiaxial loading with changing principal stress directions compared with welds with constant principal directions, e.g. simulated by out-of-phase and in-phase bending and torsion, can be described fairly well by the Effective Equivalent Stress Hypothesis (EESH). While uniaxial variable amplitude bending and variable amplitude torsion tests reveal different real damage sums ($D_{real,b}$ = 0.38, $D_{real,t}$ = 0.08) the value of the real damage sum for cumulative multiaxial fatigue assessments was found to be D_{real} = 0.35 to 0.38. However, a systematical investigation of multiaxial random fatigue, as it is presently performed **(26, 27)**, is necessary, because the obtained real damage sum cannot yet be generalised. Despite conservative recommendations how to use existing knowledge for the assessment of multiaxial random loading of welds validation tests must be carried out in critical cases.
The multiaxial fatigue research presently comprises steel welds, which have ductile material behaviour. From present knowledge on multiaxial fatigue behaviour of unwelded materials it can be concluded that aluminium welds, which are not as ductile as steel, may require the application of other stress and cumulative damage hypotheses. A new research project on the multiaxial fatigue behaviour of aluminium welds is also in progress.

Fracture mechanics concepts **(29-32)** have not yet been applied to multiaxial fatigue problems in order to predict crack propagation. The project **(27)** in progress also aims at applicable fracture mechanics concepts.

REFERENCES

(1) EUROCODE Nr. 3, (1992) Gemeinsame einheitliche Regeln für Stahlbauten, Kommission der Europäischen Gemeinschaften, Stahlbau-Verlagsgesellschaft mbH, Köln

(2) British Standard BS 5400, (1980) Steel, Concrete and Composite Bridges, Part 10, Code of Practice for Fatigue, BSI

(3) SME Boiler and Pressure Vessel Code, (1984) Section III, Division 1, Subsection NA, Article XIV-1212, Subsection NB-3352.4, New York

(4) AD-Merkblatt S 2, (1982) Berechnung gegen Schwingbeanspruchung, Beuth-Verlag, Berlin/Köln

(5) DS 804, (1996) Vorschrift für Eisenbahnbrücken und sonstige Ingenieurbauwerke (VEI), Deutsche Bahn, Munic

(6) Hobbacher, A., (1994) Recommendations on Fatigue of Welded Components, IIW Fatigue Design Recommendation, IIW-Document No. XIII-1539-94/XV 845-94

(7) Niemi, E., (1995) Stress Determination for Fatigue Analysis of Welded Components, Abington Publishing, Cambridge

(8) De Back, J., (1981) Festigkeit von Rohranschlüssen, In: Stahl in Meeresbauwerken, Internationale Konferenz 5.-8.10.19981, Paris, Hrsg.: Kommission der Europäischen Gemeinschaften, Luxemburg: EUR-Bericht No. 7347, pp. 439 – 483

(9) Haibach, E., 1968) Die Schwingfestigkeit von Schweißverbindungen aus der Sicht einer örtlichen Beanspruchungsmessung, Fraunhofer-Institut für Betriebs-festigkeit (LBF), Darmstadt, Report No. FB-77

(10) Yung J.-H. and Lawrence F.V., (1985) Analytical and graphical aids for the fatigue design of weldments, Fatigue Fract. Engng. Mat. Struc. 8, No. 3, pp. 223 – 241

(11) Anthes R.J., Köttgen V.B. and Seeger T., (1994) Einfluß der Nahtgeometrie auf die Dauerfestigkeit von Stumpf- und Doppel-T-Stößen, Schweißen und Schneiden 46, No. 9, pp. 433 – 436

(12) Lida K. and Uemura T., (1994) Stress Concentration Factor Formulas Widely Used in Japan, IIW-Document No. XIII-1530-94

(13) Radaj D., (1994) Design and Analysis of Fatigue Resistant Welded Structures Woodhead Publishing, Cambridge

(14) Yung J.Y. and Lawrence F.V. Jr., (1989) Predicting the fatigue life of welds under combined bending and torsion, In: Biaxial and Multiaxial Fatigue, EGF 3, Ed. by M.W. Brown and K.J. Miller, MEP, London, pp. 53 – 69

(15) Lawrence F.V. Jr. Mattos R.J., Higashida Y. and Burk J.D., (1978)Estimating fatigue crack initiation life of welds, In: ASTM STP 648, pp. 134 –158

(16) Siljander A., Kurath P. and Lawrence F.V. Jr., (1989) Proportional and non-proportional multiaxial fatigue of tube-to-plate weldments, University of Illinois at Urbana-Champaign, Urbana, Illinois, Report to the Welding Research Council

(17) Siljander A., (1991) Nonproportional Biaxial Fatigue of Welded Joints Dissertation Univ. Illinois

(18) Sonsino C.M., (1995) Multiaxial fatigue of welded joints under in-phase and out-of-phase local strain and stresses, Int. J. Fatigue 17, No. 1, pp. 55 – 70

(19) Sonsino C.M., (1994) Multiaxial Fatigue of Welded Flange-Tube- and Tube-Tube Connections under In- and Out-of-Phase Loading and Local Stresses IIW-Document No. XIII-1542-94

(20) Radaj D., (1996) Review of fatigue strength assessment of nonwelded and welded structures based on local parameters, Int. J. Fatigue 18, No. 3, pp. 153 –170

(21) Neuber H., (1985) Kerbspannungslehre - Theorie der Spannungskonzentration, Genaue Berechnung der Festigkeit, Springer Verlag, Berlin, 3. Edition

(22) Buxbaum O., (1992) Betriebsfestigkeit - Sichere und wirtschaftliche Bemessung schwingbruch-gefährdeter Bauteile, Verlag Stahleisen mbH, Düsseldorf, 2. Edition

(23) Haibach E., (1989) Betriebsfestigkeit - Verfahren und Daten zur Bauteilberechnung, VDI-Verlag GmbH, Düsseldorf

(24) Sonsino C.M., (1994) Über den Einfluß von Eigenspannungen, Nahtgeometrie und mehrachsigen Spannungszuständen auf die Betriebsfestigkeit geschweißter Konstruktionen aus Baustählen, Materialwissenschaft und Werkstofftechnik 25 No. 3, pp. 97 – 109

(25) Sonsino C.M. and Umbach R., (1993) Corrosion fatigue of welded and cast steel hybrid nodes under constant and variable amplitude loading, In: Proc. 12th Int. Conf. on Offshore Mechanics and Arctic Engineering, OMAE, Glasgow, Eds M.H. Salama et al. ASME, New York, Vol. III, Part B, pp. 667 674

(26) Fatigue Behaviour of Welded High Strength Components under Combined Multiaxial Variable Amplitude Loading, European Community for Steel and Coal (ECSC), Contract 7210 MC/109, Fraunhofer-Institut für Betriebsfestigkeit (LBF), Darmstadt

(27) Research Project of the Deutsche Forschungsgemeinschaft (DFG), (1995-1999) Fatigue Strength of Welded Connections under Multiaxial Loading

(28) Sonsino C.M. and Pfohl R., (1990) Multiaxial fatigue of welded shaft-flange connections of stirrers under random non-proportional torsion and bending, Int. J. Fatigue 12, No. 5, pp. 425 - 431

(29) Maddox S.J., (1991) Fatigue Strength of Welded Structures, Abington Publishing, Cambridge

(30) PD 6493, (1991) Guidance on Methods for Assessing the Acceptability of Flaws in Fusion Welded Structures, British Standards Institution - BSI, London

(31) DVS-Merkblätter 2401, (1989)Bruchmechanische Bewertung von Fehlern in Schweißverbindungen, Teil 1: Grundlagen und Vorgehensweise; Teil 2: Praktische Anwendung, Fachbuchreihe Schweißtechnik, Bd. 101, DVS-Verlag, Düsseldorf

(32) Hobbacher A., (1993) Stress intensity factors of welded joints, Eng. Fracture Mech. 46, No. 2, pp. 173 - 182

Acknowledgement

The author thanks Dr. S. Maddox (TWI) and Prof. Dr. E. Macha (TU Opole) for reviewing this paper.

A STRESS-BASED APPROACH FOR FATIGUE ASSESSMENT UNDER MULTIAXIAL VARIABLE AMPLITUDE LOADING

Bastien WEBER*, Alain CARMET*, Bienvenu KENMEUGNE**
and Jean-Louis ROBERT**

* SOLLAC – LEDEPP, 17, avenue des Tilleuls - 57191 Florange CEDEX, France
** INSA Lyon, Laboratoire de Mécanique des Solides, Bât. 304
20, avenue Albert Einstein, 69621 VILLEURBANNE CEDEX, France

ABSTRACT

This paper presents a new multiaxial variable amplitude fatigue life prediction method. The main steps of this stress-based approach are the definition of a so-called counting variable to identify and extract cycles from the multiaxial random stress history, the use of two multiaxial criteria which are extended from endurance to finite fatigue lives, in order to calculate the life of each extracted cycle by Rainflow counting and then the use of damage rules to assess the life of the material submitted to the multiaxial stress sequence. The different steps are detailed: the chosen counting variable is justified, the global approach and critical plane approach criteria are presented and both linear Miner and non linear Lemaitre & Chaboche damage laws may be used. Several biaxial variable amplitude fatigue tests that were carried out at the Technical University of Opole (Poland) allow validation the proposed method with both damage rules.

KEY WORDS

Fatigue life, multiaxial criterion, counting variable, damage law, damage indicator

NOTATION

β, $aM_0^{-\beta}$ material coefficients of the non-linear damage law
$\sigma_{-1}(N)$ fatigue strength corresponding to N cycles for a reversed tensile test (R=-1)
$\tau_{-1}(N)$ fatigue strength corresponding to N cycles for a reversed torsion test (R=-1)
$\sigma_0(N)$ fatigue strength corresponding to N cycles for zero to maximum tensile test (R= 0)
$\sigma_{ij}(t)$ stress states tensor versus time t
$\vec{h}$ unit vector normal to a physical plane P
$\sigma_{hh}(t)$ normal stress versus time t acting on the plane P
σ_{hhm} mean normal stress acting on P during the cycle

σ_{hha} amplitude of the normal stress acting on P
$\sigma_{hha}(t)$ alternate normal stress versus time t acting on P
τ_{ha} amplitude of the shear stress acting on P
$\tau_{ha}(t)$ alternate shear stress versus time t acting on P
D damage
E_h damage indicator of the plane whose unit normal vector is $\vec{h}$
E fatigue function of a multiaxial criterion
X function of damage
b slope of the material endurance diagram.

INTRODUCTION

Economical constraints, which are becoming stronger and stronger nowadays, create the need for significant of an important weight reduction of structures or components. This purpose imposes to reduce dimensions and thickness of thin walled components and, as a consequence, to submit structures to generally higher stress levels. Thus resulting in the fatigue phenomenon as a problem of great priority.
The aim of this paper is to present a stress-based approach (Kenmeugne **(1)**) which allows engineering designers to assess the fatigue life of a component under multiaxial variable amplitude loading which is the most general case of loading. The method described in this paper contains different steps as shown on the following flow chart (Fig.1).

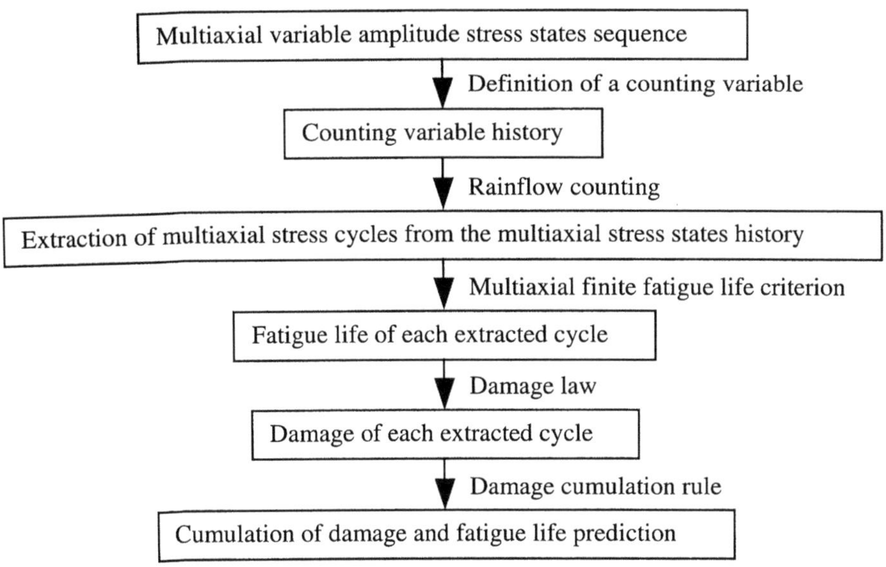

Fig.1 Flow chart of the proposed stress-based fatigue life prediction method.

The proposed stress-based approach derives from an extension of criteria from endurance (infinite lives) to finite fatigue lives (Robert **(2)**). The procedure keeps the main steps of uniaxial fatigue life prediction methods (Robert and Bahuaud **(3)**). Each step of the method is explained and justified in the following sections.

COUNTING VARIABLE DEFINITION

One of the main problems with multiaxial variable amplitude stress is the definition of a cycle. A cycle is generally defined by a hysteresis loop of the material uniaxial stress-strain response (Fig.2).
The Rainflow counting is a simple and well-established technique (Amzallag **(4)**) to identify uniaxial stress cycles represented by these hysteresis loops.

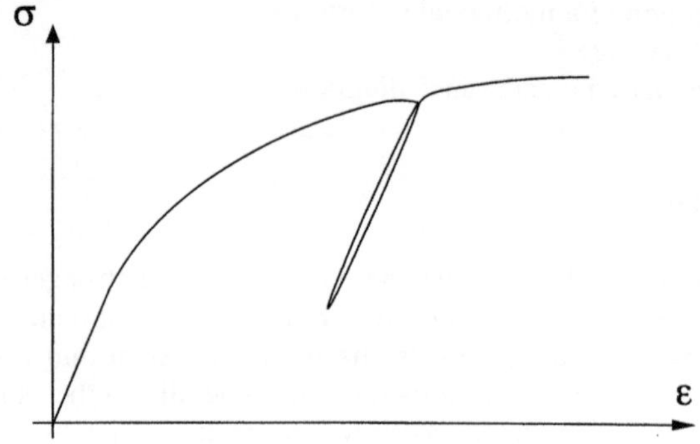

Fig.2 Hysteresis loop of the material stress-strain response.

The most complex counting problem concerns the case where any of the six different components of the stress tensor varies independently from each other. Fig.3 shows for instance that a cycle may occur on one channel (σ_{22}) but during the same time interval [t_1, t_2] none of the other channels experiences exactly a cycle.

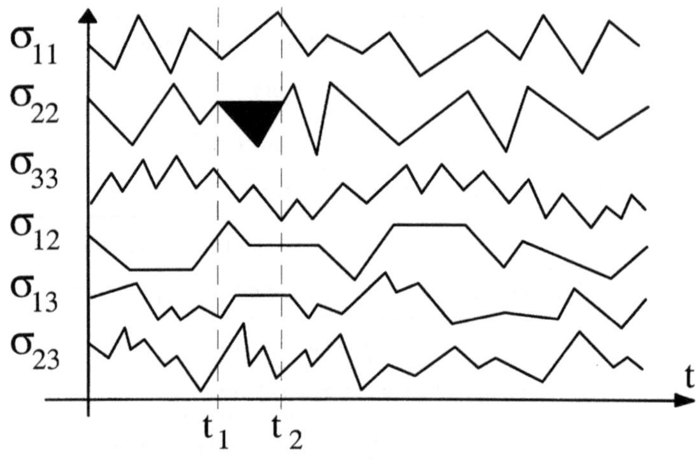

Fig.3 Problem of the identification of cycles inside a multiaxial sequence.

The proposed way to overcome this difficulty is to define a counting variable V(t), which must clearly represent both of the stress states and their evolution versus time. Cycles are then Rainflow-counted by the use of the French AFNOR recommended procedure **(5)**.
The first idea was to choose as the counting variable the projection of the octahedral shear stress onto the deviatoric plane **(2)** to build the counting variable sequence. But Kenmeugne **(1)** showed that this variable does not allow identification of the real

period of cycles in some particular cases of stress states. Then he proposed to use the normal stress acting on a fixed (relative to the body) physical plane as the counting variable (Fig.4). The physical plane is represented by its unit normal vector $\vec{h}$ and the normal stress that is acting on it is denoted σ_{hh}. This counting variable avoids the periodic mistakes of the particular cases of stress states that make the first counting variable fail. The only difficulty which occurs, is to find the right fixed plane where the more conservative fatigue life result is obtained. This point will be discussed further, when the multiaxial stress sequences is considered.

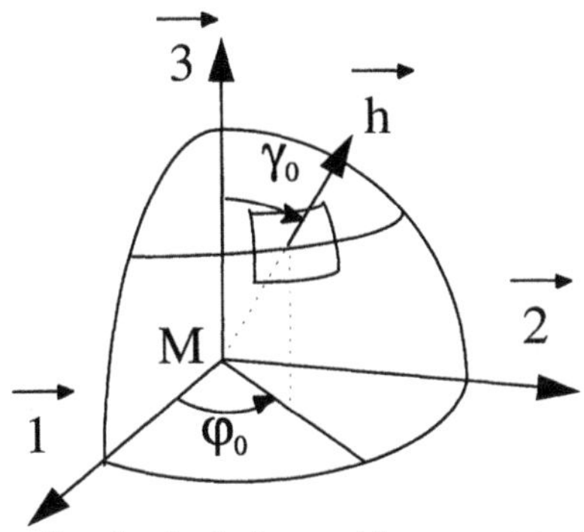

Fig.4 Location of a physical plane with respect to the body frame.

A validity condition is induced by the Rainflow counting procedure. It consists of the fact that the counting variable must not remain constant when the stress tensor is varying. A preliminary test, previous to the cycles counting, is thus realised for the choice of the counting plane.
When a cycle is identified within the counting variable sequence, the multiaxial time-corresponding stress states are extracted from the multiaxial sequence and are considered as a multiaxial stress cycle (Fig.5).

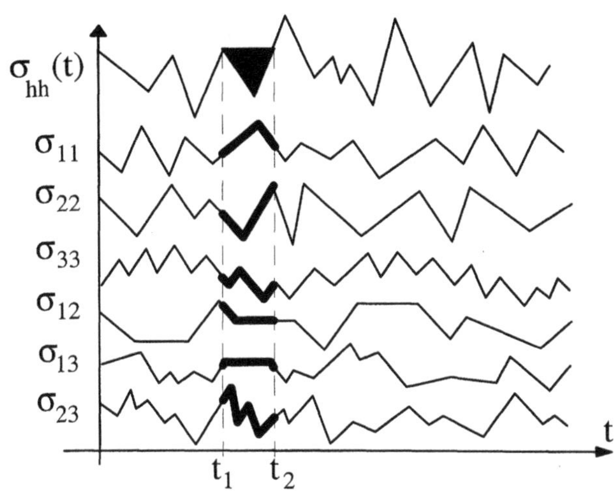

Fig.5 Extraction of a multiaxial stress cycle corresponding to a cycle experienced by $\sigma_{hh}(t)$.

FATIGUE LIFE ASSESSMENT OF MULTIAXIAL CYCLES

When a multiaxial cycle is identified and extracted from the multiaxial sequence by Rainflow application, the second main step calculates its fatigue life by way of a multiaxial fatigue criterion. Then the damage is assessed by way of a damage law.
A multiaxial criterion is a fatigue function E that allows comparison with any multiaxial stress cycle with the N cycles-fatigue strength of the material. E equals the unit value when the stress cycle reaches the fatigue strength; E is smaller (respectively greater) than the unit value if the fatigue strength is not reached (respectively is exceeded). From this point of view the applicability domain of the fatigue criteria has been extended from the infinite fatigue lives (endurance domain) to finite ones **(2, 6)**.
A criterion can be written as:

$$E\left(\sigma_{ij}(t), \sigma_{-1}(N), \tau_{-1}(N), \sigma_0(N)\right) = 1 \qquad (1)$$

The fatigue function depends on the cycle stress states tensor $\sigma_{ij}(t)$ and some material fatigue data which are three S-N curves $\sigma_{-1}(N)$, $\tau_{-1}(N)$ and $\sigma_0(N)$. These give the material fatigue strengths corresponding to N cycles for a reversed tensile test (R=-1), a reversed torsion test (R =-1) and a zero to maximum tensile test (R = 0) respectively. When E is equal to unit, the fatigue life of the cycle $\sigma_{ij}(t)$ is equal to N cycles. It is calculated directly from Eq.(1) by way of an implicit algorithm. Two formulations of criteria have been proposed by the INSA Laboratory of Solids Mechanics.
The first one proposed by Fogue **(7)** in 1987 is a Global Approach (GA). Fogue defines a fatigue indicator E_h for any physical plane P where unit normal vector is $\vec{h}$:

$$E_h = \frac{1}{\sigma_{-1}(N)}\left[a(N)\tau_{ha} + b(N)\sigma_{hha} + d(N)\sigma_{hhm}\right] \qquad (2)$$

where a(N), b(N) and d(N) are criterion parameters depending on $\sigma_{-1}(N)$, $\tau_{-1}(N)$ and $\sigma_0(N)$. They are determined by stating that the criterion is checked (E=1) for these three basic N cycles fatigue strengths.
τ_{ha} is the shear stress amplitude acting on the plane $P(\vec{u}, \vec{v})$, σ_{hha} is the normal stress amplitude, σ_{hhm} is the mean normal stress acting on P during the cycle. τ_{ha} is obtained by building the surrounding circle to the loading path, i.e. the tip of the shear stress vector acting on the plane P during the whole cycle (Fig.6). τ_{ha} is the radius of this surrounding circle.
The fatigue function (E_{GA}) of the criterion is given by the root mean square of its fatigue indicator E_h all over the possible planes through the calculation point M:

$$E_{GA} = \sqrt{\frac{1}{S}\int_S E_h^2 dS} \qquad (3)$$

where S is the area of a sphere surrounding M and which radius is equal to unit ($S = 4\pi$).

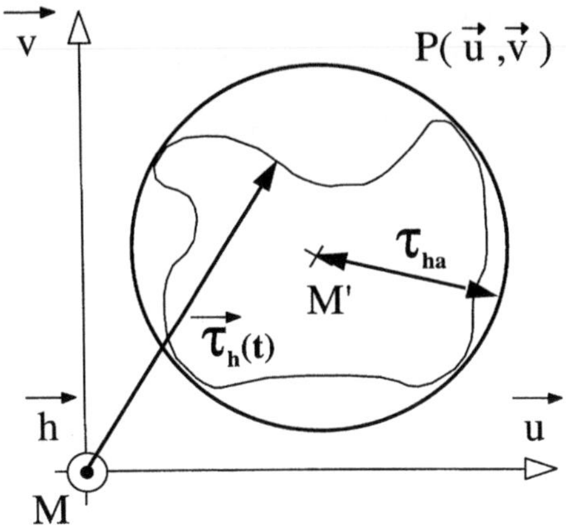

Fig.6 The surrounding circle to the loading path.

The second criterion proposed by Robert **(2)** is based on the Critical Plane Approach (CPA). The fatigue indicator $E_h(t)$ is time dependent and is a linear combination of the components of the stress vector acting at time t on the surface element which unit normal vector is $\vec{h}$:

$$E_h(t) = \|\vec{\tau}_{ha}(t)\| + \alpha(N)\sigma_{hha}(t) + \beta(N)\sigma_{hhm} \tag{4}$$

whereas its maximum value, E_h at time t on the considered plane is:

$$E_h = \frac{1}{\theta(N)} \max_t [E_h(t)] \tag{5}$$

where $\alpha(N)$, $\beta(N)$ and $\theta(N)$ are criterion parameters depending on $\sigma_{-1}(N)$, $\tau_{-1}(N)$ and $\sigma_0(N)$, $\|\vec{\tau}_{ha}(t)\|$ is the alternate shear stress versus time t acting on the plane $P(\vec{u}, \vec{v})$ (Fig.7), $\sigma_{hha}(t)$ is the alternate normal stress versus time t, σ_{hhm} is the mean normal stress during the cycle.

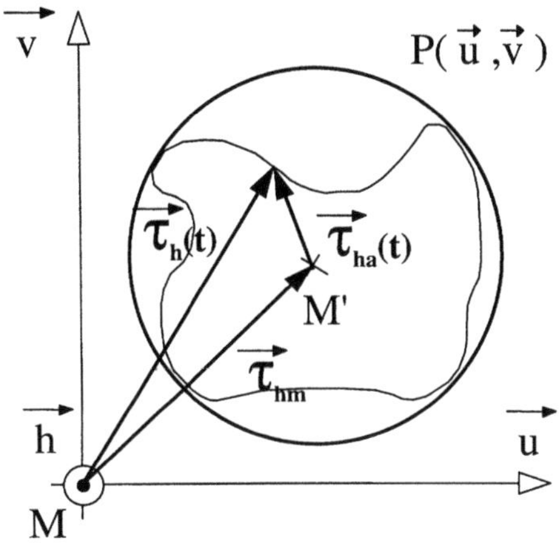

Fig.7 Definition of the alternate shear stress vector $\vec{\tau}_{ha}(t)$.

The concept of the criterion is to search the material plane where the fatigue indicator E_h is maximum. The fatigue function is hence obtained by:

$$E_{CPA} = \max_{\vec{h}}\left[E_h\right] \qquad (6)$$

A previous work (Robert, Fogue & Bahuaud **(8)**) has shown that the criterion based on the Critical Plane Approach is especially suitable when principal stress directions remain fixed during the cycle relative to the body, as the most activated slipping plane is always the same. The Global Approach gives the best description of the fatigue behaviour of the material when principal stress directions rotate during the cycle, relative to the body, because in that case several slipping plane are activated. The root mean square, which makes a quadratic average of the fatigue indicator, is a way to take into account that physical phenomenon.

Figs.8 and 9 give the distributions of the fatigue indicator E_h of the Global Approach criterion for a fixed principal stress directions cycle and a rotating principal stress directions cycle respectively. The distributions include all the possible physical planes which unit normal vector $\vec{h}$ may be defined by two angles φ and γ (see Fig.4 where these angles are denoted φ_0 and γ_0 respectively). The distributions obtained for the Critical Plane Approach criterion which are not given here have similar shapes. The most important points are that, a finite number of physical planes are critical in the first case (fixed principal directions), whereas a large set of physical planes are equally critical in the second case (rotating principal stress directions).

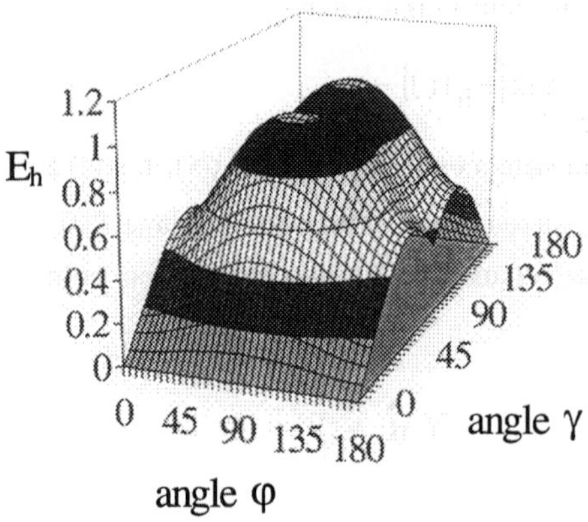

Fig.8 Distribution of the fatigue indicator E_h. Fixed principal stress directions.

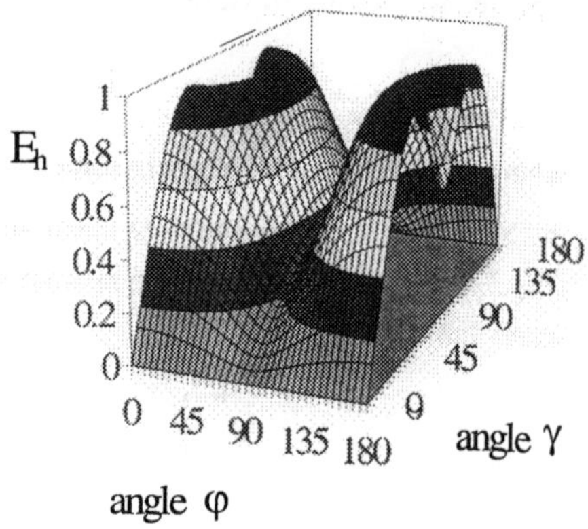

Fig.9 Distribution of the fatigue indicator E_h. Rotating principal stress directions.

LINEAR AND NON-LINEAR DAMAGE LAWS

For this third step of the method, two damage and accumulation rules will be used. The first one is the linear Miner's rule **(9)**. The second one is the non-linear Lemaitre and Chaboche damage law **(10, 11)**.

Linear Miner's rule

The damage d_k induced by a cycle extracted from the multiaxial sequence is obtained from its fatigue life N_k as:

$$d_k = \frac{1}{N_k} \tag{7}$$

The damage accumulation is defined by the summation of the damage d_k of all the cycles (Eq.(8)). The fatigue life N of the whole sequence corresponds to the number of repetitions of this stress history up to crack initiation and is obtained by Eq.(9).

$$D = \sum_k d_k \tag{8}$$

$$N = \frac{1}{D} \tag{9}$$

Non-linear Lemaitre & Chaboche's rule

This law can not be used with its initial formulation as it is designed by its authors for uniaxial stress states sequences only. Its necessary adaptation to multiaxial cycles is hereafter proposed. A multiaxial fatigue criterion gives the fatigue life N_k of any cycle and so gives the corresponding fatigue strength $\sigma_{-1}(N_k)$. The procedure defines this way a uniaxial cycle equivalent to the initial one from the fatigue life point of view. Then the non-linear law can be applied.
The Lemaitre and Chaboche's rule which is developed further is regarded as having many advantages such as taking into account the cycles occurrence order, considering cycles below and over the so-called fatigue limit in different manners, and presenting a non-linear damage accumulation. The differential expression of the law gives the increase of damage δD due to δN identical uniaxial stress cycles defined by their amplitude σ_a and their mean value σ_m, as follows:

$$\delta D = \left[1-(1-D)^{\beta+1}\right]^{\alpha}\left[\frac{\sigma_a}{M_0(1-b\sigma_m)(1-D)}\right]^{\beta}\delta N \tag{10}$$

with

$$\alpha = 1 - a\left\langle \frac{\sigma_a - \sigma_A(\sigma_m)}{R_m - \sigma_a - \sigma_m} \right\rangle \tag{11}$$

$\sigma_A(\sigma_m)$ is the fatigue limit for a non-zero mean stress σ_m, corresponding to the endurance constant life diagram that describes the dependency between σ_m and σ_a according to a linear model (Fig.10):

$$\sigma_A = \sigma_{-1}(1 - b\sigma_m) \tag{12}$$

b, β, a and M_0 are material coefficients. R_m is the ultimate tensile strength. σ_{-1} is the fatigue limit of the material for a reversed tensile test (R = -1).
The symbol $< >$ is defined as $<g>=0$ if $g\leq 0$ and $<g>=g$ if $g>0$. It gives the expressions of damage for both large amplitude cycles ($\sigma_a>\sigma_A$) and small amplitude cycles ($\sigma_a\leq\sigma_A$).

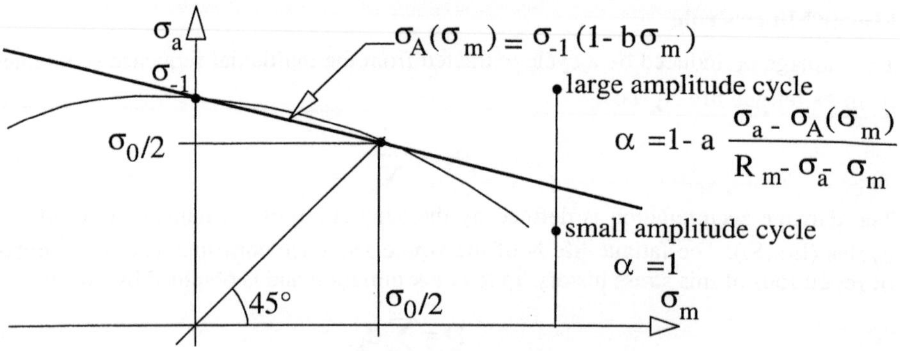

Fig.10 Endurance constant life diagram.

Equation (10) is integrated. In case of n_j successive small amplitude cycles (σ_{aj}, σ_{mj}, $\alpha = 1$), the damage increases from D_i to D_j. Stating $X_i=1-(1-D_i)^{\beta+1}$, it can be expressed as:

$$X_j = X_i e^{n_j(\beta+1)\left[\frac{\sigma_{aj}}{M_0(1-b\sigma_{mj})}\right]^\beta} \tag{13}$$

In the case of large amplitude cycles, the damage increase due to n_j cycles (σ_{aj}, σ_{mj}) is expressed as:

$$X_j^{1-\alpha_j} - X_i^{1-\alpha_j} = n_j(1-\alpha_j)(\beta+1)\left[\frac{\sigma_{aj}}{M_0(1-b\sigma_{mj})}\right]^\beta \tag{14}$$

X_i is a parameter that allows continuation of the damage evolution even if the corresponding damage D_i is not directly known. From Eq. (14) the fatigue life N_{fj} of a reversed tensile test ($\sigma_m=0$, $D_i=0$ and $D_j=1$) can be derived:

$$N_{fj} = \frac{1}{aK_j(\beta+1)}\left[\frac{M_0(1-b\sigma_{mj})}{\sigma_{aj}}\right]^\beta \quad \text{with} \quad 1-\alpha_j = aK_j \tag{15}$$

Both relations (13) and (14) are simultaneously utilised as the continuos damage function X_i^a and the damage D_i are together equal to zero when the material is not damaged and equal to the unit value when the crack occurs. Two parameters $aM_0^{-\beta}$ and β are needed to calculate X_i^a. They are obtained with the $\sigma_{-1}(N)$ S-N curve. The following algorithm (Fig.11) summarises all the possibilities of the X_i^a function increases. It stops when the damage function X_i^a reaches the unit value.
Optimisation of Lemaitre and Chaboche's law algorithm **(12)** has provided strong calculation time reduction. The non-linear law is now no more time consuming than the linear one.

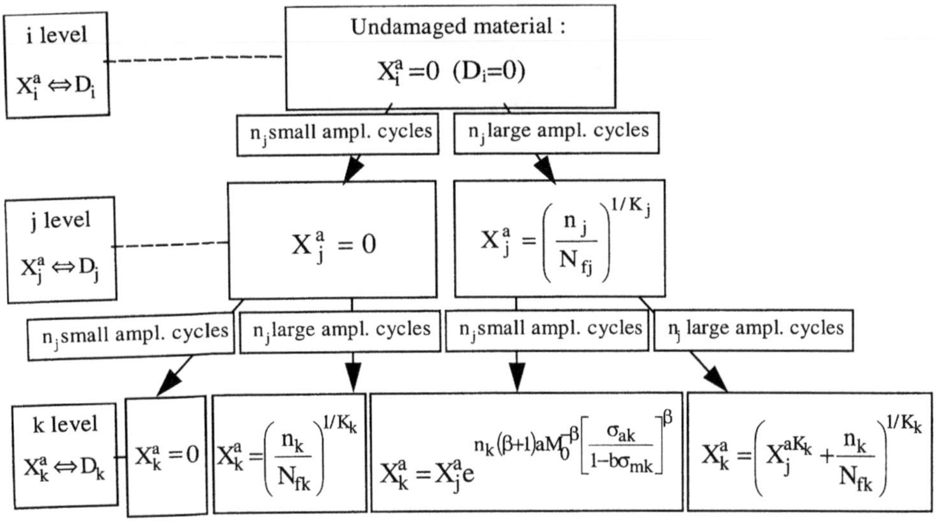

Fig.11 Algorithm of the non-linear damage function.

VALIDATION WITH BIAXIAL VARIABLE AMPLITUDE SEQUENCES

The first validation of the method has been obtained by the use of biaxial tension-compression variable amplitude tests results. Ten biaxial random stress histories are considered. They are composed of 177,000 up to 190,413 events. These tests were carried out in the laboratory of Professor Macha (Będkowski (**13**)) in Opole (Poland). The specimens have a cruciform shape as shown in Fig.12.

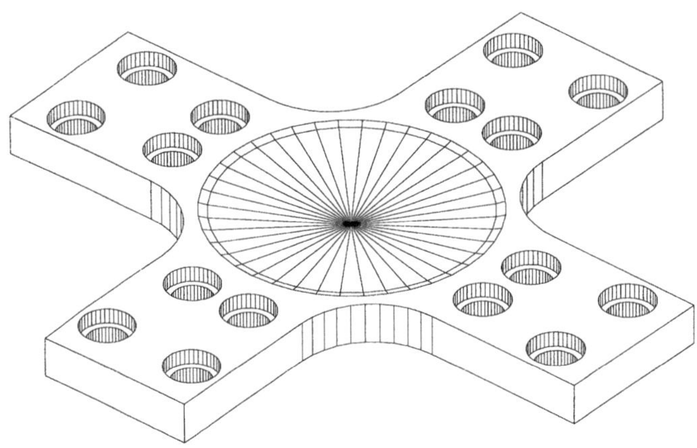

Fig.12 Cruciform specimen description.

They are made of low carbon steel 10HNAP (see chemical composition in Table 1). Mechanical static properties are presented in Table 2.

Table 1. Chemical composition

Elements	C	Mn	Si	P	S	Cr	Cu	Ni
Content [%]	0.115	0.71	0.41	0.082	0.028	0.81	0.30	0.50

Table 2. Mechanical static properties

σ_e [MPa]	R_m [MPa]	ν	E [MPa]
418	566	0.29	215 000

The required material fatigue data are the three S-N curves $\sigma_{-1}(N)$, $\tau_{-1}(N)$ and $\sigma_0(N)$. These give the material fatigue strengths versus the number N of cycles for a reversed tensile test (R= -1), a reversed torsion test (R= -1) and a zero to maximum tensile test (R= 0) respectively. Będkowski and Macha obtained the following S-N curves that are shown in Fig.13. The non-linear damage law coefficients for this material are $\beta = 4.509$ and $aM_0^{-\beta} = 2.714\times10^{-17}$.

As the biaxial variable amplitude fatigue tests involve fixed principal stress directions, the criterion based on the Critical Plane Approach is here the most suitable one for predicting fatigue lives.

Experimental lives of the ten biaxial sequences are compared with the theoretical ones (assessed by using linear and non-linear damage models). They are expressed as the number of repetitions of the sequence up to crack initiation. Table 3 summarises experimental and numerical results, which are plotted in Fig.14.

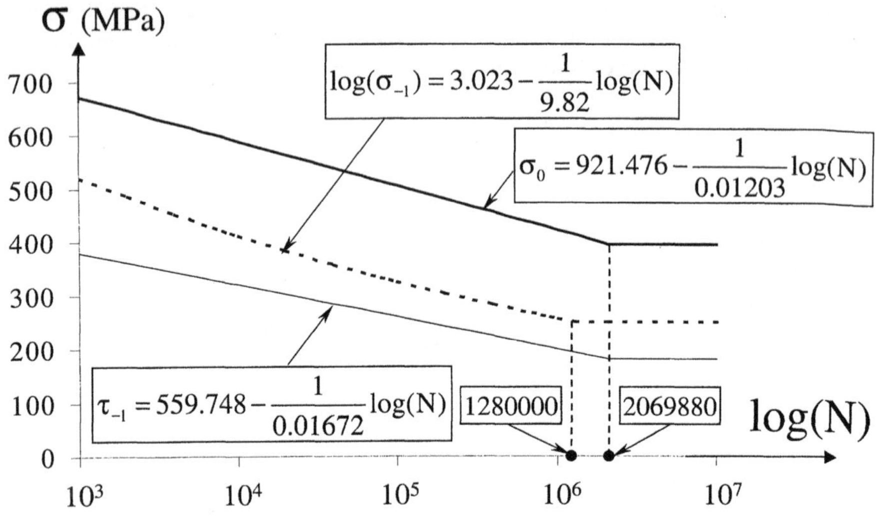

Fig.13 Reversed tensile S-N curve $\sigma_{-1}(N)$, reversed torsion S-N curve $\tau_{-1}(N)$ and zero to maximum tensile S-N curve $\sigma_0(N)$.

Table 3. Experimental and numerical lives

Sequences	Experimental fatigue life [number of sequences]	Numerical fatigue life [number of sequences]	
		Miner	Lemaitre and Chaboche
GP9302	3273	5142	650
GP9305	287	189	130
GP9307	398	652	416
GP9308	875	505	369
GP9310	1301	593	390
GP9312	2468	971	610
GP9313	1664	672	548
GP9314	848	322	239
GP9315	267	28	19
GP9619	342	308	167

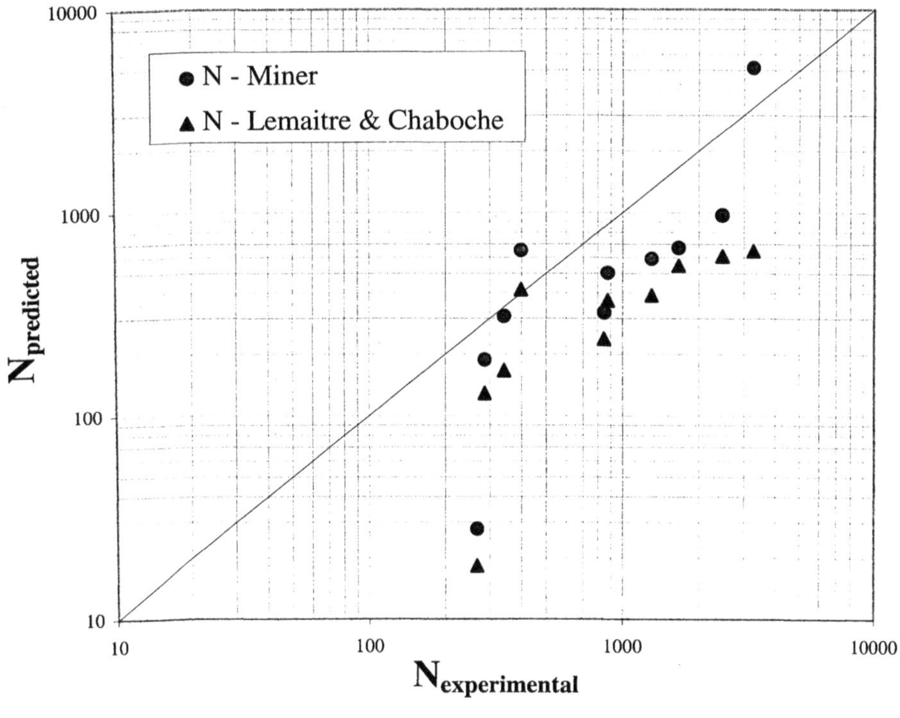

Fig.14 Assessed lives against experimental ones.

The ratio between experimental lives and predicted ones varies from 0.6 to 9.5 for the linear damage rule and 1.0 to 14.0 for the non-linear damage law. The average ratio is 2.4 and 4.0 respectively. The difference between these assessed lives is induced by the fact that small amplitude cycles contribute to the damage and make decrease the fatigue lives in the case of the non-linear law.

The lives are very conservative for one case (sequence GP9315).
The lives calculations are established by using a counting plane defined by angles $\varphi_0 = \pi/\sqrt{7}$, $\gamma_0 = \pi/\sqrt{19}$ (in radians). A study of the influence of the counting plane has been realised and has shown that the chosen plane was suitable for all the considered stress sequences as it closely corresponds to the one that gives the most conservative fatigue lives.
The problem of the determination of the most conservative counting plane is a work currently underway. The purpose is to avoid repeating the calculations for any possible counting plane. The attention is focussed on the most damaged material plane, determined by a so-called plane per plane damage accumulation **(14)**. All the material planes are not equally damaged during any stress cycle, depending on their orientation. This physical damage distribution all over the planes induced by the distribution of stresses acting on these planes, leads to the concept of the plane per plane damage accumulation. A possible way to select the counting plane is to take the most damaged plane as the reference counting one.

CONCLUSIONS AND PERSPECTIVES

A fatigue life prediction method has been proposed for multiaxial variable amplitude stress histories. It is based on a counting variable representative of the stress states and of their evolution versus time in order to identify and extract multiaxial cycles. Then finite fatigue lives criteria are used to establish the life corresponding to any multiaxial cycle. Two linear and non-linear damage laws are usable for the fatigue assessment.
Ten biaxial random stress histories issued from tests carried out on cruciform specimens allow validation the suitability of the proposed method. The ratio between experimental lives and expected ones has an average conservative value of 2.4 for Miner's rule and 4.0 for Lemaitre and Chaboche's law.
Future work will concern the determination of the most representative counting plane and the influence of the stress gradient effect on the fatigue behaviour of materials.

REFERENCES

(1) Kenmeugne B., (1996), Contribution à la modélisation du comportement en fatigue sous sollicitations multiaxiales d'amplitude variable, Thesis of the National Institute of Applied Sciences (INSA) of Lyon, Order number 96ISAL0064.

(2) Robert J.L., (1992), Contribution à l'étude de la fatigue multiaxiale sous sollicitations périodiques ou aléatoires, Thesis of the National Institute of Applied Sciences (INSA) of Lyon. Order number 92ISAL0004.

(3) Robert J.L. and Bahuaud J., (1993), Multiaxial fatigue under random loading, 5th Int. Conf. on Fatigue and Fatigue Thresholds, Montreal, Canada, pp.1515-1520.

(4) Amzallag C., Gerey J.P., Robert J.L. and Bahuaud J., (1994), Standardisation of the rainflow counting method for fatigue analysis, Int. J. Fatigue, Vol. 16, n°4, pp.287-293.

(5) Afnor, (1993), Fatigue sous sollicitations d'amplitude variable, Méthode Rainflow de comptage des cycles, Recommendation number A03-406.

(6) Vidal E., Kenmeugne B., Robert J.L. and Bahuaud J., (1996), Fatigue life prediction of components using multiaxial criteria, Multiaxial Fatigue and

Design, ESIS 21. Edited by A. Pineau, G. Cailletaud and T.C. Lindley. Mechanical Engineering Publications, London, pp.365-378.

(7) Fogue M., (1987), Critère de fatigue à longue durée de vie pour des états multiaxiaux de contraintes sinusoïdales en phase et hors phase, Thesis of the National Institute of Applied Sciences (INSA) of Lyon, Order number 87ISAL0030.

(8) Robert J.L., Fogue M. and Bahuaud J., (1994), Fatigue life prediction under periodical or random multiaxial stress states, Automation in Fatigue and Fracture: Testing and Analysis, ASTM STP 1231, (edited by Amzallag C.), American Society for Testing and Materials, Philadelphia, pp.369-387.

(9) Miner M. A., (1945), Cumulative damage in fatigue, Journal of Applied Mechanics, pp.159-164.

(10) Lemaitre J. and Chaboche J.L., (1978), Aspect phénoménologique de la rupture par endommagement, Journal of Applied Mechanics, Vol.2, n°3, pp.317-363.

(11) Chaboche J.L. and Lesne P.M., (1988), A non-linear continuous fatigue damage model. Fatigue Frac. Eng. Mat. Struct., Vol.11, n°1, pp.1-17.

(12) Weber B., Vialaton G., Carmet A. and Robert J.L., (1997), Calculation time reduction of a non linear damage rule used in variable amplitude fatigue, Third Morroco Congress of Mechanics, Tetouan, Morroco Society of Mechanics Sciences, pp.833-838.

(1) Będkowski W., (1994), Determination of the critical plane and effort criterion in fatigue life evaluation for materials under multiaxial random loading - Experimental verification based on fatigue tests of cruciform specimens, Fourth Int. Conf. on Biaxial / Multiaxial Fatigue, St Germain en Laye (France), Vol.I pp.435-447.

(14) Weber B., Clement J.C., Kenmeugne B. and Robert J.L., (1997), On a global stress-based approach for fatigue assessment under multiaxial random loading, Inter. Conf. on Engineering Against Fatigue, Sheffield (England), A.A. Balkema publishers, pp.407-414.

IV. CRACK GROWTH

A TWO DIMENSIONAL ANALYSIS OF MIXED-MODE ROLLING CONTACT FATIGUE CRACK GROWTH IN RAILS

Stanisław BOGDAŃSKI *, Jacek STUPNICKI *, Mike W. BROWN ** and Dawid F. CANNON ***

* Institute of Aeronautics and Applied Mechanics, Warsaw University of Technology, Nowowiejska 22/24, 00-665 Warsaw, Poland.
** SIRIUS, Department of Mechanical Engineering, The University of Sheffield, Mappin Street, Sheffield, S1 3JD, U.K.
***European Rail Research Institute, Utrecht, Netherlands.

ABSTRACT

In rolling contact fatigue of rails, cracks form at a shallow angle, and grow primarily in the direction of travel underneath the rail surface. Once nucleated, cracks may branch to a Mode I direction, whereas shallow cracks grow when the crack is filled with fluid. This paper attempts to model the response of shallow-angle cracks through fracture mechanics analysis, using finite element stress analysis and multiaxial fatigue tests to simulate the rolling contact history. The evolution of the history of mixed-Mode I and II stress intensity factors is derived from a 2D finite element model. Biaxial fatigue crack propagation experiments have been conducted on BS11 rail steel to investigate the effect of sequential mixed-mode loading on angled cracks. The effects of fluid in the crack, residual stress in the rail head and braking are considered to demonstrate that "squat" rolling contact fatigue cracks can only develop under prescribed loading conditions.

KEY WORDS

Rolling contact, mixed-mode, non-proportional, crack growth, fluid entrapment

INTRODUCTION

Over the last 20 years or so surface initiated rolling contact fatigue (RCF) has been increasingly observed on modern mixed traffic and high speed rail systems in countries with temperate climates. Rails made of pearlitic steels with hardness levels ranging from about 220 BHN to 380 BHN have experienced this type of fatigue damage. Once nucleated, typical cracks propagate at a shallow angle under the rail's running surface, primarily in the direction of vehicle travel. These cracks may branch either up or down; the former case leading mostly to surface spalling and the latter to predominantly Mode I driven cracks lying close to the rail's transverse plane and, in the absence of corrective maintenance action, rail fracture may occur. Shallow cracks appear to grow only when the crack is filled with fluid. It is also believed that residual stresses in the rail and wheel traction forces influence the response. This paper attempts to model the response of shallow-angle cracks through fracture mechanics analysis, using finite element stress analysis and multiaxial fatigue tests to simulate the rolling contact history. A satisfactory fatigue crack propagation prediction should model each of the above parameters.

The important role of liquid in creating the crack tip stress histories during cyclic contact loading has been discussed by Way **(1)**, Keer and Bryant **(2)**, Bower **(3)**, Kaneta and Murakami **(4)** and Bogdański et al. **(5)**. It is widely believed that liquid can influence fatigue crack growth by reducing friction between crack faces, by pressure on the crack faces as fluid flows into the crack, and by the "fluid entrapment effect" exerting hydrostatic pressure at the crack tip. The entrapment mechanism, illustrated in Fig.1, provides high levels of K_I due to the hydraulic pressure opening the crack as the wheel rolls towards the crack tip. Frequently invoked in discussions of RCF in rails **(5, 6, 7)**, fluid entrapment applies to bearings and gears also. It is assumed here that a liquid (e.g. rain water) present on the rail surface will be drawn into the crack during its initial opening, then entrapped during the subsequent phase of crack locking. Combining numerical stress analyses with experimental results provides a quantitative prediction of the fluid entrapment mechanism in terms of fatigue crack growth rate.

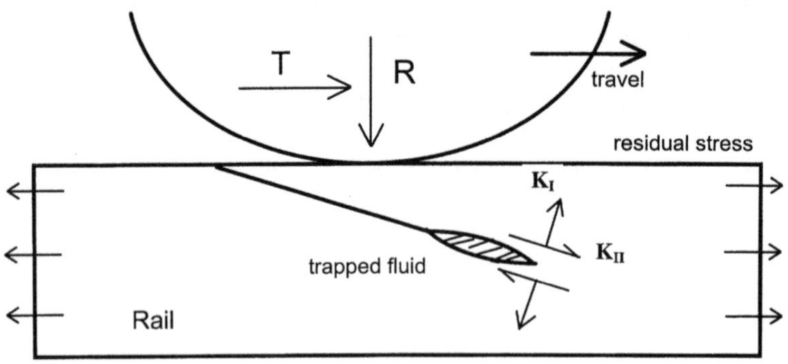

Fig.1 A schematic of the fluid entrapment mechanism for RCF in railway lines.

NUMERICAL MODELLING

The problem of rolling a wheel over a rail was modelled by three different methods. For the first method dry conditions were assumed, i.e. no liquid between the crack faces, which results in high friction. In the second and third ways liquid is present in the crack, but for the second case its only role is to lubricate the crack faces, which reduces the friction. The third way of modelling allows a certain volume of liquid to enter the crack interior in its initial phase of opening, becoming entrapped inside the crack during the phase of locking. A 2D plane strain finite element (FEM) model of a cylinder rolling over a prism (Fig.2) has been incorporated in each analysis, with contact elements at the cylinder/prism interface and along the crack faces, distributed uniformly. The crack tip is surrounded by eight special triangular elements. Boundary conditions simulating the real rail/wheel contact system were assumed, i.e. the nodes located on the bottom face of the prism are fixed in the direction perpendicular to the prism longitudinal axis, whilst the side faces of the prism are fixed in the direction parallel to this axis (Fig.2). Material properties for rail steel were modulus of elasticity E= 210 GPa and Poisson's ratio ν =0.3.

All calculations, regardless of the version of modelling were carried out with the normal load R = 90 kN, and the coefficient of friction at the cylinder/prism interface μ_s = 0.4. Various friction coefficients μ_{cr} from 0.0 to 0.4 were taken for the crack faces, under wet and dry conditions respectively. Analyses included different traction loads and residual stresses occurring in rails in practice, to investigate their influence on the values and histories of the stress intensity factors (SIFs).

It was assumed in the first method that liquid is not present in the vicinity of the crack, and hence the crack interior is free from liquid during the whole cycle of loading. Such conditions can occur in practice during sunny weather which dries the crack faces. This results in high friction between the crack faces under compressive contact loads, with an assumed coefficient of friction μ_{cr} of 0.4.

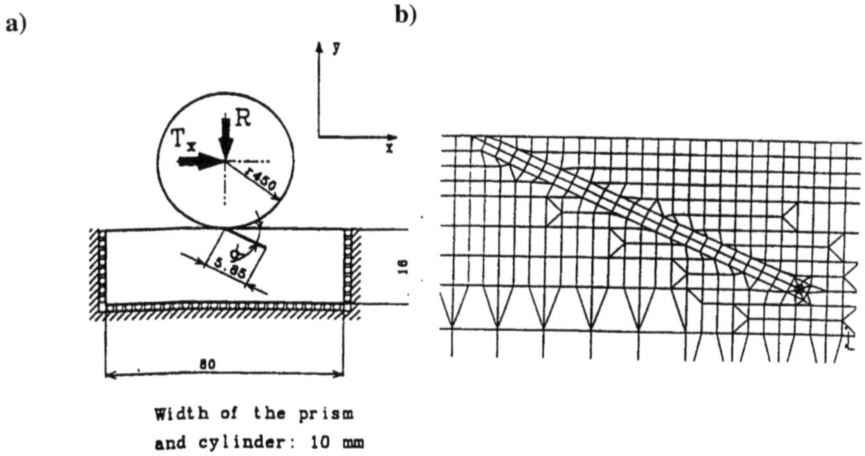

Fig.2 The contact model, a) two- dimensional model of a rail / wheel contact system, b) non-deformed finite element mesh of the prism in the vicinity of the crack, $\alpha = 25^{\circ}$.

For the second method, it was assumed that a defined volume of liquid enters the crack, reducing friction between the crack faces. No other liquid action (exertion of pressure on the crack faces, or entrapment phenomena) was taken into account here. The third method is discussed in detail in the next section.

Wet conditions - liquid entrapped in the crack interior

The entire cycle of rolling the wheel over the crack can be divided into phases; the initial phase in which the crack is open, and the second phase where the crack mouth is locked. This differs from the second analysis above only in the second phase, where liquid is entrapped inside the crack. Loading progresses as follows: the crack gradually opens as the cylinder approaches, drawing liquid into the crack interior. The opening process lasts until the cylinder touches the opposite side of the crack mouth, i.e. the apex of a wedge-shaped prism situated above the crack plane (Fig.3a).

Continued rolling of the cylinder causes the apex to bend downwards until it touches the lower crack face (Fig.3b). In the lower part of the open crack the faces remain separated from each other, creating a locked space ("bubble") which is completely filled with a volume of incompressible liquid. Further analysis is performed in steps whilst the shape of the "bubble" and the pressure inside it are determined by the equilibrium conditions. The value of the gauge pressure at the starting point, i.e. at the moment of locking the crack mouth, is taken as zero. Rolling of the cylinder causes the crack face stiction zone to spread towards the crack tip (Fig.3c), accompanied by a gradual increase of fluid pressure. In addition, it was assumed that the pressure varies linearly along the stiction zone, i.e. the gauge pressure at the first pair of nodes in contact at the crack mouth is zero, whilst the pressure at the last pair is that of the entrapped liquid.
The assumption of a linear pressure distribution relates to reality, where locking of the crack is not ideal due to the surface roughness, and hence liquid may leak slowly from the "bubble", reducing pressure along the flow path. This assumption allows estimation of upper bound SIFs, as any leakage reduces the "bubble" volume, lowering the Mode I SIF values. For high speed trains, SIFs should app-roach the upper bound.

a)
b)
c)

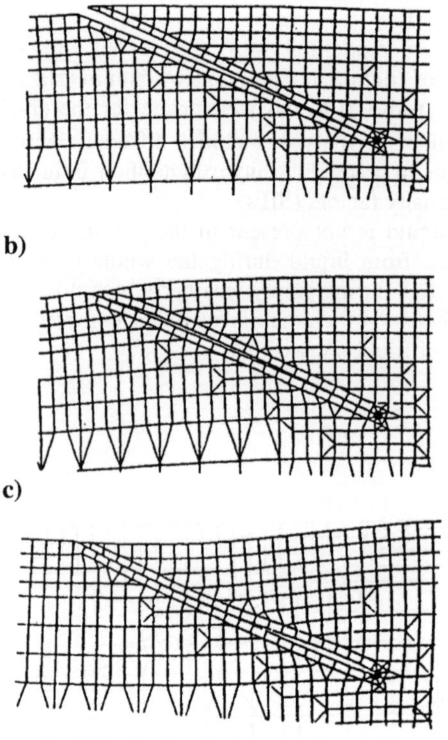

Fig.3 Typical crack shapes for selected wheel positions;
a) the and of the opening phase
b) the start of the locking phase
c) the buble moving towards the crack tip.

Mathematical model and solution

The mathematical model consists of the set of equilibrium equations which are written for every pair of contact nodes. A detailed description of the Model used in the analysis, together with the procedure of solution, was presented in **(8, 9).** Taking into account the fluid entrapment effect required some modification to the previous model. The solution to this problem leads to the sequential solution of a number of sets of equations written for the consecutive positions of the cylinder rolling over the crack, in increments of less than 0.01 mm. To ensure a reasonable accuracy of solution and stability of iteration procedure with fluid entrapment, it is necessary to apply a finer mesh in the contact zone compared to dry conditions. Before starting the main iteration procedure for solving the contact problem, the flexibility matrices for the cylinder and prism should be determined by FEM with the use of cylinder and prism finite element meshes. The iteration procedure for the analysis of the fluid entrapment effect required special care and attention, following preparatory calculations for dry conditions. The results obtained from these preliminary calculations were used as the initial data for the full analysis including the fluid entrapment effects.

Results of the SIF calculations

The purpose of this analysis was to investigate the influence of three parameters on the SIFs K_I and K_{II} and consequently on crack growth rate and direction, namely crack inclination angle, α, tangential longitudinal load, T_x (braking force), and rail residual stress, σ_r, acting separately or in combination. Plane strain calculations were carried out for dry, wet and fluid entrapment conditions for a wheel radius D/2 = 450 mm, normal load, R = 90 kN, three values of longitudinal load, T_X = 0, 9, 18 kN, and three values of residual stress, σ_r = -200, 0, +200 MPa. The numerical calculations determined the displacements, strains and stresses acting in the contact area of the cracked prism. The surface breaking cracks of length a = 5.85 mm (which is a/b = 0.87, where b = 6.7 mm is the semi-width of the contact patch) were inclined at the angles $\alpha = 15°$ and $\alpha = 25°$ to the surface of contact. The calculations were carried out for a cycle of loading covering the distance between x/b = - 2.5, and x/b = 3.3, where x is the distance of the contact patch from the crack mouth.

The comparison of the SIF histories obtained for all analysed cases yields the conclusion that the fluid entrapment mechanism has the greatest influence on the SIF ranges ΔK_{II} and ΔK_I developed during the loading cycle. Examples of the effect of this mechanism are presented in Fig.4, showing K_I and K_{II} plotted versus the location of the wheel axis in relation to the crack mouth for two angles, $\alpha = 15°$ and $\alpha = 25°$. As shown, the factor influenced most by the liquid pressure is the SIF K_I, which increased by 660% and 470% for the cracks inclined at angles of 15° and 25°, respectively. Analysing the SIF histories for the example of $\alpha = 25°$, the following should be noticed. The SIF K_I increases sharply from the moment when the locked "bubble" of liquid develops ($x \approx -6$ mm), until it reaches the value about 20.0 MPa$\sqrt{m}$ ($x \approx -1.8$ mm). Then K_I decreases to zero when $x \approx 5.6$ mm as the liquid has completely drained out from the crack. The section of the K_{II} curve for fluid entrapment with x > 5.6 mm (x/b≈ 0.8) is similar to that obtained for dry conditions. The SIF K_{II} reaches a maximum of 24.0 MPa$\sqrt{m}$ at $x \approx 2.0$ mm, and then decreases as liquid drains from the crack interior ($x \approx 5.6$ mm). The SIF K_{II} is constant and equal to 18.0 MPa$\sqrt{m}$ for the wheel positions of 5.6 mm < x < 10.0 mm, due to interfacial crack friction. This constant K_{II} value is higher than that of the case without liquid, being equal to 13.0 MPa$\sqrt{m}$.

The characteristic shape of the SIF K_{II} history can be explained by analysing the phenomena which occur during the loading cycle. With intensive liquid flow out of the crack, the crack faces are separated by the liquid film and can easily slip against each other. Similarly though not so effective, when the crack is partially locked in the stiction zone, liquid flow through this zone is possible. As a result of this flow, the tangential forces that hinder slippage of the crack faces are reduced. Finally, the shear behaviour of the crack is as if it was empty but with a reduced friction coefficient, so that the shape of the section of the K_{II} curve for -6.0 < x < 5.6 mm is similar to the curve derived for the case with no liquid and no friction $\mu_{cr} = 0$.

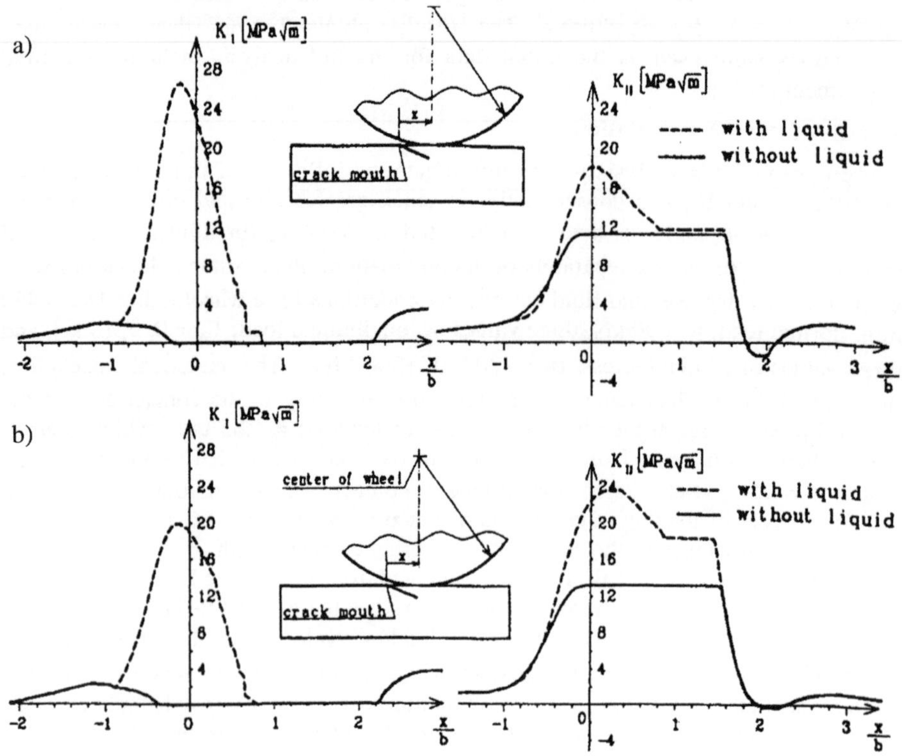

Fig.4. Comparison of K_I and K_{II} histories as the wheel rolls over the crack with/without fluid entrapment. R= 90 kN, $T_X = 0$, $\sigma_{r.}=0$, $\mu_{s.} = \mu_{cr.} =0.4$.
a) α =15°, b) α =25°.

MODELLING OF FATIGUE CRACK GROWTH

The load history experienced by a crack in rolling contact fatigue can be simulated using a biaxial cruciform specimen, with independent control of the loads on each axis. Experimental results are available for mixed-mode fatigue crack propagation in normal grade BS11 rail steel, where 45° inclined cracks in cruciform specimens subject to characteristic tensile/shear load sequences **(10, 11, 12)** emulate the histories shown in Fig.4. The results demonstrate that crack growth rate and direction are governed by the effective Mode I and Mode II SIF ranges. In rolling contact, an initial Mode I cycle is experienced as the rail bends, or as fluid is pressurised. The SIF ΔK_{Inom} is simulated by equibiaxial tension: it is followed by a Mode II cycle ΔK_{IInom} in pure shear (representing the contact patch passing over the crack tip). Generally the two waveforms may overlap. Mean stress is an additional factor that affects growth, which reflects the residual stress state in rail heads. Here ΔK_{Inom} and ΔK_{IInom} are the nominal Mode I and II SIF ranges for experiments. Details of test methods and rail steel examined are given in **(10, 11)**.

Bold **(11)** showed that when the ratio $\Delta K_{IInom} / \Delta K_{Inom}$ was greater than 2, the crack branched to follow a Mode I direction of maximum tangential stress. Thus, coplanar crack extension can be produced from this type of loading cycle with no waveform overlap under a suitable combination of ΔK_{Inom} and ΔK_{IInom}, as shown in Fig.5. (A coplanar crack on a 45° plane grows on a plane of maximum shear, whereas a Mode I crack follows a principal direction.) However when the waveforms are in-phase (180° overlap) with the Mode I and II peak SIFs occurring at the same time, initial shear mode defects in steels always deviate to become tensile Mode I fatigue cracks, because the shear mode cannot be sustained.

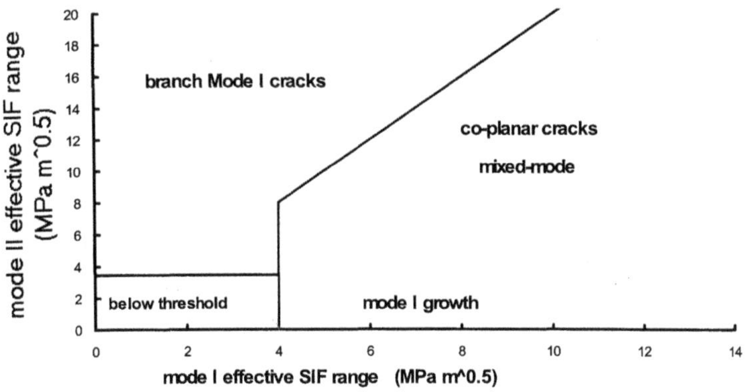

Fig.5. Three phases of fatigue crack growth under sequential mixed-mode loading.

When results are presented in terms of nominal ranges of stress intensity factor, the picture is complicated by crack closure and frictional effects. A clearer view is obtained in fatigue when effective ranges are used (Fig.5), based on values of the crack closure load. In Mode II shear, the corresponding concept is the locking up of the fracture faces due to friction and wear. Whenever crack faces are in contact, both shear and normal stresses at the crack tip are attenuated, and consequently the fatigue damage is reduced. Figure 5 indicates three regions of propagation, a) branch Mode I cracking, where rate and path are determined by the maximum tangential stress criterion, b) coplanar shear cracking on the 45^0 plane, which exhibits a characteristic very straight path due to the single slip system at the tip, and c) coplanar Mode I cracking for small ΔK_{IInom}, which shows a more tortuous path due to the operation of twin crack tip slip bands **(12).** Finally there is a threshold region, taken as a rectangular box for sequential loading after Lee **(13),** but following the curved shape for in-phase proportional stressing **(14).**

Effective measurements of the actual SIF ranges experienced by the crack tip were made in the tests. The effective Mode I and Mode II ranges, ΔK_{Ieff} and ΔK_{IIeff}, are defined as:

$$\Delta K_{Ieff} = U_I \Delta K_{Inom} = U_I Y \Delta\sigma\sqrt{\pi a} \tag{1}$$

$$\Delta K_{IIeff} = U_{II} \Delta K_{IInom} = U_{II} Y \Delta\tau\sqrt{\pi a} \tag{2}$$

where Y is the geometry factor for a central 45° crack of length 2a, and U_I and U_{II} are the proportions of applied load range or theoretical sliding displacement experienced by the crack tip, respectively.

Classical Mode I fracture mechanics tests are characterised by three parameters, the range ΔK, the mean stress *R* ratio, and the closure ratio U_I, where $R = \sigma_{min} / \sigma_{max}$ is the conventional definition for *R*. For sequential loading in RCF, values for ΔK, R and U arise for the Mode II cycle also. The waveforms overlap in Fig.4, introducing a phase angle as a seventh control parameter. (In Fig.6, the sinusoidal cycles employed always had R_I values of 0 and R_{II} values of -1, with various phase angles.) In practical situations, residual stresses may have a significant effect on RCF crack growth. Therefore, the effects of retaining some tension or compression through the shear cycle and retaining a static shear through the Mode I cycle are important. Two further load parameters are defined as S ratios, where

$$S_I = K_{Idwell} / K_{Imax} \tag{3}$$

$$S_{II} = K_{IIdwell} / K_{IImax} \tag{4}$$

Table 1 Mean stress and closure levels derived from numerical analysis.

α	fluid	T_X	σ_r					coplanar crack		branch crack	
o		kN	MPa	R_I	R_{II}	S_I	S_{II}	U_I	U_{II}	U_I	U_{II}
15	wet	0	-200	0.00	-0.83	0.00	-0.60	1.00	1.00	0.61	0.55
25	wet	0	-200	0.00	-0.75	0.00	-0.75	1.00	1.00	0.13	0.57
15	wet	0	0	0.00	-0.11	0.00	0.11	1.00	1.00	1.00	0.90
15	wet	18	0	0.00	-0.04	0.00	0.04	1.00	1.00	1.00	0.96
25	wet	0	0	0.00	-0.02	0.00	0.08	1.00	1.00	1.00	0.98
15	dry	0	-200	Mode II	-3.20	Mode II	-3.20	1.00	1.00	0.00	0.00
15	dry	18	-200	Mode II	-10.0	Mode II	-0.50	1.00	1.00	0.00	0.00
15	dry	0	0	0.00	-0.18	0.00	0.17	1.00	1.00	1.00	0.85
15	dry	18	0	0.00	0.00	0.00	0.00	1.00	1.00	1.00	1.00
25	dry	0	0	0.00	-0.04	0.00	0.17	1.00	1.00	1.00	0.96
15	dry	0	200	0.00	0.20	0.00	0.41	1.00	1.00	1.00	1.00
15	dry	18	200	0.00	0.33	0.00	0.33	1.00	1.00	1.00	1.00
25	dry	0	200	0.00	0.23	0.00	0.36	1.00	1.00	1.00	1.00

Here K_{Idwell} is the mean level of K_I during the Mode II cycle, and vice versa. The value of *S* lies between *R* and unity. Thus to define a sequential load history with sinusoidal waveforms, nine independent parameters are required, the values for ΔK, *R*, *S* and *U* for Modes I and II respectively, and the overlap angle, assuming that the cycles used are both sinusoidal. The values of characteristic parameters are given in Table 1, for the rolling contact analyses examined numerically.

EXPERIMENTAL CRACK GROWTH RESULTS FOR SEQUENTIAL MIXED-MODE LOADS

Crack growth results are presented in Fig.6 for normal grade BS11 rail steel, for the propagation of coplanar cracks. On the abscissa, the equivalent SIF plotted is $\Delta K_{eq} = \sqrt{\Delta K_I^2 + \Delta K_{II}^2}$, which is chosen purely as a convenient way to plot Mode I and II data together. Results that relate to Mode I like cracks (Fig.5) fall round the upper bound line plotted, and those which resemble the straight Mode II like cracks tend to the lower bound line. All data are contained within the factor of 2 scatter band shown as dotted lines. It is apparent that by using effective values of SIF, Eqs (1) and (2), data are bounded by the lines for the Mode I and Mode II mechanisms, irrespective of overlap angle and ratio of $\Delta K_{IInom}/\Delta K_{Inom}$, whose values are given in the key of Fig.6.

A fatigue crack growth model can be postulated for mixed-mode sequential loading, to include the three mechanisms of propagation identified in Fig.5. The Paris law describes coplanar Mode I growth with an added threshold term, assuming an intrinsic threshold of 4 MPa$\sqrt{m}$ based on data for rail steel **(15)**. Thus the upper bound line in Fig.6 becomes

$$da/dN = 0.000507(\Delta K_{eq}^{3.74} - 4^{3.74}) \qquad \text{nm/cycle} \qquad (5)$$

where the fitted constant and exponent were determined by Wong et al. **(10)** from Mode I test data. For the mixed-mode coplanar mechanism, the lower bound line gives a second crack growth relationship, which corresponds to the shear Mode law of Wong **(12)**.

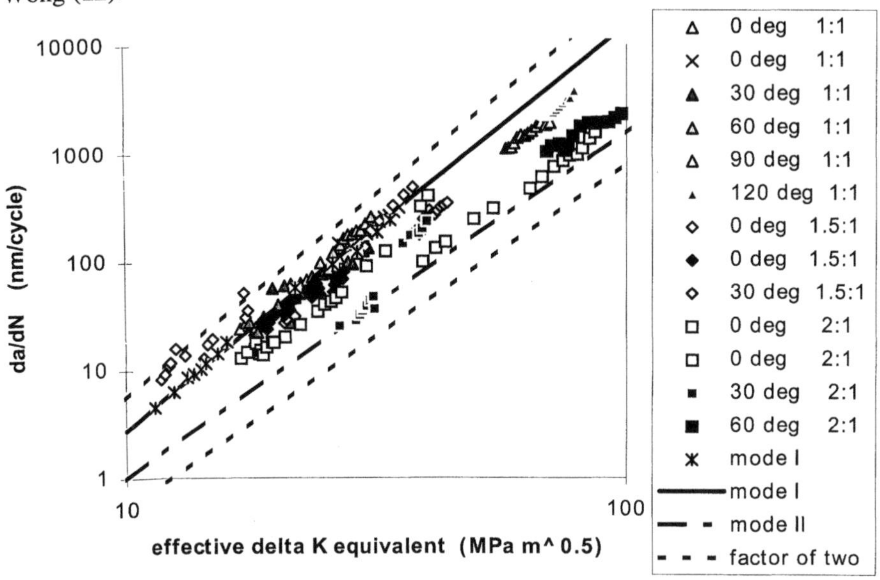

Fig.6. Coplanar fatigue crack growth under sequential mixed-mode loading for rail steel. The upper bound (solid line) is the Paris Mode I law, and the lower bound (chain dot) is a Mode II law. Key gives cycle overlap (in degrees) and ratio $\Delta K_{IInom}/\Delta K_{Inom}$.

When employing an equivalent SIF, $\Delta K_{eq} = \sqrt{\Delta K_I^2 + \Delta K_{II}^2}$, the shear Mode rule becomes

$$da/dN = 0.000614(\Delta K_{eq}^{3.21} - 4^{3.21}) \qquad \text{nm/cycle} \qquad (6)$$

A non-linear weighting is proposed to interpolate between the two bounding solutions **(16)**. Using the Mode I Paris law for fatigue crack growth in Eq (5), for coplanar propagation, an equivalent SIF range is defined such that

$$\Delta K_{eq}^2 = \Delta K_I^2 + [(614/507)\Delta K_{II}^{3.21}]^{2/3.74} \tag{7}$$

In Eqs.(5), (6) and (7), the SIFs are effective ranges. Equations (5) and (7) encompass both types of coplanar crack extension depicted in Fig.5. All available coplanar propagation data are collated in Fig.7 using the equivalent SIF of Eq.(7) in the Mode I Paris law (Eq.(5)). Results fall broadly within a factor of 2 scatter band, except where the propagation rates fall off as a crack approaches a point of branching instability. The data include tests where R_I, R_{II}, S_I and S_{II} are investigated, as well as the previous results with overlap angle and mixed-mode ratio as parameters. The degree of correlation obtained suggests that Eqs.(5) with (7) provide an effective model to predict coplanar fatigue crack growth in rolling contact.

In order to predict the onset of crack branching, Pineau's criterion of maximum crack growth rate is adopted **(17)**, as cracks follow the easiest (or fastest) path available. In torsional low cycle fatigue, cracks also adopt the path that gives the greatest speed **(18)**, either tensile or shear mode. The branch growth rate is given by Eq.(5), replacing the SIF range by the following formulae for Δk_I (the local SIF range for a branched crack tip).

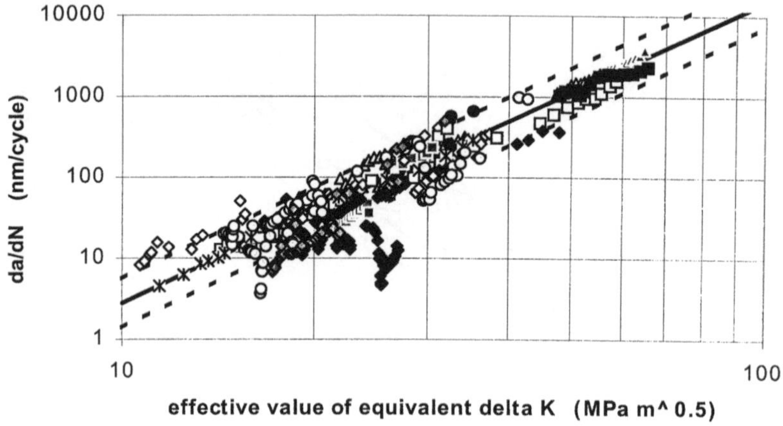

Fig.7 Correlation of coplanar fatigue crack growth data under sequential mixed-mode loading for rail steel, using the equivalent ΔK in Eq.(7).

From the maximum tangential stress criterion, for the Mode II shear cycle

$$\Delta k_I = 1.155\Delta K_{II}/(1-R_{II}) + S_I \Delta K_I/(1-R_I)$$

$$\text{or} \qquad \Delta k_I = 1.155\Delta K_{II} \qquad \text{whichever is smaller} \tag{8}$$

and for the Mode I equibiaxial cycle

$$\Delta k_I = \Delta K_I/(1-R_I) + S_{II} 1.155\Delta K_{II}/(1-R_{II})$$

$$\text{or} \qquad \Delta k_I = \Delta K_I \qquad \text{whichever is smaller} \tag{9}$$

Both SIF ranges (Eqs.(8) and (9)) contribute to the extension of branch cracks, because of the sequential nature of loading. The contributions are determined independently, then added algebraically to give overall rate, on the assumption that the increments of extension are collinear after branching occurs. (Note that the formulation of Δk_I assumes closure at zero stress, and equibiaxial Mode I stresses. Strictly collinear growth requires a zero T-stress, being the non-singular stress parallel to the crack tip.)

Prediction of rolling contact fatigue crack growth rates

Equation (5) with the equivalent SIF (Eq.(7)) furnishes a prediction model for coplanar crack speed, whereas Eq.(5) with Eqs.(8) and (9) provides a model for branch crack growth rates. These models were applied to RCF of rails by analysing the numerical results. The numerical solution takes into account the friction generated by closure and locking on the crack faces, and therefore it generates effective ranges of SIF directly, as required for the form of the Paris law in Eq.(5).

Predicted crack growth rates are listed in Table 2. The ranges of SIF were determined from histories such as Fig.4, as tabulated. Both coplanar and branch predicted rates are shown. The final two columns give the optimum propagation rate and mode of cracking, predicted by selecting the highest of the two calculated rates after Pineau's criterion.

These predictions illustrate the effect of residual stress and two coplanar crack angles on the response of wet and dry cracks. In every case, fluid entrapment gives greater growth rates, as shown in Fig.8. Coplanar or shallow angle cracks are observed in practice on high speed rail lines, showing stable growth until some set of conditions initiates a branch. In many cases stable branch growth without arrest is not sustainable, but occasionally growth persists to give a transverse crack that can lead to rail fracture. The conditions that encourage stable coplanar growth are i) a wet environment for fluid entrapment, ii) a shallow crack angle (15° here), iii) compressive residual stress along the rail, and iv) no (braking) traction. When the crack orientation is raised to 25°, branching is preferred in every case in Table 2 and Fig.8, except when the branch is suppressed by a compressive stress in the rail head.

Table 2. Predicted fatigue crack growth rates and modes.

α °	fluid	T_x kN	σ_r MPa	ΔK_I MPa$\sqrt{m}$	ΔK_{II} MPa$\sqrt{m}$	branch da/dN nm/cycle	coplanar da/dN nm/cycle	optimum da/dN nm/cycle	crack mode
15	wet	0	-200	17.0	17.6	7.1	44.6	44.6	coplanar
25	wet	0	-200	12.0	21.0	9.3	28.9	28.9	coplanar
15	wet	0	0	26.5	20.0	149.5	166.6	166.6	coplanar
15	wet	18	0	21.5	25.0	174.8	117.6	174.8	branch
25	wet	0	0	20.0	24.0	153.7	94.7	153.7	branch
15	dry	0	-200	0.0	10.5	0.1	1.1	1.1	coplanar
15	dry	18	-200	0.0	8.8	0.1	0.9	0.9	coplanar
15	dry	0	0	2.5	13.2	7.1	2.7	7.1	branch
15	dry	18	0	0.8	9.5	5.8	5.0	5.8	branch
25	dry	0	0	2.5	13.7	13.3	3.0	13.3	branch
15	dry	0	200	8.0	16.5	32.1	10.1	32.1	branch
15	dry	18	200	6.4	12.4	24.8	25.0	25.0	coplanar
25	dry	0	200	12.0	20.1	70.3	26.7	70.3	branch

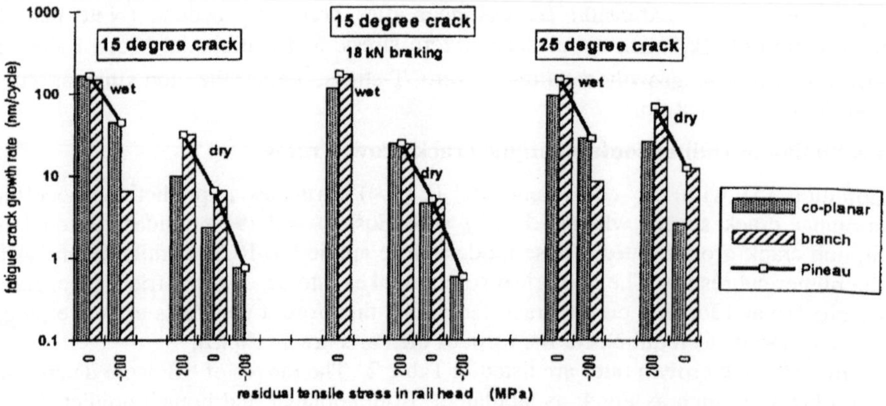

Fig.8 Prediction of growth rate and mode for various load conditions. Cracks adopt the faster rate of two calculated values, coplanar or branch (Pineau's criterion).

Traction has a small effect on the growth rates. However the largest effect on propagation is the fluid entrapment itself. The acceleration of fatigue damage is illustrated in Fig.9, where the propagation rates under wet and dry conditions are compared directly, each plotted point corresponding to fixed values of T_X, σ_r, and α. Fluid entrapment increases the growth rate obtained by an order of magnitude or more.

For the dry cracks, a friction coefficient of 0.4 was employed between the crack faces. If the sole role of the fluid was to lubricate the crack fracture surfaces, this would be best modelled by reducing the friction coefficient. An indication of the growth rates for lubricated crack faces (without the pressurisation of the fluid) was published previously by Bogdański and Brown (**16**).

Zero friction simulated the effect of full lubrication between fracture surfaces, showing that fatigue cracks were expected to branch except for shallow cracks at 25° or less. However the growth rates were 2 to 10 nm/cycle, compared to speeds approaching 200 nm/cycle in Fig.9 under fluid entrapment. As the friction was increased, the rates fell below 0.5 nm/cycle and the mode changed to coplanar.

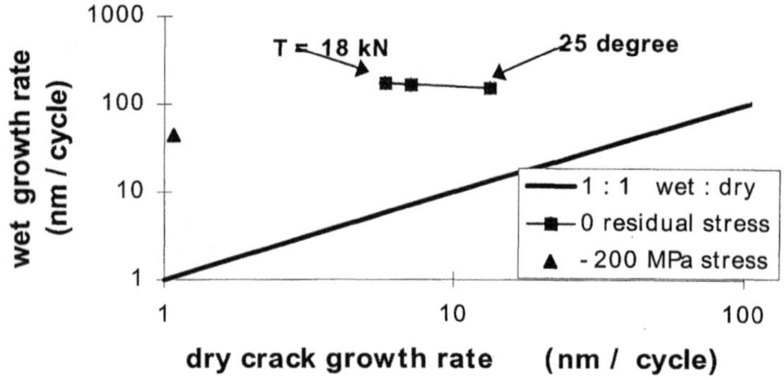

Fig.9 Comparison of wet and dry fatigue crack growth rates under sequential mixed-mode loading. Acceleration due to fluid entrapment mechanism.

DISCUSSION

The predicted results are encouraging in that they comply with experience of RCF in rails. Once nucleated, cracks may branch to a Mode I direction, either upwards to give surface spalling or downwards to produce a transverse fracture. The direction of branching is governed largely by the compressive residual stress in the rail head, which is superposed on the longitudinal tension in welded rails. Shallow cracks are able to grow when filled with fluid, due to the fluid entrapment mechanism which forces the crack faces apart, thereby permitting free shear displacement at the crack tip, which drives the mixed-mode coplanar mechanism.

As cracks dry out, they arrest or branch, and on branching they propagate slowly. Presumably with rain, a reversion to coplanar mode leaves a lot of small incipient branches in squats, which fail to extend as cracks and become wet periodically in a temperate climate. When the coplanar crack tip falls below the surface compressive layer, branching is likely if encouraged by a dry spell. However a braking force will precipitate branching with a high growth rate, to generate a stable transverse crack.

Since wheel loads and crack angles were selected as representative values for European railways, the analysis demonstrates that "squat" RCF cracks can develop under prescribed loading conditions. The integrity of rails requires that we understand the mechanisms of crack extension, and particularly the loads required for transverse cracking because such cracks can lead to fracture.

CONCLUSIONS

A model for prediction of fatigue crack growth rates in rails under rolling contact is presented. The model demonstrates that propagation of "squats" under the rail surface is feasible when a fluid entrapment mechanism is introduced, encouraging a mixed-mode shear dominated growth. However cracks may branch to a Mode I direction when the residual stress, crack inclination and braking force create favourable conditions.

REFERENCES

(1) Way S., (1935), Pitting due to rolling contact, J. Applied Mech., Trans. ASME, vol. 2, pp.49-58.

(2) Bower A.F., (1988), The influence of crack face friction and trapped fluid on surface initiated rolling contact fatigue cracks, J. Tribology, Trans. ASME, vol. 110, pp.704-711.

(3) Keer L.M. and Bryant, M.D., (1983), A pitting model for rolling contact fatigue, J. Lubr. Technol., Trans. ASME, vol. 105, pp.198-205.

(4) Kaneta M. and Murakami Y., (1991), Propagation of semi-elliptical surface cracks in lubricated rolling/sliding elliptical contacts, J. Tribology, Trans. ASME, vol. 113, pp.270-275.

(5) Bogdański, S., Olzak, M. and Stupnicki, J., (1996), Numerical stress analysis of rail rolling contact fatigue cracks, Wear, vol. 191, pp.14-24.

(6) Bogdański S., Olzak M. and Stupnicki J.,(1996), Influence of liquid interaction on propagation of rail rolling contact fatigue cracks, Proc. 2nd Mini Conference On Contact Mechanics And Wear Of Rail/Wheel Systems, (ed. I. Zobory), pp.134-143.

(7) Bogdanski S., Olzak M. and Stupnicki J., (1996), The effects of face friction and tractive force on propagation of 3D 'squat' type of rolling contact fatigue crack, Proc. 2th Mini Conference On Contact Mechanics And Wear Of Rail/Wheel Systems, (ed. I. Zobory), pp.164-173.

(8) Olzak M., Stupnicki J. and Wójcik R., (1991), Investigation of crack propagation during contact by a finite element method, Wear, vol. 146, pp.229-240.

(9) Bogdanski S., Olzak M. and Stupnicki J., (1993),. An effect of internal stress and liquid pressure on crack propagation in contact area, Proc. 6th Int. Congress on Tribology - Eurotrib '93, Budapest 30 Aug.-2 Sept. 1993, vol. 5, pp.310-315.

(10) Wong S.L., Bold P.E., Brown M.W. and Allen R.J., (1996), A branch criterion for shallow angled rolling contact fatigue cracks in rails, Wear, vol. 191, pp. 45-53.

(11) Bold P.E., Brown M.W. and Allen R.J., (1991), Shear crack growth and rolling contact fatigue, Wear, vol.144, pp.307-317.

(12) Brown M.W., Hemsworth S., Wong S.L. and Allen R.J. (1996), Rolling contact fatigue crack growth in rail steel, Proc. 2nd Mini Conference on Contact Mechanics and Wear of Rail/Wheel Systems, (ed. I. Zobory), pp.144-153.

(13) Lee S.B., (1985), A criterion for fully reversed out-of-phase torsion and bending, Multiaxial Fatigue, (eds. K.J. Miller and M.W. Brown,), ASTM STP 853, pp. 553-568.

(14) Baloch R.A. and Brown M.W., (1993), Crack closure analysis for the threshold of fatigue crack growth under mixed Mode I/II loading, Mixed Mode Fatigue and Fracture, (eds. Rossmanith, H.P. and Miller, K.J.) ESIS Publication No. 14, pp.125-137.

(15) Thompson A.W., Albert D.E. and Gray G.T., (1993), Fatigue crack growth in rail steels, Rail Quality and Maintenance for Modern Railway Operation, (eds. J.J. Kalker, D.F. Cannon and O. Orringer), pp.361-372.

(16) Bogdański S. and Brown M.W., (1997), Modelling of surface fatigue crack growth in EHD contact, Proc. Int. Conf. on Engineering Against Fatigue, Sheffield, UK, 17-21 March 1997.

(17) Hourlier F. and Pineau A., (1981), Fatigue crack propagation behaviour under complex mode loading, Advances in Fracture Research, Proc. 5th Int. Conf. on Fracture, vol. 4, pp.1841-1849.

(18) Suker D.K., A crack propagation approach to prediction of fatigue life and failure mode. Ph.D. thesis, University of Sheffield, 1994.

Acknowledgements

The authors would like to thank the British Council in Warsaw and the Polish State Committee of Scientific Research (KBN) for providing financial support for this collaborative research project. The authors are indebted to the research workers in Warsaw and Sheffield who have generated the numerical solutions and the experimental results.

STRESS INTENSITY FACTORS FOR SEMI-ELLIPTICAL SURFACE CRACKS IN ROUND BARS SUBJECTED TO MODE I (BENDING) AND MODE III (TORSION) LOADING

Manuel de FONTE *, Edgar GOMES ** and Manuel de FREITAS **

* Escola Náutica, Departamento de Máquinas Marítimas, Paço de Arcos, 2780 Oeiras, Portugal

** Instituto Superior Técnico, Dept. Eng Mecânica, Av. Rovisco Pais, 1049-001, Lisboa, Portugal

ABSTRACT

Crack growth under mixed-mode loading of rotor shafts is the reason for many failures occurring under cyclic Mode I (ΔK_I) combined with steady Mode III (K_{III}). This study presents Stress Intensity Factors calculations of semi-elliptical surface cracks in round bars subjected to bending, torsion and bending/torsion using a three-dimensional finite element model. The configuration of the semi-ellipse follows the experimentally determined equation: $b=(2s)/\pi$, where b is the crack depth and s is the semi-arc crack length. Stress Intensity Factors are obtained for bending loading and compared with available literature results in order to validate the proposed model. For Mode III, Stress Intensity Factors are obtained with the full shaft geometry due to non-symmetrical torsion loading. When the analysis is performed for combined bending and torsion, different Stress Intensity Factors are obtained at both sides where the crack front intersects the shaft surface. This result explains clearly the rotation of the crack front experimentally observed whenever a steady torsion is superimposed on the reversed or rotary bending. Results are compared with experimental crack growth measurements.

KEY WORDS

Mode I, Mode II, Mode III, Stress Intensity Factors, semi-elliptical surface cracks, round bars.

NOTATION

a,b	semi-major axis and semi-minor axis of the ellipse
A	point at maximum crack depth
B,C	interception points of crack front with free surface
B*,C*	points near points B and C
b/a	flaw aspect ratio
b/r	relative crack depth
b	crack depth, or minor semi-ellipse axis
Bs	remote bending stress

F_i	dimensionless SIF (i = I, II, III)
Ki/K_o	dimensionless SIF (i = I, II, III)
Ki	Mode I, Mode II and Mode III Stress Intensity Factors (i = I, II, III)
r	bar radius
SIF	Stress Intensity Factors
s	half crack length
2s	total arc crack length
Ts	remote torsion stress
β	interception angle between crack front and external surface
θ	position angle of crack front with external surface
ϕ	parameter angle of crack front
ν	Poisson's ratio

INTRODUCTION

Cylindrical components have many different applications in mechanical design such as shafts, bolts and screws. The behaviour of fatigue cracks in such components is dependent on the crack geometry and loading conditions. Surface cracks, with either a circumferential or a semi-elliptical shape, can cause premature failure of these structural components. These cracks occur under different types of loading: tension, bending, torsion and combined loads of tension/torsion or bending/torsion. In order to predict crack growth behaviour, the linear elastic fracture mechanics can usefully be applied since the respective Stress Intensity Factor solutions are available.

A great deal of fracture mechanics research has been devoted to Mode I crack growth. Consequently, Stress Intensity Factors solutions for a wide range of geometries in Mode I loading are reported in the literature. Concerning surface cracks in round bars, several solutions have been proposed: in Raju and Newman **(1)** Stress Intensity Factors for circumferential surface cracks in pipes and rods under tension and bending loads are presented. Shiratori et al **(2)** and Murakami et al **(3)** have presented some solutions for semi-elliptical surface cracks subjected to tension and bending. In Carpinteri **(4-7)** Stress Intensity Factors are presented for part-through cracks in round bars under cyclic combined axial and bending loading. For torsion loading of shafts with cracks, where Mode III Stress Intensity Factors is present, solutions are less common than for Mode I loading due to difficulties in obtaining analytical or numerical solutions. However, experimental studies on mixed-mode have been performed with simple geometries such as circumferentially notched round bars in cyclic bending and torsion using mainly servo-hydraulic fatigue testing machines **(8-11)**.

For semi-elliptical surface cracks in round bars subjected to Mode III loading, no solutions are available for the Stress Intensity Factors. This fact results in difficulties for the analysis of mixed-mode (I+III) fatigue crack growth in structural components where surface fatigue cracks occur under bending and steady torsion, such as power shafts. Experimental results are anyway available in Akhurst and Lindley **(12)** and Freitas et al. **(13-15)**, where fatigue crack growth tests were carried out in cylindrical specimens subjected to bending and torsion.

The aim of the present study is to obtain Stress Intensity Factors (K_I, K_{II}, K_{III}) for semi-elliptical surface cracks in round bars, subjected to bending (Mode I) and torsion (Mode III), having in view the analysis of the steady torsion on crack growth rate when combined with cyclic bending loading.

STRESS INTENSITY FACTOR CALCULATIONS

A three-dimensional finite-element analysis was performed to obtain the Mode I, Mode II and Mode III Stress Intensity Factors along the crack front for a semi-elliptical surface crack in a round bar subjected to bending, torsion and bending/torsion. The elliptical shape of the crack front was adopted with a constant arc crack length/depth ratio, according to the experimental equation obtained in previous studies (13,14), b=(2s)/π, and illustrated in Fig.1.

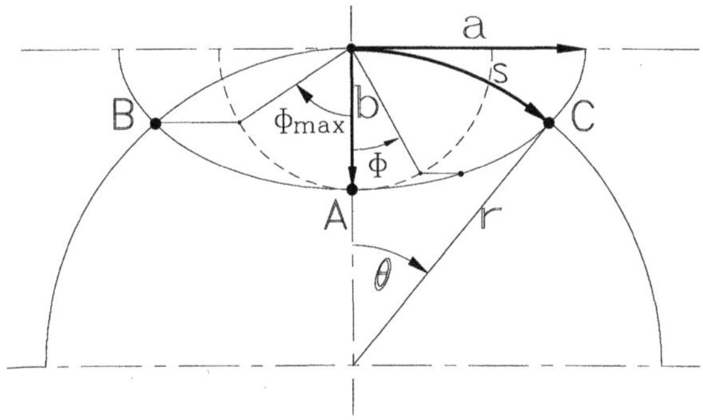

Fig.1 Geometry of a semi-elliptical crack.

Table 1 Geometric parameters of the ellipses.

Nº	θ	b/r	b/a	b/s
1	10º	0.111	0.6338	0.6366
2	20º	0.222	0.6253	0.6366
3	30º	0.333	0.6104	0.6366
4	40º	0.444	0.5878	0.6366
5	50º	0.555	0.5555	0.6366
6	60º	0.666	0.5091	0.6366
7	70º	0.777	0.4413	0.6366
8	80º	0.888	0.3325	0.6366

Table 1 presents the geometric parameters for the eight ellipses considered. The major semi-axis of the ellipse is obtained through the equation:

$$a = \frac{r.\sin\theta}{\sqrt{1 - \frac{r^2}{b^2}(1-\cos\theta)^2}} \qquad (1)$$

In this study a numerical analysis is carried out using the three-dimensional finite element method through the commercial COSMOS/M (version 1.75A) program **(16)**, for the Stress Intensity Factor calculation of semi-elliptical surface cracks in round bars subjected to bending and torsion. As a first step, Stress Intensity Factors are computed for Mode I and compared with available results **(1-4)** in order to validate the proposed model; finally, Stress Intensity Factors due to torsion (Mode III) and mixed-mode (I+III) loading are computed.

Symmetry conditions could only be used to obtain the Stress Intensity Factors in bending, as was performed in previous studies of Raju and Newman **(1)** or a quarter of the geometry as made by Carpinteri **(4)**. In the present case, the aim of the study is to obtain Stress Intensity Factors under torsion and bending/torsion. Due to both torsion loading and a semi-elliptical surface crack, it is necessary to model all the geometry of the shaft with the semi-elliptical crack placed at the mid-distance of the applied loading, as is shown in Fig.2 and Fig.3. The round bar diameter is equal to 80 mm, whereas the total length is 120 mm. A previous study was carried out where for one particular case; three lengths were considered respectively 2x30, 2x45, 2x60 mm. The difference in the calculated K values between the 2x30 mm and 2x45 mm was around 20% and between the 2x45 mm and the 2x60 mm lengths is about 5%. Larger lengths would lead to impractical computer time calculations. This geometry was performed for a round bar of 80 mm in diameter and 120 mm in total length.

The three-dimensional finite element analysis is carried out by employing 20-node isoparametric solid elements. The stress square-root singularity is modelled by shifting the finite element mid-side nodes near the crack front to quarter-point positions **(17,18)**. Elastic material properties are assumed to be equal to E = 207 GPa and 0.3 Poisson's ratio. The present finite element mesh was developed with a total of 3000 elements and 14000 nodes approximately, depending on the sizes of the considered ellipse.

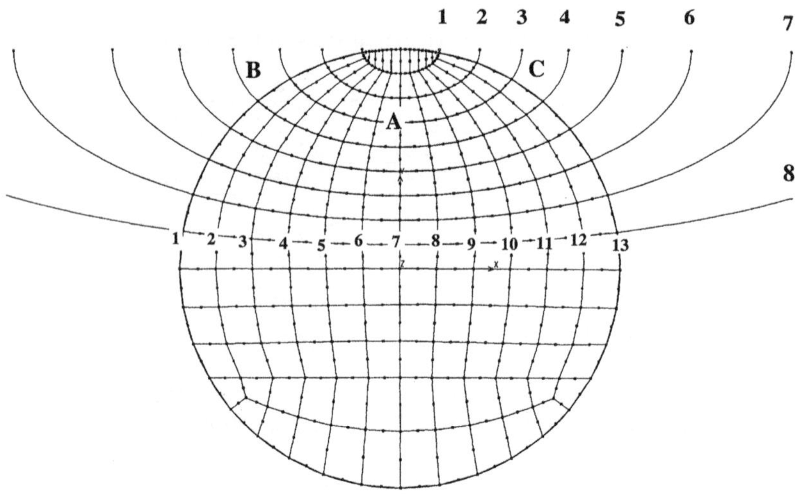

Fig.2 Eight possible positions of a semi-elliptical surface crack.

In order to compare the proposed model and the available results reported in the literature, a first series of tests were performed for pure bending load. Several solutions are available in Raju and Newman **(1)**, Shiratori **(2)**, Murakami **(3)** and Carpinteri **(4)** where the Stress Intensity Factors were obtained by different techniques (finite elements, body forces) and different parameters for the measurement of the crack, either crack depth (b) or arc crack length (s), were used.

The arc crack length (2s) has been chosen for the present study because it is the experimentally determined parameter in fatigue crack growth tests. Therefore the

Mode I Stress Intensity Factor K_I for any point along the surface crack is obtained from the following equation:

$$K_I = F_I\left(\frac{b}{a}, \frac{b}{r}, \phi\right) B_s \sqrt{\pi s} \tag{2}$$

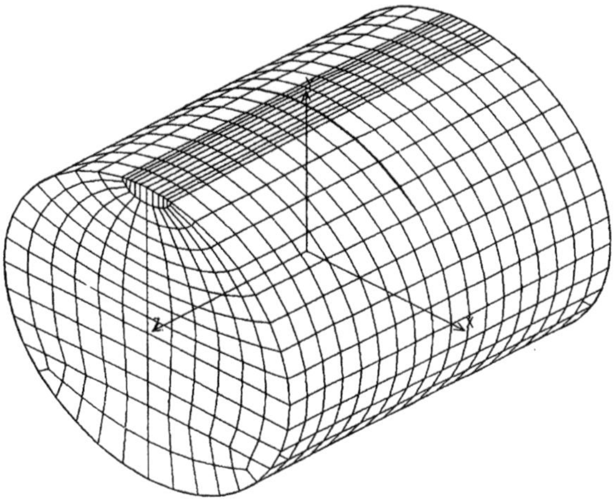

Fig.3 Finite element mesh.

where $\boldsymbol{B_S}$ is the remote applied bending stress, $\boldsymbol{s}$ is the half arc crack length and $\boldsymbol{F_I}$ is the boundary correction factor for Mode I, which is a function of the elliptical crack shape (b/a), the shaft radius r and the position along the crack front, determined through the parametric angle ϕ, see Fig.1.

Assuming that the crack always has an elliptical shape, the points on the crack front at the deepest crack depth (A) and where the crack intersects the outer side of the shaft surface (B and C) are the most important ones. Therefore the comparison will only be presented at these particular points.

It is known that when analysing a crack in 3D geometry, numerical problems may be present at free boundaries, as was shown by Carpinteri and Brighenti **(6,7)** and Pook **(19)**. It has been shown theoretically, from fracture energy considerations, that the singularity power at border (points B and C of crack front) depends on the material Poisson's ratio ν and on the intersection angle β between crack front and external surface. In the present study β=87° and it is constant for each ellipse. For Mode I loading and ν=0.3 the necessary singularity of ½ is obtained for about β=100.4°, but for Mode II and III loading β=67° as is referred in **(6,7)**.

The use of quarter point finite element (square-root singularity) does not generally produce reliable results in a boundary layer at points B and C. This effect is confined only to a small zone and within this region it is only possible to define K_I, K_{II} and K_{III} in an asymptotic sense **(19)**. Corner points values of K_i obtained from finite elements analysis are extrapolations whose precise values depend on details of method used, such element size Then we can use the nearest nodes in the mesh for points B and C in order to obtain the approximate results for K_I, K_{II} and K_{III}, and this is a procedure that is usually made. Therefore the considered results are only presented for these particular points, A, B* and C*.

Mode II and Mode III displacements and K_{II} and K_{III} values cannot exist in isolation in the vicinity of a corner point: the presence of one mode always induces the other in a fixed ratio **(19)**. As referred to above, no Stress Intensity Factors results are available for these type of semi-elliptical surface cracks in round bars subjected to torsion stress. Therefore the Stress Intensity Factors K_{II} and K_{III} are presented by employing the same geometric parameters used for Mode I (bending) obtained as follows:

$$K_{II} = F_{II}\left(\frac{b}{a}, \frac{b}{r}, \phi\right) T_s \sqrt{\pi s} \tag{3}$$

$$K_{III} = F_{III}\left(\frac{b}{a}, \frac{b}{r}, \phi\right) T_s \sqrt{\pi s} \tag{4}$$

where T_S is the remote applied torsion stress in the shaft, *s* is the half crack length and F_{II} and F_{III} are the boundary correction factor for Mode II and Mode III, respectively, which are a function of the shape of the elliptical crack (b/a), the shaft radius r and the position along the crack front, determined through the parametric angle ϕ. Once again, it is assumed that the crack always has an elliptical shape. The results for points A, B* and C* are presented in the following paragraphs.

A further study is performed in order to obtain the Stress Intensity Factors along the crack front under both remote bending and remote torsion stresses. According to linear elastic fracture mechanics, this mixed loading presents mixed-mode Stress Intensity Factors at the crack front and it is a linear function of the previous single loading cases.

RESULTS AND DISCUSSION

Bending case

Stress Intensity Factors were calculated for the geometry shown in Fig.2 and Fig.3 subjected to a remote bending stress. Eight semi-elliptical surface cracks are designed with a constant ratio b/s, whereas the angle θ ranges from 10° to 80°, according to Fig.1 and Table 1.

In order to check the accuracy of the present results, the boundary correction factor F_I defined for bending stress, was calculated for the eight semi-elliptical surface cracks and compared with the same correction factors available in the literature. Note that some of the literature results were recalculated as a function of the half arc crack length s. The comparison between the results reported in **(1-4)** and those of the present study, at the deepest point (A) and points B* and C* on the crack front is shown in Table 2 and illustrated in Fig.4 (a) and (b), respectively.

For Bs=200 MPa and the semi-ellipse number 3, the results of the Stress Intensity Factors (K_I, K_{III}, K_{III}) for the three points (A, B*, C*) along the crack front are shown in Table 3 (bending case).

The correlation between present results and the literature mean values range from 2% and 5% for ellipse 1 and 12% and 15 % for ellipse 8, for both points (A, B^* and C^*) respectively. Therefore, a satisfactory agreement is observed between such results although they were obtained according to different approaches, as mentioned above.

Table 2. Dimensionless Stress Intensity Factors F_I for the eight semi-elliptical surface crack

ELLIPSE N°	Points	Present	Shiratori	Newman	Murakami	Carpinteri
1	**A**	***0.602***	***0.603***	***0.627***	***0.618***	
b/r=0.111	B* - C*	0.568	0.530	0.569		
2	**A**	***0.599***	***0.577***	***0.601***	***0.588***	
b/r=0.222	B* - C*	0.525	0.543	*0.567*		
3	**A**	***0.564***	***0.562***	***0.590***	***0.564***	
b/r=0.333	B* - C*	0.500	0.542	0.557		
4	**A**	***0.546***	***0.559***	***0.593***	***0.542***	
b/r=0.444	B* - C*	0.485	0.540	0.573		
5	**A**	***0.555***	***0.575***	***0.611***	***0.553***	***0.596***
b/r=0.555	B* - C*	0.484	0.552	0.579		0.598
6	**A**	***0.564***	***0.613***	***0.615***		***0.635***
b/r=0.666	B* - C*	0.500	0.551	0.561		0.613
7	**A**	***0.611***	***0.676***	***0.626***		***0.691***
b/r=0.777	B* - C*	0.524	0.559	0.551		0.648
8	**A**	***0.681***	***0.767***			***0.799***
b/r=0.888	B* - C*	0.568	0.615			0.713

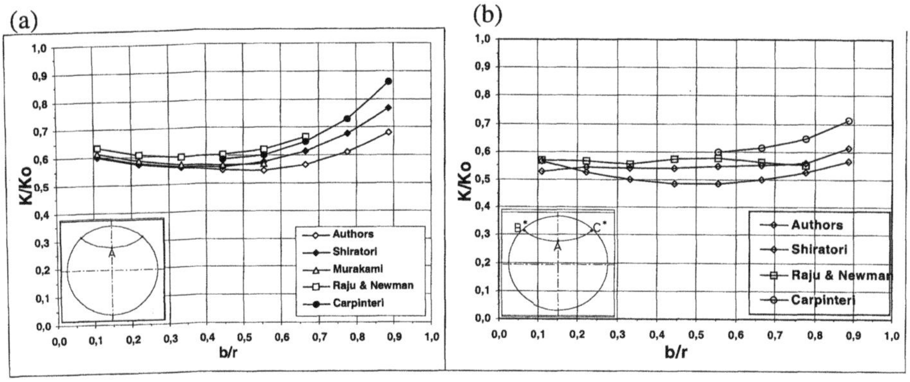

Fig.4 Dimensionless SIF (F_I=K_I/Ko) at deepest point A (a) and points B*, C* (b) for pure bending.

At points B^* and C^* identical Stress Intensity Factors are obtained which confirms the symmetry of the present geometry when subjected to a remote bending stress.
The dimensionless Stress Intensity Factors F_I are illustrated in Fig.5 along the crack front n° 3 (a) and crack front n° 8 (b).

Torsion case

Stress Intensity Factors K_I, K_{II} and K_{III}, were calculated for applied remote torsion stress, according to the same procedure used under bending stress and eight semi-elliptical surface cracks as shown in Fig.2. In order to observe and compare the effect of the applied torsion, concerning Mode I, Mode II and Mode III, a torsion stress was applied: Ts = 200 MPa.

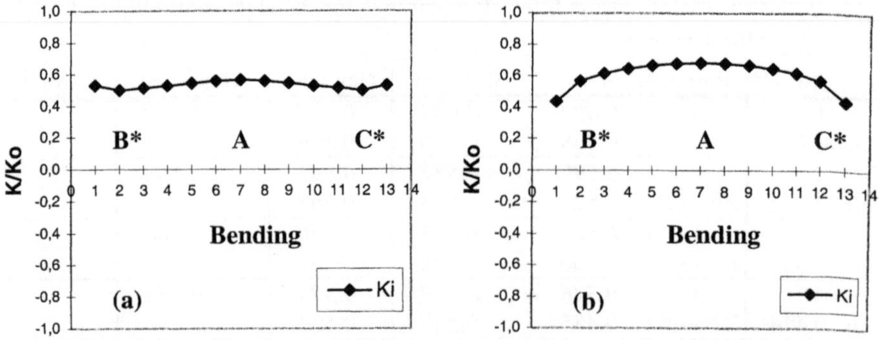

Fig.5 Dimensions SIF ($F_I = K_I/Ko$) along the crack front n° 3 (a) and n° 8 (b).

In this situation we should note an important property of circular shafts: when a circular shaft is subjected to torsion, every cross section remains plane and undistorted. In other words, while the various cross sections along the shaft rotate through different amounts, each cross section rotates as a solid rigid slab. Otherwise, when a bar of square cross section is subjected to torsion, its various cross sections are warped and do not remain plane **(20)**. When a round bar has a semi-elliptical surface crack, a non-circular section exists in the crack plane. Therefore the cross section of the crack plane warps, creating a partial helical surface and consequently a Mode I is induced by the applied torsion between point A and points B and C. Results have clearly shown that Mode I induced by the helical crack plane, due to the applied torsion, increases with the crack depth b under the effect of torsion loading, and has negligible value when the crack depth is close to zero, i. e. when the section at the crack plane is nearly circular.

Results are presented in Table 3 (torsion case) for the semi-elliptical surface crack n° 3 and show that at the point of maximum crack depth (A) a pure state of Mode III ($K_I=K_{II}=0$) is observed. For all the other points at the crack front a value of K_{II} is obtained. This is due to the elliptical crack front and the applied torsion. But a numerical value of K_I is also obtained for the points at the crack front between A and B and C. This K_I is obviously positive and is zero at point A and it is asymptotically zero close to points B and C. This K_I is a consequence of the warping of the crack plane, which increases with the crack depth as shown in Fig.6 (a) and (b), respectively for the crack front n°3 and n° 8. For this case no comparisons can be performed since, as far as we know, there are no available calculations in the literature.

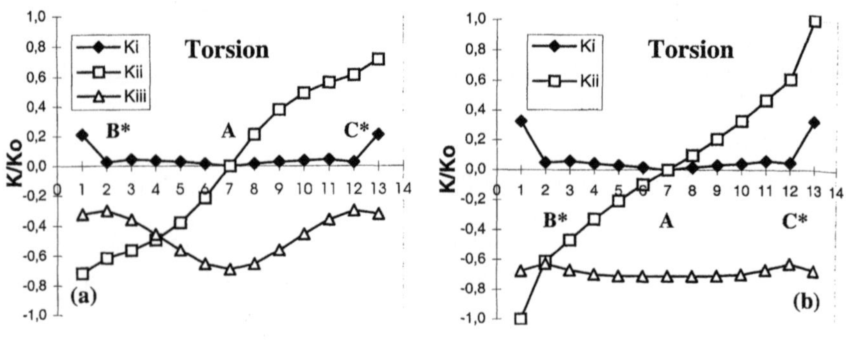

Fig.6 Dimensionless SIF ($F_i = K_i/Ko$) along the crack front n° 3 (a) and n° 8 (b).

Bending and torsion case

Finally, the shaft with a semi-elliptical surface crack is subjected to both bending and torsion stresses. The SIF, K_I, K_{II} and K_{III}, are calculated according to the same procedure as in the previous cases. Table 3 shows the results of the Stress Intensity Factors obtained for the semi-elliptical surface crack number 3 and points A, B^* and C^*, respectively for bending, torsion and combined bending and torsion. As was expected, a linear elastic behaviour is observed with the exception of the induced Mode I by torsion. At point A (maximum crack depth) the contribution to K_I is given by the bending stress while the contribution to K_{III} is obtained only from the torsion stress. At points B^* and C^* a different behaviour is observed. The numerical K_{II} and K_{III} values are the respective numerical values obtained for the torsion load cases, which results in non-symmetrical numerical values of K_{II}.

Table 3. Stress Intensity Factors K_I, K_{II} and K_{III} for semi-elliptical surface crack n° 3

ELLIPSE N° 3	K_I (MPa $\sqrt{m}$)			K_{II} (MPa $\sqrt{m}$)			K_{III} (MPa $\sqrt{m}$)		
Points	B*	A	C*	B*	A	C*	B*	A	C*
Mode I, Mode III:									
Bending (200 MPa)	25.64	**28.93**	25.64	0.00	**0.00**	0.00	0.00	**0.00**	0.00
Torsion (200 MPa)	1.39	**0.00**	1.39	-31.53	**0.00**	31.53	-15.30	**-35.48**	-15.30
Mode (I+III):									
Bending + Torsion	27.03	**28.93**	24.25	-31.53	**0.00**	31.53	-15.30	**-35.48**	-15.30

Similar behaviour for K_I would be expected, and should be symmetric under combined bending and torsion. However, K_I (positive and symmetric at these points for the case of torsion) becomes non-symmetric in the presence of a bending stress. This is due to the different helical angles of the crack plane when subjected to torsion what that not apparent for torsion case only, because of the constraint of both crack planes.

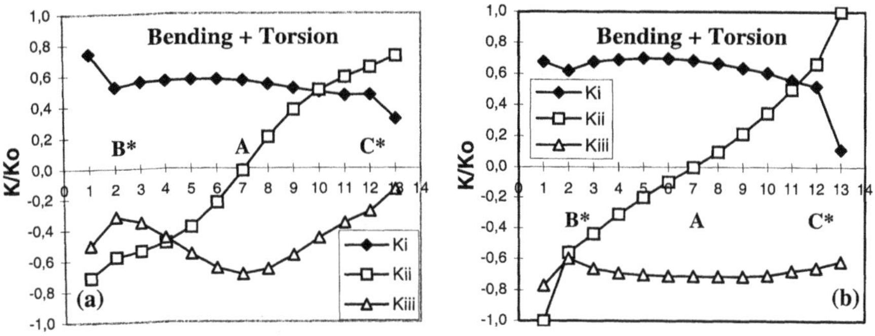

Fig.7 Dimensionless SIF ($F_i = K_i/Ko$) along the crack front n° 3 (a) and n° 8 (b) for Mode (I+III).

The asymmetry of K_I increases with the crack depth as shown in Fig.7 (a) and (b), respectively for the elliptical crack front n° 3 and n° 8.

Fig.8 shows the deformed mesh of the cracked round bar under bending, torsion and bending plus torsion. It clearly shows the opening mode for the bending case, the helical crack front induced by the torsion loading and finally for both bending and torsion loading, the different angles of the helical crack fronts from both sides of the crack, which induces the supplementary Mode I.

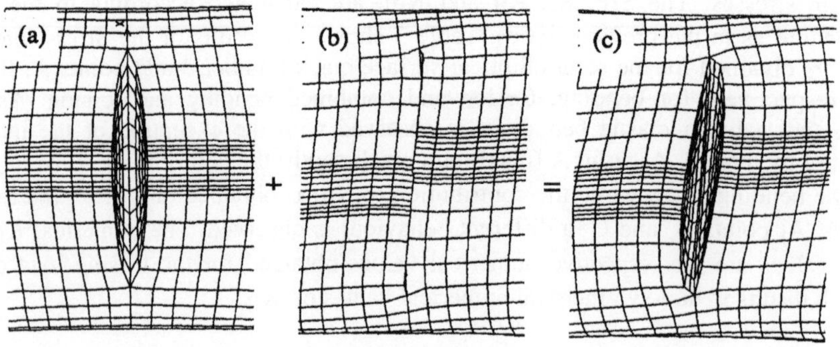

Fig.8 Deformed mesh for the crack front n° 5 in Mode I (a), Mode III (b) and mixed mode (I+III) (c).

In **(13,14)**, the authors carried out fatigue crack growth tests on cylindrical specimens of Ck45K steel, pre-cracked and subjected to reversed bending with static torsion. Some results represented in Fig.9 (a) show the expected behaviour: two crack growth rates at each side of the symmetry axis (fast and slow crack growth) for Bs=200 MPa and Ts=140 MPa. In Fig.9 (b) the fracture surface of a specimen subjected to reversed bending with steady torsion is shown, which confirms the results obtained from the finite element analysis and the non-symmetric crack growth.

Then, when performing fatigue crack growth tests under reversed bending and static torsion it should be expected that different crack growth rates would be observed at both sides of the crack front, resulting from different ΔK_I at each point near points B and C. At maximum crack depth, point A, ΔK_I results only from the Stress Intensity Factor obtained from the remote bending stress; therefore it is not affected by the static torsion stress. But with increasing torsion stresses and mainly with increasing crack depths, resulting in more pronounced non-circular sections at the crack plane; ΔK_I is different at both sides of the crack front. It should be expected that as the crack depth is larger more significant is the difference in crack growth.

CONCLUSIONS

Stress Intensity Factors for semi-elliptical surface cracks in round bars have been obtained by a three-dimensional finite element analysis. The round bar was subjected to either remote bending or remote torsion, as well as to combined remote bending and torsion loading. The semi-elliptical crack shape was modelled with a constant ratio between the arc crack length and the crack depth. Stress Intensity Factors have been obtained for several crack depths and arc crack lengths. For the bending stress loading, results have been compared with other ones available in the literature, obtained through other methods and the agreement is quite good. For the remote applied torsion new results have been obtained: at the point of maximum crack depth (A), a pure Mode III exists, while at the points where the crack intersects the free surface a mixed-mode state is observed. For combined bending and torsion, results show that, at the points of both sides where the crack intersects the free surface, the Mode I Stress Intensity Factor at point B is different from that at point C. This fact helps us to clarify experimental results obtained in fatigue crack growth tests where two different crack growth rates are observed at points B and C, respectively.

(a)

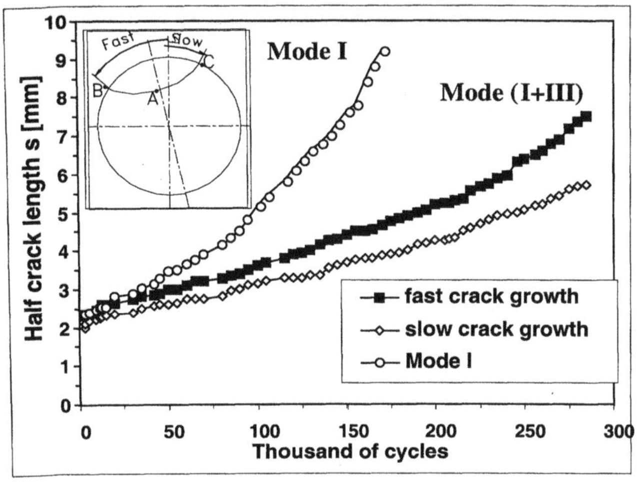

(b)

Fig.9 Mode I and Mode (I+III) crack growth (a) and crack front surface of a specimen subjected to mixed-mode ($\Delta K_I + K_{III}$) (b).

REFERENCES

(1) Raju I.S. and Newman J.S. (1986), Stress intensity factors for circumferential surface cracks in pipes and rods, Fracture Mechanics ASTM STP 905, vol.16, pp. 789-805.

(2) Shiratori M., Miyoshi T., Sakai Y. and Zhang G., (1987), Analysis and application of influence coefficients for round bar with a semi-elliptical surface crack, Stress Intensity Factors Handbook, Ed. by Y. Murakami, Pergamon Press, Oxford, vol. II, pp. 659-665.

(3) Murakami Y. and Tsuru H., (1987), Stress-intensity factor equations for semi-elliptical surface crack in a shaft under bending, Stress Intensity Factors Handbook, Ed. by Y. Murakami, Pergamon Press, Oxford, vol.II, pp.657-658.

(4) Carpinteri A., (1992), Elliptical-arc surface cracks in round bars, Fatigue Fract. Engng. Mater. Struct., vol. 15, pp. 1141-1153.

(5) Carpinteri A. (1993), Shape change of surface cracks in round bars under cyclic axial loading, Int. J. Fatigue, vol. 15, pp. 21-26.

(6) Carpinteri A. and Brighenti R. (1996), Part-through cracks in round bars under cyclic combined axial and bending loading, Int. J. Fatigue, 18, pp. 33-39.

(7) Carpinteri A. and Brighenti R. (1996), Fatigue propagation of surface flaws in round bars: a three-parameter theoretical model, Fatigue Fract. Engng Mater. Structures, 19, pp. 1471-1480.

(8) Harris D., (1967), Stress intensity factors for hollow circumferential notched round bars, J. Basic Engng. Trans. ASME, vol. 88, pp. 49-54.

(9) Nayeb-Hashemi H., Mcclintock F. A. and Ritchie R. O., (1983), Micro-mechanical modelling of Mode III fatigue crack growth in rotors steels, Int. Journal of Fracture, vol. 23, pp. 163-185.

(10) Yates J. and Miller, K., (1989), Mixed-Mode (I+III) fatigue thresholds in a forging steel, Fatigue Engng. Mater. Struct., vol. 12, n° 3, pp. 259-270.

(11) Tschegg E., Stanzl S., Mayer H. and Czegley M., (1983), Crack face interactions near threshold fatigue crack growth, Fatigue Fract. Engng. Mater. Struct. vol. 16, pp. 71-83

(12) Akhurst K., Lindley T. and Nix K., (1983), The effect of Mode III loading on fatigue crack growth in a rotating shaft, Fatigue Fract. Engng. Mater. Struct. , vol. 6, n° 4, pp. 345-348.

(13) Freitas M. and François D., (1995), Analysis of fatigue crack growth in rotary bend specimens and railway axles, Fatigue Fract. Engng. Mater. Struct., vol. 18, pp. 171-178.

(14) Fonte M. and Freitas M., (1997), Semi-elliptical fatigue crack growth under rotating or reversed bending combined with steady torsion, Fatigue Fract. Engng. Mater. Struct., 20 (6), pp. 895-906.

(15) Fonte M., (1997) "Fatigue crack growth analysis of semi-elliptical surface cracks in rotor shafts under bending and torsion", PhD. Thesis (in Portuguese), IST, Lisbon.

(16) Cosmos/M, Version 1.75, (1995), Finite Element Analysis System, Structural Research & Analysis Corporation.

(17) Barsoum R., (1976), On the use of isoparametric finite elements in linear fracture mechanics, Int. J. Num. Meth. Engng., vol. 10, pp. 25-37.

(18) Henshell R. and Shaw K., (1975), Crack-tip finite elements are unnecessary, Int. J. Num. Meth. Engng. Vol. 9, pp. 445-507.

(19) Pook L. (1993), A finite element analysis of the angle crack specimen, mixed-mode fatigue and fracture, ESIS 14 Ed. by Rossmanith and Miller, Mechanical Engineering Publications, London, pp. 285-302.

(20) Beer F. and Johnston E. (1992), Mechanics of Materials, Int. Student Edition, pp. 115-162.

CALCULATION OF STRESS INTENSITY FACTORS FOR CRACKS SUBJECTED TO ARBITRARY NON-LINEAR STRESS FIELDS

Hieronim JAKUBCZAK*, and Grzegorz GLINKA**

* Warsaw University of Technol., Inst. of Heavy Machinery Eng., Narbutta 84, 02-524 Warsaw, Poland

**Department of Mechanical Eng., University of Waterloo, Waterloo, Ontario, Canada N2L 3G1

ABSTRACT

Fatigue cracks in shot peened and case hardened notched machine components are subjected to the notch tip stress field induced by the load and the residual stress resulting from the surface treatment. Both stress fields are highly non-linear and appropriate handbook stress intensity solutions are unavailable for such configurations, especially in the case of planar surface breaking cracks. The method presented in the paper is based on the generalized weight function technique enabling the stress intensity factors to be calculated for any Mode I loading. Both the general weight functions and the calculated stress intensity factors are validated against various numerical and analytical data. The numerical procedure for calculating stress intensity factors for arbitrary non-linear stress distributions is briefly discussed as well. The method is particularly suitable for modeling fatigue crack growth of single buried elliptical, surface semi-elliptical and multiple cracks

KEY WORDS

Stress intensity factor, weight function

NOTATION

a - depth of a semi-elliptical, elliptical (minor semi-axis) or edge crack
A - the deepest point of surface, semi-elliptical crack
B - the surface point of semi-elliptical crack
c - half length of semi-elliptical or elliptical crack (major semi-axis)
K_I - mode I stress intensity factor (general)
K_{IA} - mode I stress intensity factor at the deepest point A
K_{IB} - mode I stress intensity factor at the surface point B
M_i - coefficients of weight functions (i= 1, 2, 3)
M_{iA} - coefficients of the weight functions for the deepest point A (i= 1, 2, 3)
M_{iB} - coefficients of the weight functions for the surface point B (i= 1, 2, 3)
$m(x,a)$ - weight function (general)
$m_A(x,a)$ - weight function for the deepest point A of a semi-elliptical surface crack
$m_B(x,a)$ - weight function for the surface point B of a semi-elliptical crack

Q - elliptical crack shape factor
S - external (applied) load
s - shortest distance between the point load and the crack front
t - thickness
Φ - angle co-ordinate for parametric representation of an ellipse
Ω - crack area
ρ - distance between the point load and any point A on the crack front
ρ_1, ρ_2, ρ_3, ρ_4 - geometrical parameters of a planar crack
$\sigma(x)$ - a stress distribution over the crack surfaces
σ_0 - nominal or reference stress (usually the maximum value of s(x))
x - the local, through the thickness co-ordinate
Y - geometric stress intensity correction factor

INTRODUCTION

Fatigue durability, damage tolerance and strength evaluation of notched and cracked structural elements require calculation of stress intensity factors for cracks located in regions characterized by complex stress fields. This is particularly true for cracks emanating from notches or other stress concentration regions that are frequently found in mechanical and structural components. In the case of engine components, complex stress distributions are often due to temperature, geometry and surface finish resulting in superposition of applied, thermal and residual stresses. In the case of welded or riveted structural components, it is often necessary to deal with cracked components repaired by overlapping patches. Such components require fatigue analysis of cracks propagating through a variety of interacting stress fields. Moreover, these are often planar two-dimensional surface or buried cracks with irregular shapes. The existing handbook stress intensity factor solutions are not sufficient in such cases due to the fact that most of them have been derived for simple geometry and load configurations. The variety of notch and crack configurations, and the complexity of stress fields occurring in engineering components require more versatile tools for calculating stress intensity factors than the currently available ready made solutions, obtained for a range of specific geometry and load combinations. Therefore, a method for calculating stress intensity factors for one- and two-dimensional cracks subjected to two-dimensional stress fields is discussed below. The method is based on the use of the weight function technique.

STRESS INTENSITY FACTORS AND WEIGHT FUNCTIONS

Most of the existing methods of calculating stress intensity factors require separate analysis of each load and geometry configuration. Fortunately, the weight function method developed by Bueckner **(1)** and Rice **(2)** simplifies considerably the determination of stress intensity factors. The important feature of the weight function is that it depends only on the geometry of the cracked body. If the weight function is known for a given cracked body, the stress intensity factor due to any load system applied to the body can be determined by using the same weight function. The success of the weight function technique for calculating stress intensity factors lies in the possibility of using superposition. It can be shown, **(3)**, that the stress intensity factor for a cracked body (Fig.1) subjected to the external loading, S, is the same as the stress intensity factor in a geometrically identical body with the local stress field σ(x) applied to the crack faces.

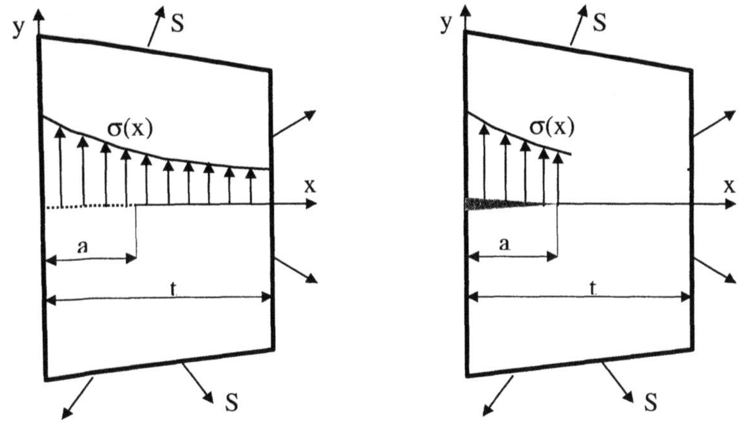

Fig.1 Nomenclature and the concept of superposition.

The local stress field, σ(x), induced in the prospective crack plane by the external load, S, is determined for *uncracked* body which makes the stress analysis relatively simple. Therefore, if the weight function is known there is no need to derive ready made stress intensity factor expressions for each load system and associated internal stress distribution.

The stress intensity factor for a one dimensional crack can be obtained by multiplying the weight function, m(x,a), and the internal stress distribution, σ(x), in the prospective crack plane, and integrating the product along the crack length 'a'.

$$K = \int_0^a \sigma(x) m(x,a) dx \tag{1}$$

The weight function, m(x,a), can be interpreted (Fig.2) as the stress intensity factor that results from a pair of splitting forces, P, applied to the crack face at position x.

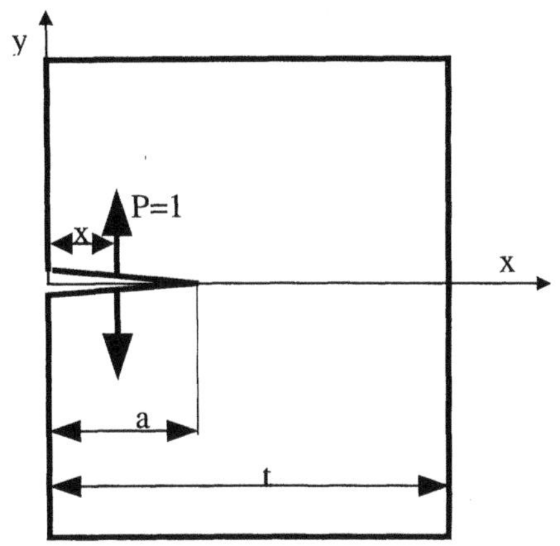

Fig.2 Weight function for an edge crack in a finite width plate; nomenclature.

Since the stress intensity factors are linearly dependent on the applied loads, the contributions from multiple splitting forces applied along the crack surface can be superposed and the resultant stress intensity factor can be calculated as the sum of all individual load contributions. This results in the integral, (1), of the product of the weight function, m(x,a), and the stress function, σ(x), for a continuously distributed stress field. A variety of one dimensional (line-load) weight functions can be found in references **(4,5,6)**. However, their mathematical forms vary from case to case and therefore they are not easy to use. Therefore, Shen and Glinka **(7)** have proposed one general weight function form which can be used for a wide variety of Mode I cracks.

UNIVERSAL WEIGHT FUNCTIONS FOR ONE-DIMENSIONAL STRESS FIELDS

The weight function is dependent on the geometry only and in principle should be derived individually for each geometrical configuration. However, Glinka and Shen **(7)** have found that one general weight function expression can be used to approximate weight functions for a variety of geometrical crack configurations subjected to one-dimensional stress fields of Mode I.

$$m(x,a)=\frac{2}{\sqrt{2\pi(a-x)}}\left[1+M_1\left(1-\frac{x}{a}\right)^{\frac{1}{2}}+M_2\left(1-\frac{x}{a}\right)^{1}+M_3\left(1-\frac{x}{a}\right)^{\frac{3}{2}}\right] \qquad (2)$$

As an example the system of coordinates and the notation for an edge crack are given in Fig.2. In order to determine the weight function, m(x,a), for a particular cracked body, it is sufficient to determine, **(8)**, the three parameters M_1, M_2, and M_3 in expression (2). Because the mathematical form of the weight function, (2), is the same for all cracks, the same methods can be used for the determination of parameters M_1, M_2, and M_3 and the integration routine for calculating stress intensity factors from Eq.(1). The method of finding the M_i parameters has been discussed in reference **(8)**. Moreover, it has been found that only a limited number of generic weight functions is needed to enable the calculation of stress intensity factors for a large number of load and geometry configurations. In the case of 2-D cracks such as the surface breaking semi-elliptical crack in a finite thickness plate or cylinder, the stress intensity factor changes along the crack front.

However, in many practical cases the deepest point, A, and the surface point, B, are associated (Fig.3) with the highest and the lowest value of the stress intensity factor respectively. Therefore, weight functions for the points A and B of a semi-elliptical crack have been derived, **(9)**, analogously to the universal weight function of Eq.(2).

- For point A (Fig.3)

$$m_A(x,a,a/c,a/t)=\frac{2}{\sqrt{2\pi(a-x)}}\left[1+M_{1A}\left(1-\frac{x}{a}\right)^{\frac{1}{2}}+M_{2A}\left(1-\frac{x}{a}\right)^{1}+M_{3A}\left(1-\frac{x}{a}\right)^{\frac{3}{2}}\right] \qquad (3)$$

- For point B (Fig.3)

$$m_B(x,a,a/c,a/t)=\frac{2}{\sqrt{\pi x}}\left[1+M_{1B}\left(\frac{x}{a}\right)^{\frac{1}{2}}+M_{2B}\left(\frac{x}{a}\right)^{1}+M_{3B}\left(\frac{x}{a}\right)^{\frac{3}{2}}\right] \qquad (4)$$

The weight functions, $m_A(x,a)$ and $m_B(x,a)$, for the deepest and the surface points, A and B, respectively have been derived for the crack face unit *line loading* making it possible to analyze one-dimensional stress fields (Fig.3), dependent on one variable, x, only.

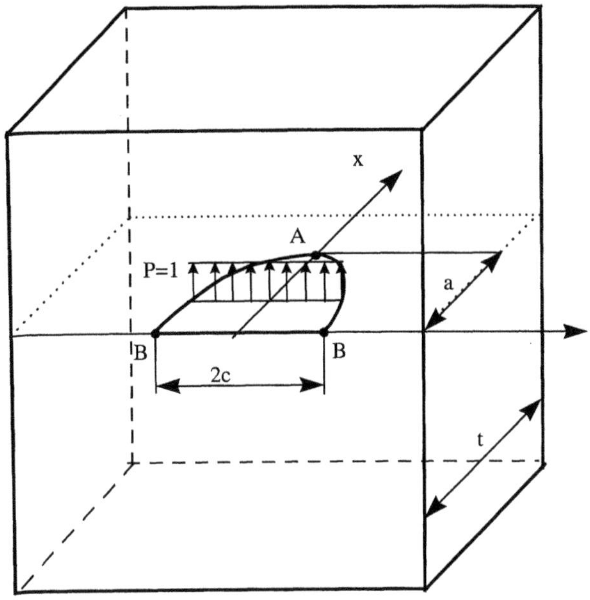

Fig.3 Semi-elliptical surface crack under the unit line load; weight function notations.

A variety of universal line load weight functions **(9-13)** have been derived and published already. The M_i parameters for the edge (Fig.2) and the semi-elliptical surface crack (Fig.3) in a finite thickness plate are given in the Appendix.

SEQUENCE OF STEPS FOR CALCULATING STRESS INTENSITY FACTORS USING WEIGHT FUNCTIONS

In order to calculate stress intensity factors using the weight function technique the following tasks need to be carried out:

- Determine stress distribution, $\sigma(x)$, in the prospective crack plane using linear elastic analysis of uncracked body (Fig.1a), i.e. perform the stress analysis ignoring the crack and determine the stress distribution $\sigma(x) = \sigma_0 f(S,x)$;
- Apply the "uncracked" stress distribution, $\sigma(x)$, to the crack surfaces (Fig.1b) as traction
- Choose appropriate generic weight function
- Integrate the product of the stress function $\sigma(x)$ and the weight function, $m(x,a)$, over the entire crack length or crack surface, Eq.(1).

The weight function (3) for the deepest point A was used to calculate stress intensity factor for the non-linear stress field (5) acting in the crack plane.

$$\sigma(x) = \sigma_0\left(1-\frac{x}{a}\right)^2 \tag{5}$$

The M_{iA} parameters for the weight function (3) are given in the Appendix (Eq A4-A6). The accuracy and the versatility of the weight function (3) for the semi-elliptical crack in a finite plate (Fig.3) is illustrated in Figs. 4 and 5, showing the comparison with the Finite Element results of Wang **(12)** and Shiratori **(14)**.

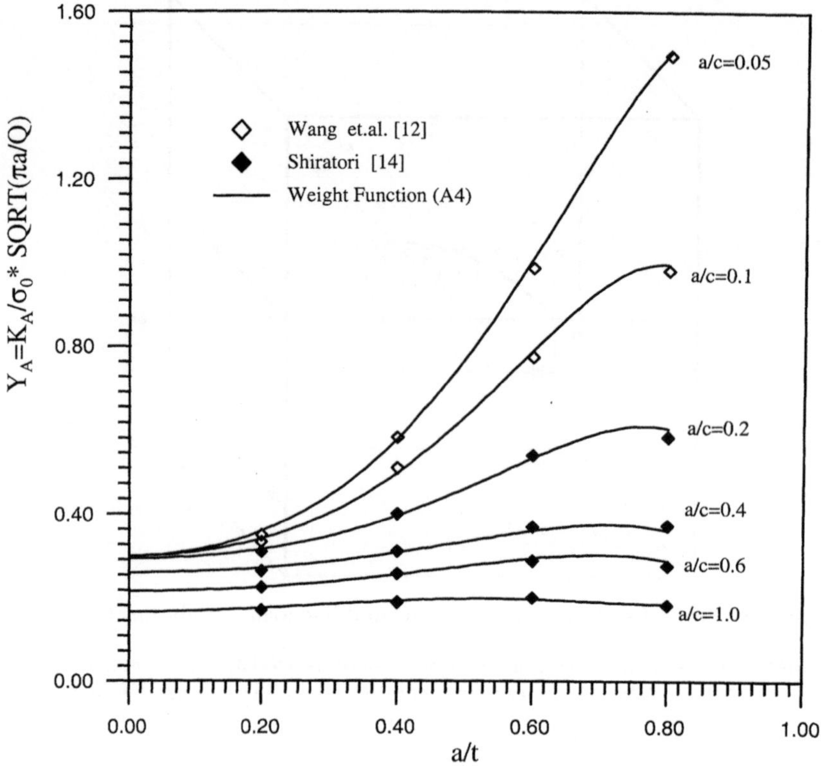

Fig.4 Comparison of the weight function based stress intensity factor and FEM data **(14)** for quadratic stress distribution; the deepest point A.

It can be seen that the agreement is good over the entire range of parameters for which the weight function parameters (A4-A6) have been derived. The parameters M_i of the weight function (3) were derived using the reference data for $0<a/t<0.9$ and $0<a/c<1$. The accuracy of the weight function while compared with the finite element data was better than 3%.

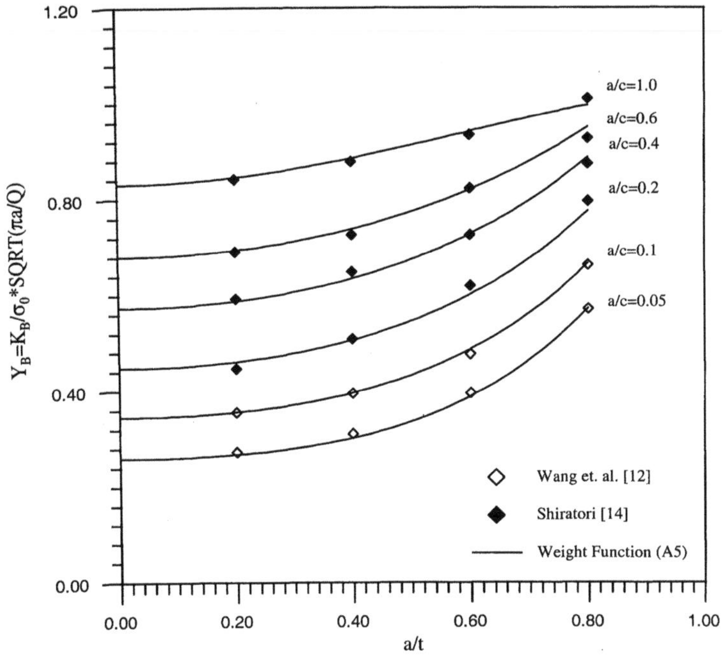

Fig.5 Comparison of the weight function based stress intensity factor and FEM data **(12, 14)**.

WEIGHT FUNCTIONS FOR TWO-DIMENSIONAL STRESS FIELDS

In spite of the efficiency and great usefulness of the line load weight functions (Figs. 2 & 3), they cannot be used in many practical cases where the stress fields is of two-dimensional nature, i.e. where the stress field $\sigma(x,y)$ in the crack plane depends on the x and y coordinates. Therefore, a weight function for the unit point load (Fig.6) is needed in order to calculate stress intensity factors for cracks subjected to two-dimensional stress fields.

A two-dimensional point-load weight function, $m_A(x,y)$, represents the stress intensity factor at point, A, on the crack front (Fig.6), induced by a pair of unit forces attached to the crack surface at point P(x,y). If the weight function is given in a closed mathematical form, it allows to calculation of the stress intensity factor at any point along the crack front. In order to determine the stress intensity factor at any point on the crack front induced by a two-dimensional stress field, $\sigma(x,y)$, the product of the stress and the weight function needs to be integrated over the entire crack surface area Ω.

$$K_A = \int_\Omega \sigma(x,y) m_A(x,y,P) dxdy \tag{6}$$

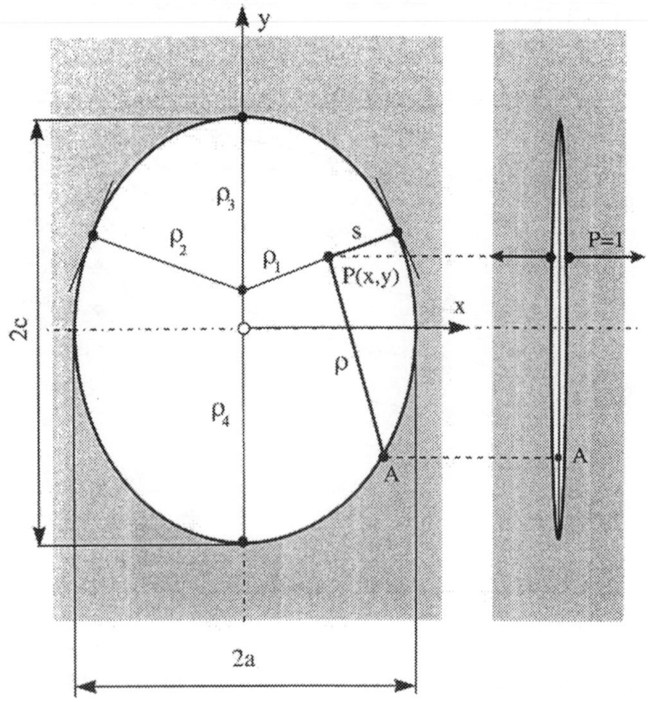

Fig.6 Elliptical crack in infinite three-dimensional body: point load weight function notation.

There are only a few point load (2-D) weight functions available. Among them the weight function for an embedded elliptical crack in an infinite body is the most often discussed in the literature **(15-17)**. Unfortunately, most of the existing point-load weight functions are given in the form of complex mathematical expressions difficult to use in practice. Therefore an attempt was made to present all the existing point-load weight functions in an uniform form making them easier for comparisons and numerical analysis.

It has been found that all the existing point load weight functions for cracks in an infinite three dimensional bodies can be expressed using the s, ρ, ρ_1, ρ_2, ρ_3 and ρ_4 parameters shown in Fig.6. It was found that the weight function for an elliptical crack in an infinite 3-D body can be written in the form of Eq.(7).

$$m_A(x,y,P) = \frac{P\sqrt{s}}{\pi^{3/2}\rho^2}\sqrt{2 - \frac{s}{4\rho_1} - \frac{s}{4\rho_2} - \frac{s}{4\rho_3} - \frac{s}{4\rho_4}} \qquad (7)$$

The weight function (7) was verified **(18)** by comparison of the weight function (7) based stress intensity factors with the semi-analytical data of Shah and Kobayashi **(19)**. The stress intensity factor results shown in Fig.7 were calculated using Eqs.(6) and (7) and three different stress fields given by expressions (8-10).

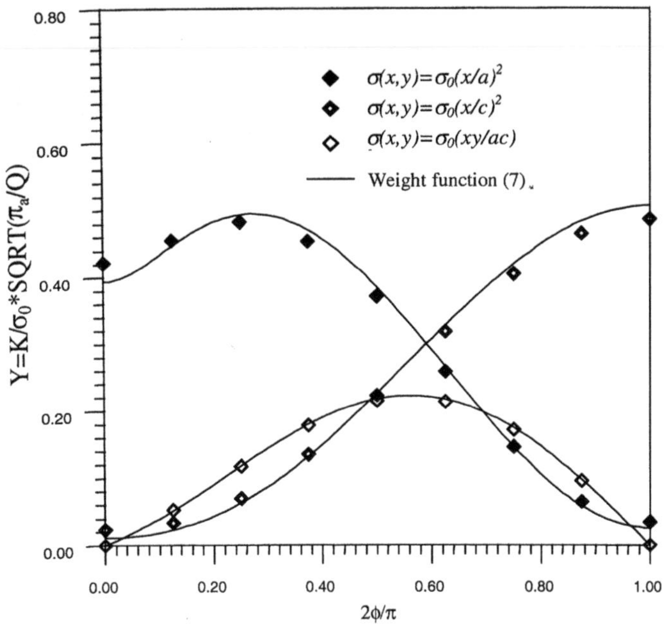

Fig.7 Comparison of the analytical **(19)** and the weight function **(18)** based stress intensity factors for an elliptical embedded crack subjected to nonlinear stress fields; a/c=0.2.

$$\sigma(x,y) = \sigma_0\left(\frac{x}{a}\right)^2 \tag{8}$$

$$\sigma(x,y) = \sigma_0\left(\frac{y}{c}\right)^2 \tag{9}$$

$$\sigma(x,y) = \sigma_0\frac{xy}{ac} \tag{10}$$

The agreement between the weight function based calculations **(18)** and the data obtained by Shah and Kobayashi **(19)** was very good for a wide range of ellipse aspect ratios a/c. The data shown in Fig.7 were obtained for a/c=0.2.

The weight function (7) can be used to derive weight functions already known in the literature. By assuming that parameters ρ_1, ρ_2, ρ_3 and ρ_4 tend to infinity, the weight function (11) for an infinite edge crack in an infinite **(20)** body can be derived (Fig.8)

$$m_A(x,y,P) = \frac{P\sqrt{s}}{\pi^{3/2}\rho^2}\sqrt{2} \tag{11}$$

By setting all the parameters $\rho_1=\rho_2=\rho_3=\rho_4=a$ the well known weight function (12) for a penny shape crack **(21)** in infinite body (Fig.9) is derived.

$$m_A(x,y,P) = \frac{P\sqrt{s}}{\pi^{3/2}\rho^2}\sqrt{2-\frac{s}{a}} \tag{12}$$

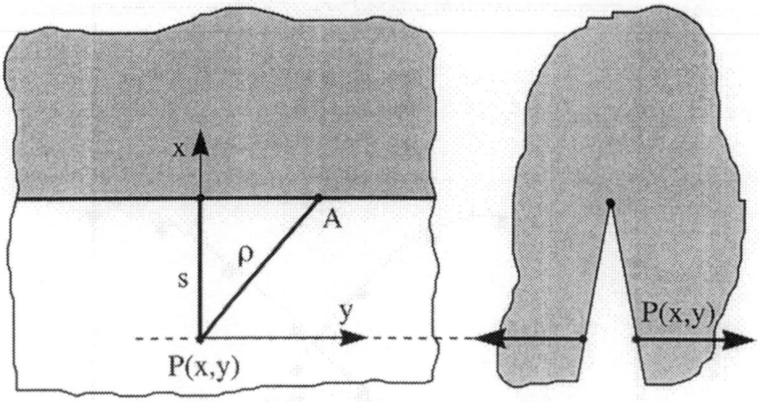

Fig.8. Infinite edge crack in an infinite body

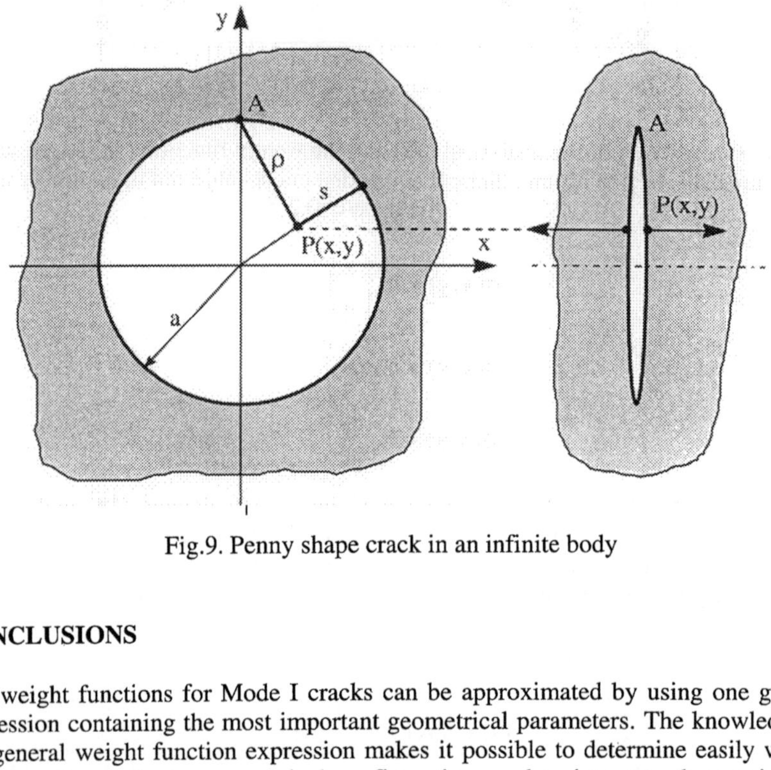

Fig.9. Penny shape crack in an infinite body

CONCLUSIONS

The weight functions for Mode I cracks can be approximated by using one general expression containing the most important geometrical parameters. The knowledge of the general weight function expression makes it possible to determine easily weight functions for particular geometrical configurations and to integrate them using the same numerical procedure. It has been found that the approximate weight functions gave accurate estimation of stress intensity factors for a variety of non-linear stress fields.

REFERENCES

(1) Bueckner H. F., (1970), A novel principle for the computation of stress intensity factors. Zeitschrift fur Angewandte Mathematik und Mechanik, vol. 50, pp.529-546

(2) Rice J. R., (1972), Some remarks on elastic crack-tip stress field, International Journal of Solids and Structures, vol. 8, pp. 751-758

(3) Broek D.,(1988), The Practical Use of Fracture Mechanics, Amsterdam,Kluwer

(4) Tada H., Paris P. and Irwin G., (1985), The Stress Analysis of Cracks Handbook, 2nd Edn, Paris Production Inc., St. Louis, Missouri, USA

(5) Wu X.R. and Carlsson A.J., (1991), Weight Functions and Stress Intensity Factor Solutions, Pergamon Press, Oxford, UK

(6) Fett T. and Munz D., (1994), Stress Intensity Factors and Weight Functions for One-dimensional Cracks, Report No.KfK 5290, Kernforschungszentrum Karlsruhe, Institut fuer Materialforschung, Dezember 1994, Karlsruhe, Germany

(7) Glinka G. and Shen G., (1991), Universal features of weight functions for cracks in mode I, Engineering Fracture Mechanics, vol. 40, pp. 1135-1146

(8) Shen G. and Glinka G., (1991), Determination of weight functions from reference stress intensity factors, Theoretical and Applied Fracture Mechanics, vol. 15, pp. 237-245

(9) Shen G. and Glinka G., (1991), Weight Functions for a Surface Semi-Elliptical Crack in a Finite Thickness Plate, Theor. Appl. Fract. Mech., Vol. 15, No. 2, pp. 247-255

(10) Zheng X.J., Glinka G. and Dubey R., (1995), Calculation of stress intensity factors for semi-elliptical cracks in a thick-wall cylinder, Int. J. Pres. Ves, & Piping, vol. 62, pp 249- 258

(11) Zheng X.J., Glinka G. and Dubey R., (1997), Stress intensity factors and weight functions for a corner crack in a finite thickness plate, Engineering Fracture Mechanics (to be published)

(12) Wang X. and Lambert S. B., (1995), Stress intensity factors for low aspect ratio semi-elliptical surface cracks in finite-thickness plates subjected to non-uniform stresses, Engineering Fracture Mechanics, vol. 51, pp.517-532

(13) Wang X. and Lambert S. B., (1997), Stress intensity factors and weight functions for high aspect ratio semi-elliptical surface cracks in finite-thickness plates, Engineering Fracure Mechanics, (to be published)

(14) Shiratori M., Miyoshi T. and Tanikawa K., (1986), Analysis of stress intensity factors for surface cracks subjected to arbitrary distributed surface stresses, Transactions of Japan Society for Mechanical Engineers, vol. 52, pp. 390-398

(15) Oore M and Burns D.J.,(1980), Estimation of stress intensity factors for embedded irregular cracks subjected to arbitrary normal stress fields, J. of Pressure Vessel Technology, ASME, vol.102, pp 202-211

(16) Vainshtok V. A., (1991), Application of the weight function method to solving multiparametric three-dimensional fracture mechanics problems, International Journal of Fracture, vol. 47, pp. 201-212

(17) Dhont G., (1995), Green functions for a half circular crack in half infinite space under normal loading, Int. Jrnl of Solids and Strucutures, vol.32, pp.1807-1857

(18) Wang X., Lambert S.B. and Glinka G., (1997), Approximate weight functions for embedded elliptical cracks, Engineering Fracture Mechanics, (to be published)

(19) Shah R.C. and Kobayashi A.C., (1971), Stress intensity factor for an elliptical crack under arbitrary normal loading, Engineering Fracture Mechanics, vol. 15, pp. 71-96

(20) Sih G. C., (1968), Mathematical theories of brittle fracture, in Treatice on Fracture, Ed. H. Liebowitz, vol. II, p. 69, Academic Press, New York

(21) Gallin L. A., (1953), Contact Problems in the Theory of Elasticity, Gostechizdat, Moscow (in Russian)

APPENDIX

a) Parameters M_1 (A1), M_2 (A2) and M_3 (A3) of weight function (2) for an edge crack in a finite thickness plate (Fig.3)

$$M_1 = \frac{-0.029207+\frac{a}{t}(0.213074+\frac{a}{t}(-3.029553+\frac{a}{t}(5.901933+\frac{a}{t}(-2.657820))))}{1.0+\frac{a}{t}(-1.259723+\frac{a}{t}(-0.048475+\frac{a}{t}(0.481250+\frac{a}{t}(-0.526796+\frac{a}{t}(0.345012)))))} \tag{A.1}$$

$$M_1 = \frac{0.451116+\frac{a}{t}(3.462425+\frac{a}{t}(-1.078459+\frac{a}{t}(3.558573+\frac{a}{t}(-7.553533))))}{1.0+\frac{a}{t}(-1.496612+\frac{a}{t}(0.764586+\frac{a}{t}(-0.659316+\frac{a}{t}(0.258506+\frac{a}{t}(0.114568)))))} \tag{A.2}$$

$$M_3 = \frac{0.427195+\frac{a}{t}(-3.730114+\frac{a}{t}(16.276333+\frac{a}{t}(-18.799956+\frac{a}{t}(14.112118))))}{1.0+\frac{a}{t}(-1.129189+\frac{a}{t}(0.033758+\frac{a}{t}(0.192114+\frac{a}{t}(-0.658242+\frac{a}{t}(0.554666)))))} \tag{A.3}$$

Valid for $0 < a/t < 0.9$

b) Parameters M_{iA} of weight function (3) for a semi-elliptical surface crack in a finite thickness plate (Fig.4) - the deepest point A

$$M_{1A} = \frac{\pi}{\sqrt{2Q}}(4Y_0 - 6Y_1) - \frac{24}{5} \tag{A4}$$

$$M_{2A} = 3 \tag{A5}$$

$$M_{3A} = 2(\frac{\pi}{\sqrt{2Q}}Y_0 - M_{1A} - 4) \tag{A6}$$

where:

$$Q = 1.0 + 1.464\left(\frac{a}{c}\right)^{1.65}$$

$$Y_0 = B_0 + B_1\left(\frac{a}{t}\right)^2 + B_2\left(\frac{a}{t}\right)^4 + B_3\left(\frac{a}{t}\right)^6$$

$$B_0 = 1.0929 + 0.2581\left(\frac{a}{c}\right) - 0.7703\left(\frac{a}{c}\right)^2 + 0.4394\left(\frac{a}{c}\right)^3$$

$$B_1 = 0.456 - 3.045\left(\frac{a}{c}\right) + 2.007\left(\frac{a}{c}\right)^2 + \frac{1.0}{0.147 + \left(\frac{a}{c}\right)^{0.688}}$$

$$B_2 = 0.995 - \frac{1.0}{0.027 + \frac{a}{c}} + 22.0\left(1 - \frac{a}{c}\right)^{9.953}$$

$$B_3 = -1.459 + \frac{1.0}{0.014 + \frac{a}{c}} - 24.211\left(1 - \frac{a}{c}\right)^{8.071}$$

and

$$Y_1 = A_0 + A_1\left(\frac{a}{t}\right)^2 + A_2\left(\frac{a}{t}\right)^4 + A_3\left(\frac{a}{t}\right)^6$$

$$A_0 = 0.4537 + 0.1231\left(\frac{a}{c}\right) - 0.7412\left(\frac{a}{c}\right)^2 + 0.4600\left(\frac{a}{c}\right)^3$$

$$A_1 = -1.652 + 1.665\left(\frac{a}{c}\right) - 0.534\left(\frac{a}{c}\right)^2 + \frac{1.0}{0.198 + \left(\frac{a}{c}\right)^{0.846}}$$

$$A_2 = 3.418 - 3.126\left(\frac{a}{c}\right) - \frac{1.0}{0.041 + \left(\frac{a}{c}\right)} + 17.259\left(1 - \frac{a}{c}\right)^{9.286}$$

$$A_3 = -4.228 + 3.643\left(\frac{a}{c}\right) + \frac{1.0}{0.020 + \frac{a}{c}} - 21.924\left(1 - \frac{a}{c}\right)^{9.203}$$

c) Parameters M_{iB} of weight function (4) for a semi-elliptical surface crack in a finite thickness plate (Fig.4) - the surface point B

$$M_{1B} = \frac{\pi}{\sqrt{4Q}}(30F_1 - 18F_0) - 8 \qquad \text{(A7)}$$

$$M_{2B} = \frac{\pi}{\sqrt{4Q}}(60F_0 - 90F_1) + 15 \qquad \text{(A8)}$$

$$M_{3B} = -(1 + M_{1B} + M_{1B}) \qquad \text{(A9)}$$

where:

$$F_0 = \left[C_0 + C_1\left(\frac{a}{t}\right)^2 + C_2\left(\frac{a}{t}\right)^4\right]\sqrt{\frac{a}{c}}$$

$$C_0 = 1.2972 - 0.1548\left(\frac{a}{c}\right) - 0.0185\left(\frac{a}{c}\right)^2$$

$$C_1 = 1.5083 - 1.3219\left(\frac{a}{c}\right) + 0.5128\left(\frac{a}{c}\right)^2$$

$$C_2 = -1.101 + \frac{0.879}{0.157 + \frac{a}{c}}$$

and

$$F_1 = \left[D_0 + D_1\left(\frac{a}{t}\right)^2 + D_2\left(\frac{a}{t}\right)^4\right]\sqrt{\frac{a}{c}}$$

$$D_0 = 1.2687 - 1.0642\left(\frac{a}{c}\right) + 1.4646\left(\frac{a}{c}\right)^2 - 0.7250\left(\frac{a}{c}\right)^3$$

$$D_1 = 1.1207 - 1.2289\left(\frac{a}{c}\right) + 0.5876\left(\frac{a}{c}\right)^2$$

$$D_2 = 0.190 - 0.608\left(\frac{a}{c}\right) + \frac{0.199}{0.035 + \frac{a}{c}}$$

AUTHOR INDEX

SUBJECT INDEX

www.ingramcontent.com/pod-product-compliance
Ingram Content Group UK Ltd.
Pitfield, Milton Keynes, MK11 3LW, UK
UKHW050946290926
15470UKWH00005B/1658